Conquering RAS

From Biology to Cancer Therapy

Conquering RAS

From Biology to Cancer Therapy

Edited by

Asfar S. Azmi
Department of Oncology, Wayne State University School of Medicine,
Karmanos Cancer Institute, Detroit, MI, USA

AMSTERDAM • BOSTON • HEIDELBERG • LONDON
NEW YORK • OXFORD • PARIS • SAN DIEGO
SAN FRANCISCO • SINGAPORE • SYDNEY • TOKYO

Academic Press is an imprint of Elsevier

Academic Press is an imprint of Elsevier
125 London Wall, London EC2Y 5AS, United Kingdom
525 B Street, Suite 1800, San Diego, CA 92101-4495, United States
50 Hampshire Street, 5th Floor, Cambridge, MA 02139, United States
The Boulevard, Langford Lane, Kidlington, Oxford OX5 1GB, United Kingdom

Library of Congress Cataloging-in-Publication Data
A catalog record for this book is available from the Library of Congress

British Library Cataloguing-in-Publication Data
A catalogue record for this book is available from the British Library

ISBN: 978-0-12-803505-4

For information on all Academic Press publications
visit our website at https://www.elsevier.com/

Publisher: Mica Haley
Acquisition Editor: Peter Linsley
Editorial Project Manager: Lisa Eppich
Production Project Manager: Edward Taylor
Designer: Mark Rogers

Typeset by TNQ Books and Journals

Dedication

Dedicated to Sheila Sky Kasselman—a pancreatic cancer survivor, inspirational personality and hope for those fighting the disease.

Contents

List of Contributors

A.S. Azmi Wayne State University, Detroit, MI, United States
W. Bai University of South Florida, Tampa, FL, United States
G. Bepler Karmanos Cancer Institute, Detroit, MI, United States
M. Bian Eastern Virginia Medical School, Norfolk, VA, United States
J.K. Bruflat Cellular and Molecular Immunology Laboratory, Rochester, MN, United States
H.-H. Chang David Geffen School of Medicine at UCLA, Los Angeles, CA, United States
C.R. Chow Northwestern University, Chicago, IL, United States
G.J. Clark University of Louisville, Louisville, KY, United States
K. Ebine Northwestern University, Chicago, IL, United States
G. Eibl David Geffen School of Medicine at UCLA, Los Angeles, CA, United States
J.L. Eisner Eastern Virginia Medical School, Norfolk, VA, United States
N.S. Gray Dana Farber Cancer Institute, Boston, MA, United States
H.Z. Hattaway Northwestern University, Chicago, IL, United States
J.C. Hunter The University of Texas Southwestern Medical Center at Dallas, Dallas, TX, United States
J.A.E. Irving Newcastle University, Newcastle upon Tyne, United Kingdom
A.J. Isbell Eastern Virginia Medical School, Norfolk, VA, United States
D.F. Kashatus The University of Virginia School of Medicine, Charlottesville, VA, United States
A.B. Keeton University of South Alabama Mitchell Cancer Institute, Mobile, AL, United States; ADT Pharmaceuticals, Inc., Orange Beach, AL, United States
K. Kumar Northwestern University, Chicago, IL, United States; Jesse Brown VA Medical Center, Chicago, IL, United States
G.A. McArthur Peter MacCallum Cancer Centre, East Melbourne, VIC, Australia; University of Melbourne, Parkville, VIC, Australia
J.N. Mezzanotte University of Louisville, Louisville, KY, United States
H.G. Munshi Northwestern University, Chicago, IL, United States; Jesse Brown VA Medical Center, Chicago, IL, United States
M.M. Njogu Eastern Virginia Medical School, Norfolk, VA, United States
E. O'Neill University of Oxford, Oxford, United Kingdom
J.J. Odanga Eastern Virginia Medical School, Norfolk, VA, United States
P.A. Philip Wayne State University, Detroit, MI, United States

G.A. Piazza University of South Alabama Mitchell Cancer Institute, Mobile, AL, United States; ADT Pharmaceuticals, Inc., Orange Beach, AL, United States

A.D. Rao Peter MacCallum Cancer Centre, East Melbourne, VIC, Australia; University of Melbourne, Parkville, VIC, Australia

A.A. Samatar TheraMet Biosciences LLC, Princeton Junction, NJ, United States

A. Schmidt David Geffen School of Medicine at UCLA, Los Angeles, CA, United States; Universitätsklinikum Freiburg, Freiburg, Germany

R.L. Schmidt Upper Iowa University, Fayette, IA, United States

L.L. Siewertsz van Reesema Eastern Virginia Medical School, Norfolk, VA, United States

D.D. Stuart Novartis Institutes for Biomedical Research, Cambridge, MA, United States

E. Svyatova Eastern Virginia Medical School, Norfolk, VA, United States

A.H. Tang Eastern Virginia Medical School, Norfolk, VA, United States

A.M. Tang-Tan Princess Anne High School, Virginia Beach, VA, United States

R.E. Van Sciver Eastern Virginia Medical School, Norfolk, VA, United States

K.D. Westover The University of Texas Southwestern Medical Center at Dallas, Dallas, TX, United States

S. Xiang University of South Florida, Tampa, FL, United States

X. Zhang Karmanos Cancer Institute, Detroit, MI, United States

V. Zheleva Eastern Virginia Medical School, Norfolk, VA, United States

Acknowledgments

I would like to especially thank the editorial team at Elsevier for their help in the publication process. Special thanks to the peer reviewers and technical staff who helped improve the chapters. The efforts of Lisa Eppich are gratefully acknowledged. Her help was invaluable during the entire production process. Last but not least, all of the co-authors are acknowledged for their important contributions to this book.

Introduction

Ras gene mutations are observed in more than 30% of all cancers and are more prevalent in some of the difficult-to-treat malignancies, such as >90% in pancreatic cancer and also in lung and colon cancers. Ras proteins (N-Ras, H-Ras, and K-Ras) act as molecular switches that when activated, through binding of GTP, initiate a cascade of signaling events controlling important cellular processes such as proliferation and cell division. A precise and recurring cycling of GTP to GDP (inactive state) occurs through the intrinsic GTPase activity of ras. However, mutations in ras result in the loss of this intrinsic GTPase activity rendering the protein in a constantly activated state. In this scenario, a continuous signaling from ras results in cells growing uncontrollably, evading cell death mechanisms and also becoming resistant to therapies. These facts are well known for the past 30 years; nevertheless, till date strategies to block ras mutation-driven signaling remain futile. The reasons for such failures have been attributed to many factors. Chief among them is the lack of any possible druggable pocket within the ras structure for optimal attachment of small molecule drugs. In addition, the inherent affinity of ras to GTP (in the picomolar range) restricts the design of high-affinity drugs to displace GTP. Researchers have evaluated the benefits of targeting important upstream (EGFR and IGFR) and downstream (RAF, MEK, and AKT) and other signaling molecules. With few exceptions, sadly, none of these targets have proven to be effectual in the clinical setting. The redundancies and cross talk within the associated pathways pose additional hurdles making the ras fortress impenetrable. Collectively, these multitude number of challenges have led to a sort of consensus that the ras protein itself in un-druggable.

Thanks to the National Cancer Institute (NCI) ras Initiative, there is a renewed spark in the field of ras research. Ras Initiative is a concerted and broad-spectrum approach to ras biology with a single goal and that is to develop effective therapies against this important master cancer regulator. When such an initiative is underway, a book specifically focused on ras biology becomes an important resource for researchers working in the field. This is especially important given that the literature on ras is distributed in the web of knowledge sometimes out of the reach of the most avid researchers working in the field. With this goal in mind, I have designed this book to bring forward some of the newer topics in the field of ras biology under one volume. The book is divided into two parts. Part I deals with ras biology and in Part II many novel therapeutic approaches are highlighted. Each chapter carries a comprehensive list of up-to-date references that are surely going to find their way in the libraries of ras researchers. Unlike before, in this book, an attempt has been made to accommodate some of the most burning topics such as ras metabolic vulnerabilities, impact on microenvironment, role in stemness, effect of post-translational mechanisms, and the biology of effectors. On the therapeutic side, some very new targets and novel agents have been presented that will surely make for an interesting reading.

It is recognized that aside from the contributors of this book there are many additional groups working in the field. Therefore, every effort has been made to include the vast library of important references extracted from major contributions across a wide spectrum of related research papers. It was my pleasure collecting these novel ideas from so many experts working in the field who have a single goal and that is to conquer ras.

Asfar S. Azmi, PhD
Department of Oncology
Wayne State University School of Medicine
Karmanos Cancer Institute
Detroit, MI 48201, USA

1

SECTION

RAS Cancer Biology

CHAPTER 1

Ras and RASSF Effector Proteins

J.N. Mezzanotte, G.J. Clark
University of Louisville, Louisville, KY, United States

INTRODUCTION

Ras is the most frequently activated oncoprotein in human cancer. When we consider the prevalence of activating point mutations in Ras in tumors combined with the frequent inactivation of GTPase-activating proteins, it seems likely that the majority of human cancers use Ras activation as a driving force [1]. In contrast, the RASSF1A tumor suppressor appears to be the most frequently inactivated tumor suppressor in human cancer [2]. Not only is RASSF1A subjected to frequent epigenetic inactivation in human tumors but also it can be inactivated by protein degradation or point mutation at a significant frequency. RASSF1A contains a Ras-association (RA) domain and can bind directly to Ras [2]. Thus the most frequently activated oncoprotein in human cancer forms a complex with the most frequently inactivated tumor suppressor.

Although the mechanistic basis of the potent transforming effects of activated Ras is well documented, it has also become apparent that Ras activation can stimulate signaling pathways that suppress growth and survival [3]. For example, in primary cells, introduction of an activated Ras gene tends to promote apoptosis or senescence, not transformation [3–5]. The signaling pathways involved in these anti-transformation events are only now being understood. Many of them appear to involve the RASSF family of Ras effector/tumor suppressors. Thus RASSF proteins may serve as Ras death effectors, and their inactivation may enable Ras-dependent tumors to progress to malignancy.

The RASSF Family of Proteins

The RASSF family of proteins consists of 10 members, all of which contain a Ras-association, or RA, domain, hence the term RASSF: Ras-association domain family. RASSF1 through RASSF6 have their RA domain toward the C-terminus, whereas RASSF7 through RASSF10 all have an N-terminal RA domain. The C-terminal RASSF proteins have been more widely studied and have shown

CONTENTS

Conquering RAS. http://dx.doi.org/10.1016/B978-0-12-803505-4.00001-1

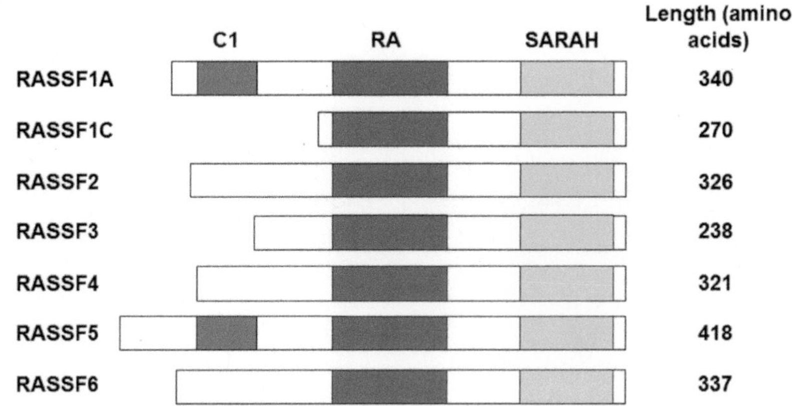

FIGURE 1.1 RASSF protein structure.
Protein structures for the C-terminal RASSF members are shown. *C1*, zinc finger domain; *RA*, Ras association domain; *SARAH*, Salvador/RASSF/Hippo domain.

extensive epigenetic inactivation in numerous cancers, thus they will be the focus of this discussion.

A general feature of the RASSF proteins is that they do not appear to have any enzymatic activity; instead, they appear to act as scaffolding molecules, facilitating the growth and survival suppressing effects of Ras by scaffolding it to various pro-apoptotic or pro-senescent signaling pathway proteins. All of the C-terminal RASSF proteins also contain a Salvador/RASSF/Hippo, or SARAH, domain, which directly binds the mammalian sterile 20 like (MST) kinases, connecting them to the Hippo signaling pathway [6,7]. Relevant structural domains of the C-terminal RASSF proteins are highlighted in Fig. 1.1.

Another unique feature of RASSF proteins is their high rate of epigenetic inactivation in numerous cancers. Epigenetics refers to changes in gene expression that are not due to changes in the DNA sequence itself, and in the case of RASSF proteins, they commonly experience methylation of CpG islands in their promoter regions, leading to the loss of expression of RASSF proteins in the cell. Suppressing RASSF proteins experimentally can enhance Ras transformation and disconnect Ras from apoptotic and senescent pathways [8,9]. Thus loss of RASSF protein expression facilitates Ras transformation. The known relevance of each of the C-terminal RASSF proteins to Ras function will be summarized in the following sections.

RAS AND RASSF1

The RASSF1 gene was identified serendipitously in a two-hybrid screen for proteins that interact with the DNA repair protein xeroderma pigmentosum group A-complementing protein (XPA) [10]. It was shown to produce two main transcripts, RASSF1A and RASSF1C, both of which contain an RA domain. Several other isoforms appear to exist, but because there is little or no data available on them, they will not be considered here.

The RASSF1 proteins were shown to bind to activated Ras and promote Ras-dependent apoptosis [9,11]. Initially, some controversy arose as to the physiological nature of the interaction between Ras and RASSF1A, with some groups suggesting the interaction did not occur or was indirect [12]. However, multiple groups have now reported that activated Ras forms an endogenous complex with RASSF1A and that the interaction is likely to be direct [9,11,13]. A possible explanation for the confusion was identified when we observed that RASSF1A preferentially associates with K-Ras and fails to bind non-farnesylated Ras [2]. Consequently, experiments using recombinant H-Ras protein from bacteria or non-farnesylated Ras mutants in yeast would not be expected to give positive results.

Early work showed that RASSF1A was frequently down-regulated in human tumor cells and could act to suppress the tumorigenic phenotype in vitro [9,10]. RASSF1A has no apparent enzymatic activity, and we hypothesize that it acts as a scaffolding protein under the control of K-Ras. This allows K-Ras to control multiple tumor suppressing pathways.

The first biological properties of RASSF1A that were characterized were that RASSF1A can promote both G1 and G2/M cell cycle arrest [6,9,14]. The G2/M arrest can be explained by the powerful effects of RASSF1A over-expression on microtubule polymerization. RASSF1A directly binds multiple microtubule-associated proteins (Maps), which themselves directly bind to tubulin [15,16]. Maps modulate microtubule polymerization. We have found RASSF1A associating with most forms of tubulin, including gamma tubulin at the spindle poles, and this may explain the ability of RASSF1A to suppress K-Ras-induced genetic instability [16]. A more technically sophisticated study suggested that in interphase cells, RASSF1A preferentially associates with a subset of microtubules at the Golgi to promote correct cell polarity and Golgi orientation [17]. In addition to a microtubular localization, we can identify endogenous RASSF1A in the nuclear compartment, and the protein has also been reported to associate with mitochondria [18].

In addition to its effects on microtubules, RASSF1A has also been shown to connect Ras to two major pro-apoptotic signaling pathways: the Bax pathway and the Hippo pathway. Bax is a pro-apoptotic protein that contains a Bcl-2 homology, or BH, domain and is critical for most forms of apoptosis in the cell. In 2005, two studies by independent groups identified RASSF1A as a critical mediator of Bax activation through its ability to directly interact with the protein modulator of apoptosis-1 (MOAP-1) [19,20]. MOAP-1 in turn directly binds and activates Bax. Activated K-Ras enhances the interaction of RASSF1A and MOAP-1 to stimulate Bax activation and translocation to the mitochondria. Suppressing RASSF1A impairs the ability of Ras to activate Bax in tumor cells [20].

RASSF1A also connects Ras to another major pro-apoptotic signaling pathway, the Hippo pathway. The major splice variants of RASSF proteins 1–6 all contain a C-terminal SARAH motif. This serves to bind to the Hippo kinases MST1 and MST2 [6,7,11,21]. The MST kinases can in turn phosphorylate and activate the large tumor suppressors (LATs) kinases in a kinase cascade. LATs kinases have several targets, but the most important targets appear to be the transcriptional co-activators yes-associated protein (YAP) and tafazzin (TAZ) [22]. Phosphorylation of YAP/TAZ by LATs promotes their exclusion from the nucleus and leads to their proteosomal degradation. YAP acts as a potent oncogene and pro-survival factor, so its suppression by the Hippo pathway can lead to apoptosis or senescence [22]. The Hippo pathway plays a key role in normal cellular homeostasis, and it is commonly dysregulated in human cancers, leading to YAP activation and pro-growth effects (reviewed in Ref. [23]). RASSF1A serves to connect Ras to the control of the Hippo pathway as the interaction of Ras with RASSF1A promotes MST kinase stability and activation [24,25]. Thus the loss of RASSF1A uncouples Ras from the activation of the Hippo pathway, suppressing apoptotic signaling. However, this story may be even more complex. In an in vivo system using a point mutant of RASSF1A that specifically fails to bind the MST kinases, we found that cardiomyocytes and cardiac fibroblasts react quite differently to RASSF1A/Hippo signaling [24]. Thus there may be a strong cell type specificity associated with the net result of this pathway.

RASSF1A has also been shown to complex with the mouse double minute 2 homolog (MDM2) ubiquitin ligase, which can degrade both p53 and Rb [26]. By promoting the degradation of MDM2, RASSF1A may stabilize p53 and potentially Rb, thereby activating them. This connection with p53 may explain why RASSF1A/p53 dual heterozygous knockout mice exhibit synergistic tumor formation [27]. The role of Ras in this process remains unclear.

Furthermore, RASSF1A has now been shown to play an important role in both the DNA damage response and the DNA repair process itself [28–30]. The O'Neill group showed that RASSF1A is involved in activating MST2 and LATS1 upon DNA damage, leading to the stabilization of the pro-apoptotic protein p73 [29]. Therefore, RASSF1A-defective cells fail to activate apoptotic processes when they are subjected to DNA damage, thus promoting the survival of cells carrying mutations, which can lead to cancer.

Donninger et al. showed that RASSF1A-defective cells not only fail to induce the DNA damage apoptotic response, but also fail to repair that DNA damage. They found that the original yeast two-hybrid observation that RASSF1A might bind the DNA repair protein XPA was in fact correct [31]. Moreover, they found that RASSF1A-negative cells were defective for proper XPA regulation and, as a result, were less able to repair DNA damage due to UV radiation. This observation was confirmed in vivo by enhanced tumor formation in UV-irradiated mice heterozygous for both RASSF1A and XPA [31]. Intriguingly, a single nucleotide polymorphism (SNP) variant of RASSF1A that exhibits an alteration in the consensus phosphorylation site for the DNA damage kinases ataxia telangiectasia mutated protein/ataxia telangiectasia and Rad3-related protein has been identified. This variant was found to be defective for both the DNA damage response and for supporting DNA repair after damage [29,31]. Thus SNP carriers are both less able to respond to DNA damage by inducing cell death and less able to repair DNA damage. These observations explain the reported enhanced cancer predisposition of SNP carriers but fail to explain why the SNP is so common in European populations (~22%) but very rare in African populations (~2%) [32–34]. We speculate that the attenuated apoptotic properties of the SNP variant may give some biological advantage to carriers. RASSF1A has been shown to play an important role in cell death during cardiac hypertrophy [35]. Perhaps, the attenuated apoptotic response due to the SNP variant may suppress cardiac disease.

Further investigation showed us that the mechanism by which RASSF1A appears to control DNA repair involves regulating the acetylation status of DNA repair proteins via the SIRT1 deacetylase [31]. It has been reported that RASSF1A can form a complex with the deacetylase HDAC6 [36]. Thus RASSF1A can link K-Ras to the control of protein acetylation by multiple deacetylases. Loss of RASSF1A may induce general defects in the acetylome. As acetylation may be an even more widespread post-translational modification in the cell than phosphorylation [37], this effect could be of profound importance to Ras-driven tumor development and to tumor response to acetyl transferase inhibitors.

Transgenic mouse studies have confirmed a tumor suppressor role for RASSF1A in vivo as knockout mice developed a modest increase in spontaneous tumor development with age or carcinogen treatment [38]. However, the results were subtle and curious in that the heterozygous knockout mice developed more tumors than the homozygous knockout mice. This hints that the cell may require a minimal, reduced RASSF1A expression for survival. Indeed, potent suppression of many tumor suppressors, such as breast cancer early onset 1 or von hippel-lindau, can lead to the reduced growth of target cells, and the same appears to be true for RASSF1A [39]. This may explain why RASSF1A is so seldom completely deleted in human cancer. Studies examining the loss of both RASSF1A and p53 showed a synergistic effect, with RASSF1A-null, p53-null mice showing a large amount of spontaneous tumor formation at a young age [27]. It will be revealing to examine the results of RASSF1A suppression and Ras activation in mouse models.

The main alternative splice form of RASSF1 is RASSF1C, a shorter form that lacks the N-terminus of RASSF1A. RASSF1C can complex with K-Ras and has apoptotic properties [9]. The RASSF1C promoter has not been reported to suffer epigenetic inactivation in tumors, so the protein is not regarded as a tumor suppressor. However, we have found that RASSF1C protein expression is lost in some tumor cell lines. Indeed, it is lost in some cases where the RASSF1A protein expression is retained (unpublished data). This implies that RASSF1C may be regulated at a post-transcriptional level and may also act as a tumor suppressor, at least in some cell types.

Contradictory roles for RASSF1C have also been reported. In our hands, it appears to behave rather like a weaker form of RASSF1A, polymerizing microtubules and promoting apoptotic cell death [16]. It has also been shown to play a role in ovarian cancer cell death and in the activation of the apoptotic jun N-terminal kinase (JNK) pathway after DNA damage [40,41]. Other groups have found that RASSF1C can have a mild stimulatory effect on tumor cell growth and may up-regulate the β-catenin oncoprotein [42–44]. Consequently, the physiological functions of this isoform remain unclear.

RAS AND RASSF5 (NORE1)

The second best-studied RASSF family member is RASSF5. The RASSF5 gene produces two main protein isoforms, RASSF5A, also known as NORE1A, and NORE1B, or RAPL. NORE1A is broadly expressed in tissue, whereas NORE1B seems mostly restricted to the lymphoid compartment. NORE1A was originally identified as a Ras-binding protein in a two-hybrid screen [45]. RAPL/NORE1B was identified as a Rap binder in a similar screen. NORE1A binds to Ras via the effector domain in a GTP-dependent manner [46]. It can be found in an endogenous complex with Ras, so it meets the definition of a Ras effector. Unlike RASSF1A, it readily binds H-Ras [2].

NORE1A is often down-regulated in tumors by epigenetic mechanisms [2]. It can also be down-regulated at a protein level in tumor cells by calpains and by ubiquitination [47,48]. In liver cancer, more malignant primary tumor samples expressed less NORE1A [13]. Moreover, a human family with a translocation that inactivates the *NORE1A* gene suffers from a hereditary cancer syndrome [49]. Thus the evidence that NORE1A is a tumor suppressor is strong.

Ras can use NORE1A as a pro-apoptotic effector [7]. Like RASSF1A, NORE1A binds the MST kinases and has the potential to modulate the pro-apoptotic Hippo pathway. However, deletion mutagenesis has shown that the canonical Hippo pathway is not essential to the growth suppressing function of NORE1A, and it is unclear if NORE1A can stimulate the canonical Hippo kinase cascade [21,50].

We find that NORE1A is a highly potent senescence effector of Ras [8]. Overexpression of NORE1A induces senescence at levels comparable with those induced by activated Ras, whereas suppression of NORE1A severely impairs the senescence response and enhances Ras-mediated transformation [8]. Moreover, the expression of NORE1A in primary human tumors correlates well with the expression of senescence markers [51]. Specifically, NORE1A can form a Ras-regulated complex with p53 and scaffolds it to the kinase HIPK2, so in the presence of NORE1A, we can detect Ras in a complex with p53 [8]. Thus the long-known but poorly understood link between Ras and p53 may be solved by NORE1A. HIPK2 can phosphorylate p53 at residue S46 to promote apoptosis; however, it can also recruit acetyl transferases to acetylate p53 to promote its pro-senescence effects (reviewed in Ref. [52]). NORE1A acts to suppress S46 phosphorylation of p53 and enhances p53 acetylation at residues 382 and 320, thus driving p53-dependent senescence [8]. Once again, this is evidence that RASSF proteins may be able to couple Ras to the control of protein acetylation.

In addition, NORE1A can associate with MDM2, a negative regulator of p53, but it can use this association to induce the ubiquitination and degradation of the oncoprotein HIPK1 by scaffolding the two proteins together, highlighting the functionality of NORE1A as a scaffolding molecule [53]. In our hands, the NORE1A/MDM2 interaction appears to be Ras regulated, adding a further level to the Ras/NORE1A/p53 relationship.

Another major component of Ras/NORE1A signaling that has been identified is the β-catenin protein [54]. β-catenin is an adherens junction protein and transcription co-factor that serves as the terminal executor of the Wnt signaling pathway. A multi-protein complex normally phosphorylates β-catenin, allowing it to bind the SCFβ-TrCP ubiquitin ligase complex [55]. In the absence of Wnt stimulation, this process results in β-catenin degradation. When Wnt is activated, un-phosphorylated β-catenin translocates to the nucleus, where it activates the transcription of growth-promoting, pro-survival genes. Thus β-catenin, when either dysregulated or mutated, can act as an oncogene in cancer, and β-TrCP, the SCFβ-TrCP substrate recognition component, can act as a tumor suppressor because of its influence on β-catenin degradation. Schmidt et al. found that NORE1A plays a role in this signaling process by directly binding β-TrCP in an interaction that is enhanced by activated Ras, allowing Ras to specifically stimulate SCFβ-TrCP-mediated degradation of β-catenin [54]. Thus in cancers where NORE1A expression is lost, negative Ras regulation of β-catenin is disrupted, revealing another crucial barrier that the RASSF proteins present to uncontrolled cell growth and tumorigenesis [54].

In vivo studies investigating the role of NORE1A as a potential tumor suppressor found that NORE1A knockout mice appear overtly normal, yet mouse embryonic fibroblasts (MEFs) from the animal can be transformed by activated Ras in a single step [56]. Normal MEFs require the addition of other oncogenic events such as mutant p53 or the activation of other oncogenes to suppress oncogene-induced senescence to transform [5], so the loss of NORE1A increases the susceptibility to transformation. In addition, the same study revealed that NORE1A helps mediate tumor necrosis factor (TNF)-α–mediated, or death receptor mediated, apoptosis, most likely through its interaction with MST1, providing further evidence that NORE1A is a tumor suppressor [56]. In human tumors, few studies have examined the results of NORE1A inactivation and Ras activation, but one study of hepatocellular carcinomas with activated Ras signaling found that NORE1A promoter methylation was only found in a subset of tumors with poor patient survival, indicating the potential impact of NORE1A loss on human tumors [13]. Studies of NORE1A-null, Ras-positive mice will provide more insight into the full inhibitory effect that NORE1A plays in cancer.

A lesser-studied NORE1 isoform, NORE1B has mostly been shown to play a role in immune cells, and it seems to be expressed more in the lymphoid compartment compared with the fairly ubiquitous expression of other RASSF proteins. Specifically, NORE1B plays an important role in lymphocyte and dendritic cell adhesion and migration and was shown to be a crucial part of immunosurveillance. The loss of NORE1B expression in a mouse model resulted in immune dysfunction that included the inability of lymphocytes and dendritic cells to migrate to tissues and impaired the maturation of B cells [57]. Like NORE1A, NORE1B can associate with MST1, and the two form a synergistic relationship, negatively regulating T cell proliferation when the T cell antigen receptor is stimulated [58]. In addition, NORE1B has been shown to work with Ras to regulate T cell signaling; NORE1B recruits activated Ras to T cell synapses to promote Ras signaling in immune cells [59].

Unlike the other C-terminal RASSF proteins, very little evidence exists to show that NORE1B can act as a tumor suppressor or that it experiences epigenetic inactivation, but it can associate with Ras and Ras-related proteins [60]. One study revealed that NORE1B experiences a high percentage (62%) of promoter methylation in hepatocellular carcinoma, indicating that NORE1B loss may enhance tumorigenesis in some systems [61]. The same group found that NORE1B interacts with RASSF1A and that the two are frequently lost together in hepatocellular carcinomas, leading to the hypothesis that NORE1B and RASSF1A work together to prevent hepatocellular carcinoma formation [62]. Further work on this subject is needed to elucidate the mechanism of action of NORE1B.

RAS AND RASSF2

RASSF2 forms an endogenous complex with activated K-Ras via the Ras effector domain. It appears to be specific for K-Ras as its interaction with H-Ras is very weak [63]. RASSF2 is expressed at particularly high levels in the brain but is expressed in many other tissues at lower levels. Like other RASSF proteins, RASSF2 has been found to undergo epigenetic inactivation in a large number of cancers, and its inactivation was found to increase oncogenic transformation induced by K-Ras in colorectal cancer and in lung cancer cells [64,65]. RASSF2 has also been identified as a potential metastasis suppressor [66]. Loss of RASSF2 expression enhances cell growth, disrupts adhesion, and leads to the up-regulation of phosphorylated AKT [65]. Curiously, RASSF2 knockout mice die soon after birth. Thus RASSF2 may actually be the most critical member of the RASSF family as it is the only one that is essential to life.

Of all the tumor types screened, RASSF2 shows the most intense level of aberrant promoter methylation in prostate tumors, with up to 95% promoter methylation observed in one study [67]. Moreover, this methylation correlated well with the frequent loss of protein expression in primary prostate cancers. RASSF2 may prove to be a highly effective diagnostic marker for prostate cancer, as it is so commonly methylated, and RASSF2 promoter methylated DNA can be detected in urine samples by a sensitive polymerase chain reaction assay [68]. Indeed, this was found to be a more predictive biomarker for prostate cancer than the prostate-specific antigen test. RASSF2 has also been proposed as a potential biomarker for gliomas [69].

RASSF2 is pro-apoptotic and acts, in part, by binding the prostate apoptosis response protein (PAR-4). This interaction is K-Ras regulated, and activated K-Ras promotes translocation of PAR-4 to the nucleus via RASSF2. There, PAR-4 can interact with the TNF-related apoptosis-inducing ligand to induce apoptosis [67]. RASSF2 can also bind the MST kinases and may modulate their stability. However, like NORE1A, RASSF2 can induce apoptosis independently of MST [70,71]. RASSF2 can also modulate both nuclear factor κb signaling and the JNK pathway, but the precise mechanisms of these effects and if they are coupled to Ras remain to be elucidated [70,72].

RAS AND RASSF3

We have found that RASSF3 can bind activated K-Ras in over-expression systems, but an endogenous complex between Ras and RASSF3 has yet to be confirmed. We also found that K-Ras and RASSF3 co-operate to induce cell death. We have not found much evidence of RASSF3 methylation or protein down-regulation in tumors in our studies; however, RASSF3 may be deleted in some colorectal cancers and in patients with relapsed acute lymphoblastic leukemia [73,74]. Moreover, several RASSF3 SNPs have been reported to be associated with an enhanced risk of head and neck cancer [75]. In one tumor type, somatotroph adenomas, high-frequency promoter methylation has been reported [76]. Thus, although there is evidence for RASSF3 loss of function in human cancer, this is a much less frequent event than that observed for some other family members.

Experimentally, RASSF3 inactivation has been reported to lead to defects in DNA repair and enhanced genomic instability as well as enhanced transformation of lung tumor systems [77,78]. RASSF3 can also bind MST1 but does not appear to activate it [7]. Instead, RASSF3 is involved in p53-mediated apoptosis and has been shown to modulate p53 via binding and promoting MDM2 degradation [76,78].

RAS AND RASSF4 (AD037)

RASSF4 (originally designated AD037) can bind activated K-Ras via the effector domain [79], but the lack of good antibodies for RASSF4 has prevented the confirmation of endogenous complex formation between the two. Over-expressed RASSF4 promotes a Ras-dependent apoptosis, and RASSF4 expression is down-regulated by promoter methylation in some tumor cells, including nasopharyngeal carcinoma [79,80]. RASSF4 down-regulation has also been linked to the maintenance of cancer stem cells in oral squamous cell cancer stem-like cells [81].

Although RASSF4 is down-regulated in some tumors, we did observe that some primary human breast cancers exhibited enhanced RASSF4 expression [79]. Moreover, RASSF4 has been implicated as pro-oncogenic by binding to MST1 and inhibiting the Hippo pathway in alveolar rhabdomyosarcoma systems, resulting in increased YAP expression and increased cell growth and senescence evasion [82]. Thus RASSF4 may have different biological effects in different cell systems.

RAS AND RASSF6

RASSF6 can bind activated K-Ras via the effector domain in over-expression studies but has yet to be confirmed in an endogenous complex with Ras. RASSF6 can induce apoptosis, and suppression of RASSF6 can enhance the transformed phenotype of tumor cell lines [83,84]. RASSF6 is epigenetically down-regulated in primary cancers although less so than RASSF1A or RASSF5. It has been found to be specifically inactivated in neuroblastoma, childhood leukemias, and melanoma and melanoma metastases [85–87].

RASSF6 was the first RASSF family member to be identified that binds and inhibits, rather than activates, the MST kinases to suppress the Hippo pathway [88]. This is the same effect as observed with the single *Drosophila* RASSF protein. Therefore, RASSF6 must induce apoptosis independently of Hippo signaling. As with other RASSF family members, RASSF6 interacts with the MDM2 protein to modulate p53, apoptosis, and the cell cycle [89]. Like RASSF1A, RASSF6 can form a Ras-regulated complex with the Bax activating protein MOAP-1 [83,84]. In addition, over-expressed RASSF6 has been shown to increase the association of the inhibitory kinase MST1 and mutant B-Raf to suppress the mitogen activated protein kinase (MAPK) pathway in melanoma cells [87]. Thus RASSF6 has multiple mechanisms that could be used by Ras to suppress tumorigenesis. However, we have observed occasional strong up-regulation of RASSF6 in some primary tumor samples, such as ovarian. Perhaps, like the Hippo-inhibiting RASSF4 protein, in some circumstances, RASSF6 may be pro-tumorigenic.

EFFECTS OF RASSF PROTEINS ON MITOGENIC RAS EFFECTORS

RASSF proteins bind Ras and can be considered as Ras death effectors connecting Ras to multiple signaling pathways that can mediate apoptosis or senescence. However, the role of RASSF proteins in Ras biology may be more complex and subtle. In addition to their own signaling pathways, RASSF proteins may be able to modulate the activity of the classic mitogenic signaling pathways used by Ras.

The MST2 kinase not only binds RASSF1A but it can also bind Raf-1, where it serves to inhibit Raf kinase activity [25,90]. RASSF1A acts to compete with Raf-1 for MST2 binding, so down-regulating RASSF1A can increase Raf-1/MST2 binding, suppressing the MAPK pathway. Thus Ras modulates Raf-1 directly by binding to it and indirectly via RASSF1A/MST2. RASSF1A has also been reported to suppress AKT activity by a mechanism that remains unclear [91]. AKT is a component of the Ras/phosphoinositide 3-kinase pathway. In a further twist, AKT can phosphorylate MST2 to promote its binding to Raf-1, inhibiting the kinase activity of MST2 [11,25]. We have found that RASSF6 can also modulate the interaction of MST1 with activated B-Raf in a melanoma cell line to suppress the MAPK pathway [87]. Thus RASSF proteins may have a more complex role in mediating Ras biology than simply controlling their own, separate signaling modalities. They may be able to integrate the regulation of pro-mitotic and pro-death Ras pathways.

THERAPEUTIC RAMIFICATIONS

RASSF1A is epigenetically inactivated in a broad range of primary human tumors [92,93]. For reviews of cancers known to experience epigenetic inactivation of RASSF1A, see Refs. [2,94]. Loss of RASSF1A uncouples Ras from multiple growth suppressive pathways, so it would seem reasonable that Ras tumors would often show RASSF1A down-regulation.

Although some studies have shown a correlation between Ras point mutations and RASSF1A promoter methylation, the majority have not. However, in addition to inactivation by promoter methylation, RASSF1A can be inactivated by point mutations at a significant frequency [94,95]. Moreover, a SNP variant of RASSF1A has been identified that is defective for some apoptotic responses and predisposes carriers to cancer development [32,33]. Thus many RASSF1A "positive tumors," as measured by promoter methylation, may actually be RASSF1A negative. Moreover, Ras is frequently activated in the absence of point mutations by defects in upstream activators (eg, human epidermal growth factor receptor 2) or negative regulators such as neurofibromatosis 1, Ras GTPase activating like protein, or DAB2 interacting protein [96–98]. Therefore, assays performed at a protein level will be required to definitively answer the question of the relationship between Ras activation and RASSF inactivation in human tumorigenesis.

One large-scale study that stratified non-small cell lung cancer tumors and measured Ras mutation and RASSF1A promoter methylation has been reported [99]. It showed that, although there was no general correlation, tumors with both K-Ras mutations and RASSF1A methylation had a much poorer prognosis and lower overall patient survival than other tumors of the same stage [99]. A similar result was reported for hepatocellular carcinoma tumors with NORE1A promoter methylation and Ras activation [13]. This implies that most tumors with Ras activation and RASSF methylation have the potential to be more aggressive than tumors without RASSF methylation, data that certainly correlate well with studies examining the loss of RASSF proteins in K-Ras–positive cancer cell lines. For example, the loss of RASSF2 expression was shown to enhance proliferation and invasion of K-Ras–positive lung cancer cells, and it also conferred resistance to chemotherapy in those cells [65]. Similar results were obtained for RASSF3 in non-small cell lung cancer, and our group showed that reintroduction of RASSF6 expression in a metastatic melanoma cell line that had lost RASSF6 expression was sufficient to alter mutant B-Raf signaling and decrease the invasiveness of those cells [77,87]. In addition, RASSF1A knockdown cells exhibit resistance to DNA damage-induced apoptosis and to treatment with cisplatin [29]. These observations suggest that epigenetic therapy designed to restore RASSF protein expression might be a plausible strategy to help treat aggressive Ras-positive, RASSF-negative tumors. This might be a particularly attractive approach as targeting Ras directly has so far not been successful [100].

DNA methylation is a reversible process, making it a potential target for cancer therapy. DNA is methylated by DNA methyltransferase (DNMT) proteins, and a class of drugs called DNA methyltransferase inhibitors can be used to prevent DNMTs from methylating DNA. The most commonly used DNA methyltransferase inhibitors are 5-azacytidine and decitabine, both of which are nucleoside analogs that cause DNA methyltransferases to be inactivated in a protein–DNA adduct [101]. Both these drugs have been approved for the treatment of myelodysplastic syndrome and for low-blast count acute myeloid leukemia, but their efficacy in the treatment of solid tumors is limited, potentially because of the higher doses needed that lead to unwanted side effects, like myelosuppression and nausea [102].

The aberrant methylation of the RASSF1A promoter appears to be mediated primarily by one enzyme: the DNA methyltransferase DNMT3B [103]. This protein can be up-regulated by K-Ras, so it is possible that Ras mutations actually promote RASSF1A inactivation [104]. A quinone-based antibiotic, nanaomycin A, was identified as a specific DNMT3B inhibitor. Nanaomycin A treatment was able to result in specific re-expression of RASSF1A in lung cancer cells, and the tumorigenic phenotype in those cells was effectively suppressed [105]. Similar results were also shown in a melanoma cell line, in which treatment with nanaomycin A resulted in re-expression of RASSF6 [87]. Nanaomycin A has also been shown to eradicate melanoma stem cell–like cancer cells [106]. These results imply that nanaomycin A treatment, or drugs with a similar action, has considerable potential to result in re-expression of all RASSF proteins in Ras-driven tumor cells. Such a treatment might have therapeutic potential in many different cancers without the side effects associated with the use of less specific nucleoside analogs. No clinical trials that examine the utility of nanaomycin A as an anti-cancer agent have currently been reported.

Another aspect of clinical relevance for RASSF proteins is their use as biomarkers in human tumors. RASSF1A down-regulation, thought to be one of the most common events in human cancer, is widely associated with more aggressive cancer phenotypes, and methylation of RASSF1A DNA promoter regions can be detected in sputum, serum, and urine analyses [93]. RASSF2 methylation can also be detected in the urine of patients with prostate cancer [68]. Overall, examining the methylation status of RASSF proteins is a non-invasive process that could provide insight into the overall cancer phenotype, and this information could be used to develop a more personalized treatment plan for patients with cancer.

In addition, an RASSF1A SNP variant has been discovered that renders cells less sensitive to cell death by DNA damaging agents [29]. Patients with this polymorphism could thus experience different responses to certain chemotherapeutic regimens, meaning that not only RASSF1A expression status but also RASSF1A mutation status could play a role in developing personalized therapies for patients with cancer.

CONCLUSION

Ras activation is likely to be the most common single event in the development of human cancer. When the proper checkpoints are in place, however, Ras activation does not lead to cancer. Instead, activated Ras can promote cell death and/or cell growth arrest. This surprising function of a widely known oncogene is facilitated by the RASSF protein family. Members of this family have no enzymatic activity and instead work by scaffolding Ras to various

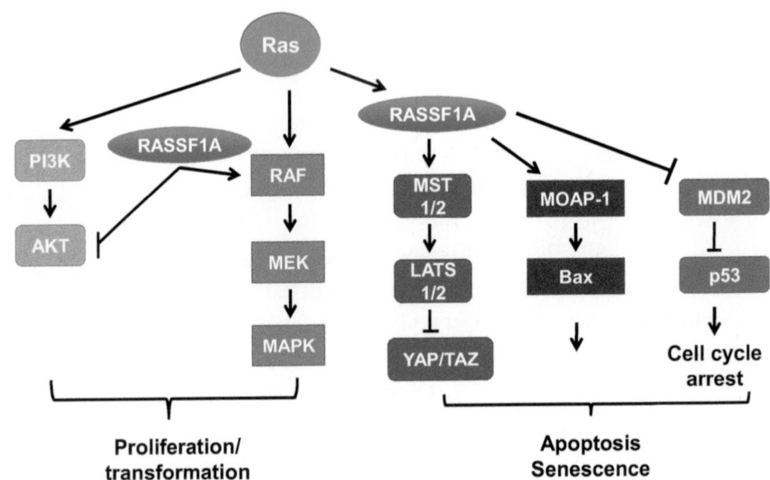

FIGURE 1.2 Signaling pathway involvement of RASSF1A.
RASSF1A is involved in a wide variety of signaling pathways, several of which are outlined here. RASSF1A can act to stimulate apoptosis upon Ras activation, but it can also act independently of Ras to inhibit Akt signaling through several mechanisms described in this discussion.

pro-apoptotic and pro-senescent signaling pathways. In addition, RASSF family members can impact the activity of other mitogenic Ras effectors or effector pathways. A summary of RASSF protein involvement in Ras signaling pathways is shown in Fig. 1.2, using RASSF1A as an example to show the wide variety of processes in which RASSF signaling plays a major role.

Further studies are needed to elucidate the full significance of losing RASSF proteins in Ras-driven tumors. For example, in the case of RASSF1A, RASSF1A-null mice show an increased susceptibility to tumors, yet the dual effect of RASSF1A loss and Ras activation remains to be examined [38]. Therapeutic approaches aimed at reactivating the expression of these proteins in cancer have the potential to offer novel, personalized treatments for patients with RASSF promoter methylation. Overall, RASSF proteins showcase an interesting and paradoxical side of Ras, and their potential clinical utility could provide a useful tool for targeting a large subset of the most intractable Ras-driven tumors in the future.

Glossary

Apoptosis Programmed cell death.

Epigenetics The study of changes that are caused by the modification of gene expression, not by alteration of DNA itself.

Kinase An enzyme that catalyzes the transfer of a phosphate group from a molecule of ATP to another molecule.

Methylation The addition of a methyl group (CH_3) to a DNA base. Methylation of cytosines in CpG island regions of DNA can lead to gene suppression.

Mitogenic A growth-promoting (mitosis-inducing) substance or signal.
Senescence A state of permanent cell cycle arrest, in which the cell is alive but not actively growing.

List of Acronyms and Abbreviations

ALL Acute lymphoblastic leukemia
ATM/ATR Ataxia telangiectasia mutated protein/ataxia telangiectasia and Rad3-related protein
BRCA1 Breast cancer early onset 1
DAB2IP DAB2 interacting protein
DNMT DNA methyltransferase
GAPs GTPase-activating proteins
HDAC6 Histone deacetylase protein 6
HER2 Human epidermal growth factor receptor 2
JNK Jun N-terminal kinase
LAT Large tumor suppressor kinase
MAPK Mitogen activated protein kinase
Maps Microtubule-associated proteins
MDM2 Mouse double minute 2 homolog
MOAP-1 Modulator of apoptosis-1
MST Mammalian sterile 20 like
NF1 Neurofibromatosis 1
PAR-4 Prostate apoptosis response protein 4
RA domain Ras-association domain
RASAL Ras GTPase activating like protein
RASSF Ras-association domain family
SARAH domain Salvador/RASSF/Hippo domain
SIRT1 NAD-dependent protein deacetylase sirtuin-1
TAZ Tafazzin
TRAIL TNF-related apoptosis-inducing ligand
VHL Von hippel-lindau
XPA Xeroderma pigmentosum group A-complementing protein
YAP Yes-associated protein

References

[1] Downward J. Targeting RAS signalling pathways in cancer therapy. Nat Rev Cancer 2003;3(1):11–22.

[2] Donninger H, Vos MD, Clark GJ. The RASSF1A tumor suppressor. J Cell Sci 2007;120 (Pt 18):3163–72.

[3] Overmeyer JH, Maltese WA. Death pathways triggered by activated Ras in cancer cells. Front Biosci (Landmark Ed) 2011;16:1693–713.

[4] Franza Jr BR, Maruyama K, Garrels JI, Ruley HE. In vitro establishment is not a sufficient prerequisite for transformation by activated ras oncogenes. Cell 1986;44(3):409–18.

[5] Land H, Parada LF, Weinberg RA. Tumorigenic conversion of primary embryo fibroblasts requires at least two cooperating oncogenes. Nature 1983;304(5927):596–602.

[6] Guo C, Tommasi S, Liu L, Yee JK, Dammann R, Pfeifer GP. RASSF1A is part of a complex similar to the Drosophila Hippo/Salvador/Lats tumor-suppressor network. Curr Biol 2007;17(8):700–5.

[7] Khokhlatchev A, Rabizadeh S, Xavier R, Nedwidek M, Chen T, Zhang XF, et al. Identification of a novel Ras-regulated proapoptotic pathway. Curr Biol 2002;12(4):253–65.

[8] Donninger H, Calvisi DF, Barnoud T, Clark J, Schmidt ML, Vos MD, et al. NORE1A is a Ras senescence effector that controls the apoptotic/senescent balance of p53 via HIPK2. J Cell Biol 2015;208(6):777–89.

[9] Vos MD, Ellis CA, Bell A, Birrer MJ, Clark GJ. Ras uses the novel tumor suppressor RASSF1 as an effector to mediate apoptosis. J Biol Chem 2000;275(46):35669–72.

[10] Dammann R, Li C, Yoon JH, Chin PL, Bates S, Pfeifer GP. Epigenetic inactivation of a RAS association domain family protein from the lung tumour suppressor locus 3p21.3. Nat Genet 2000;25(3):315–9.

[11] Matallanas D, Romano D, Yee K, Meissl K, Kucerova L, Piazzolla D, et al. RASSF1A elicits apoptosis through an MST2 pathway directing proapoptotic transcription by the p73 tumor suppressor protein. Mol Cell 2007;27(6):962–75.

[12] Ortiz-Vega S, Khokhlatchev A, Nedwidek M, Zhang XF, Dammann R, Pfeifer GP, et al. The putative tumor suppressor RASSF1A homodimerizes and heterodimerizes with the Ras-GTP binding protein Nore1. Oncogene 2002;21(9):1381–90.

[13] Calvisi DF, Ladu S, Gorden A, Farina M, Conner EA, Conner EA, et al. Ubiquitous activation of Ras and Jak/Stat pathways in human HCC. Gastroenterology 2006;130(4):1117–28.

[14] Shivakumar L, Minna J, Sakamaki T, Pestell R, White MA. The RASSF1A tumor suppressor blocks cell cycle progression and inhibits cyclin D1 accumulation. Mol Cell Biol 2002;22(12):4309–18.

[15] Dallol A, Agathanggelou A, Fenton SL, Ahmed-Choudhury J, Hesson L, Vos MD, et al. RASSF1A interacts with microtubule-associated proteins and modulates microtubule dynamics. Cancer Res 2004;64(12):4112–6.

[16] Vos MD, Martinez A, Elam C, Dallol A, Taylor BJ, Latif F, et al. A role for the RASSF1A tumor suppressor in the regulation of tubulin polymerization and genomic stability. Cancer Res 2004;64(12):4244–50.

[17] Arnette C, Efimova N, Zhu X, Clark GJ, Kaverina I. Microtubule segment stabilization by RASSF1A is required for proper microtubule dynamics and golgi integrity. Mol Biol Cell 2014;25(6):800–10.

[18] Liu L, Vo A, Liu G, McKeehan WL. Distinct structural domains within C19ORF5 support association with stabilized microtubules and mitochondrial aggregation and genome destruction. Cancer Res 2005;65(10):4191–201.

[19] Baksh S, Tommasi S, Fenton S, Yu VC, Martins LM, Pfeifer GP, et al. The tumor suppressor RASSF1A and MAP-1 link death receptor signaling to Bax conformational change and cell death. Mol Cell 2005;18(6):637–50.

[20] Vos MD, Dallol A, Eckfeld K, Allen NP, Donninger H, Hesson LB, et al. The RASSF1A tumor suppressor activates Bax via MOAP-1. J Biol Chem 2006;281(8):4557–63.

[21] Praskova M, Khoklatchev A, Ortiz-Vega S, Avruch J. Regulation of the MST1 kinase by autophosphorylation, by the growth inhibitory proteins, RASSF1 and NORE1, and by Ras. Biochem J 2004;381(Pt 2):453–62.

[22] Pan D. The hippo signaling pathway in development and cancer. Dev Cell 2010;19(4):491–505.

[23] Harvey KF, Zhang X, Thomas DM. The Hippo pathway and human cancer. Nat Rev Cancer 2013;13(4):246–57.

[24] Del Re DP, Matsuda T, Zhai P, Gao S, Clark GJ, Can Der Weyden L, et al. Proapoptotic Rassf1A/Mst1 signaling in cardiac fibroblasts is protective against pressure overload in mice. J Clin Invest 2010;120(10):3555–67.

[25] Romano D, Matallanas D, Weitsman G, Preisinger C, Ng T, Kolch W. Proapoptotic kinase MST2 coordinates signaling crosstalk between RASSF1A, Raf-1, and Akt. Cancer Res 2010;70(3):1195–203.

[26] Song MS, Song SJ, Kim SY, Oh HJ, Lim DS. The tumour suppressor RASSF1A promotes MDM2 self-ubiquitination by disrupting the MDM2-DAXX-HAUSP complex. EMBO J 2008;27(13):1863–74.

[27] Tommasi S, Besaratinia A, Wilczynski SP, Pfeifer GP. Loss of RASSF1A enhances p53-mediated tumor predisposition and accelerates progression to aneuploidy. Oncogene 2011;30(6):690–700.

[28] Pefani DE, Latusek R, Pires I, Grawenda AM, Yee KS, Hamilton G, et al. RASSF1A-LATS1 signalling stabilizes replication forks by restricting CDK2-mediated phosphorylation of BRCA2. Nat Cell Biol 2014;16(10):962–71. 1–8.

[29] Hamilton G, Yee KS, Scarce S, O'Neill E. ATM regulates a RASSF1A-dependent DNA damage response. Curr Biol 2009;19(23):2020–5.

[30] Yee KS, Grochola L, Hamilton G, Grawenda A, Bond EE, Taubert H, et al. A RASSF1A polymorphism restricts p53/p73 activation and associates with poor survival and accelerated age of onset of soft tissue sarcoma. Cancer Res 2012;72(9):2206–17.

[31] Donninger H, Clark J, Rinaldo F, Nelson N, Barnoud T, Schmidt ML, et al. The RASSF1A tumor suppressor regulates XPA-mediated DNA repair. Mol Cell Biol 2015;35(1):277–87.

[32] Gao B, Xie XJ, Huang C, Shames DS, Chen TT, Lewis CM, et al. RASSF1A polymorphism A133S is associated with early onset breast cancer in BRCA1/2 mutation carriers. Cancer Res 2008;68(1):22–5.

[33] Schagdarsurengin U, Seidel C, Ulbrich EJ, Kolbl H, Dittmer J, Dammann R. A polymorphism at codon 133 of the tumor suppressor RASSF1A is associated with tumorous alteration of the breast. Int J Oncol 2005;27(1):185–91.

[34] Donninger H, Barnoud T, Nelson N, Kassler S, Clark J, Cummins TD, et al. RASSF1A and the rs2073498 cancer associated SNP. Front Oncol 2011;1:54.

[35] Oceandy D, Pickard A, Prehar S, Zi M, Mohamed TM, Stanley PJ, et al. Tumor suppressor Ras-association domain family 1 isoform A is a novel regulator of cardiac hypertrophy. Circulation 2009;120(7):607–16.

[36] Jung HY, Jung JS, Whang YM, Kim YH. RASSF1A suppresses cell migration through inactivation of HDAC6 and increase of acetylated alpha-tubulin. Cancer Res Treat 2013;45(2):134–44.

[37] Kouzarides T. Acetylation: a regulatory modification to rival phosphorylation? EMBO J 2000;19(6):1176–9.

[38] Tommasi S, Dammann R, Zhang Z, Wang Y, Liu L, Tsark WM, et al. Tumor susceptibility of RASSF1A knockout mice. Cancer Res 2005;65(1):92–8.

[39] Ram RR, Mendiratta S, Bodemann BO, Torres MJ, Eskiocak U, White MA. RASSF1A inactivation unleashes a tumor suppressor/oncogene cascade with context-dependent consequences on cell cycle progression. Mol Cell Biol 2014;34(12):2350–8.

[40] Kitagawa D, Kajiho H, Negishi T, Ura S, Wantanabe T, Wada Y, et al. Release of RASSF1C from the nucleus by Daxx degradation links DNA damage and SAPK/JNK activation. EMBO J 2006;25(14):3286–97.

[41] Lorenzato A, Martino C, Dani N, Oligschlager Y, Ferrero AM, Biglia N, et al. The cellular apoptosis susceptibility CAS/CSE1L gene protects ovarian cancer cells from death by suppressing RASSF1C. FASEB J 2012;26(6):2446–56.

[42] Estrabaud E, Lassot I, Blot G, Le Rouzic E, Tanchou V, Quemeneur E, et al. RASSF1C, an isoform of the tumor suppressor RASSF1A, promotes the accumulation of beta-catenin by interacting with betaTrCP. Cancer Res 2007;67(3):1054–61.

[43] Reeves ME, Baldwin SW, Baldwin ML, Chen ST, Moretz JM, Aragon RJ, et al. Ras-association domain family 1C protein promotes breast cancer cell migration and attenuates apoptosis. BMC Cancer 2010;10:562.

[44] Reeves ME, Firek M, Chen ST, Amaar Y. The RASSF1 gene and the opposing effects of the RASSF1A and RASSF1C isoforms on cell proliferation and apoptosis. Mol Biol Int 2013;2013:145096.

[45] Vavvas D, Li X, Avruch J, Zhang XF. Identification of Nore1 as a potential Ras effector. J Biol Chem 1998;273(10):5439–42.

[46] Wohlgemuth S, Kiel C, Kramer A, Serrano L, Wittinghofer F, Herrmann C. Recognizing and defining true Ras binding domains I: biochemical analysis. J Mol Biol 2005;348(3):741–58.

[47] Kuznetsov S, Khokhlatchev AV. The growth and tumor suppressors NORE1A and RASSF1A are targets for calpain-mediated proteolysis. PLoS One 2008;3(12):e3997.

[48] Suryaraja R, Anitha M, Anbarasu K, Kumari G, Mahalingam S. The E3 ubiquitin ligase itch regulates tumor suppressor protein RASSF5/NORE1 stability in an acetylation-dependent manner. Cell Death Dis 2013;4:e565.

[49] Chen J, Liu WO, Vos MD, Clark GJ, Takahashi M, Schoumans J, et al. The t(1;3) breakpoint-spanning genes LSAMP and NORE1 are involved in clear cell renal cell carcinomas. Cancer Cell 2003;4(5):405–13.

[50] Aoyama Y, Avruch J, Zhang XF. Nore1 inhibits tumor cell growth independent of Ras or the MST1/2 kinases. Oncogene 2004;23(19):3426–33.

[51] Calvisi DF, Donninger H, Vos MD, Birrer MJ, Gordon L, Leaner V, et al. NORE1A tumor suppressor candidate modulates p21CIP1 via p53. Cancer Res 2009;69(11):4629–37.

[52] Puca R, Nardinocchi L, Givol D, D'Orazi G. Regulation of p53 activity by HIPK2: molecular mechanisms and therapeutical implications in human cancer cells. Oncogene 2010;29(31):4378–87.

[53] Lee D, Park SJ, Sung KS, Park J, Lee SB, Park SY, et al. MDM2 associates with Ras effector NORE1 to induce the degradation of oncoprotein HIPK1. EMBO Rep 2012;13(2):163–9.

[54] Schmidt ML, Donninger H, Clark GJ. Ras regulates SCF(beta-TrCP) protein activity and specificity via its effector protein NORE1A. J Biol Chem 2014;289(45):31102–10.

[55] Orford K, Crockett C, Jensen JP, Weissman AM, Byers SW. Serine phosphorylation-regulated ubiquitination and degradation of beta-catenin. J Biol Chem 1997;272(40):24735–8.

[56] Park J, Kang SI, Lee SY, Zhang XF, Kim MS, Beers LF, et al. Tumor suppressor ras association domain family 5 (RASSF5/NORE1) mediates death receptor ligand-induced apoptosis. J Biol Chem 2010;285(45):35029–38.

[57] Katagiri K, Ohnishi N, Kabashima K, Iyoda Y, Takeda N, Shinkai Y, et al. Crucial functions of the Rap1 effector molecule RAPL in lymphocyte and dendritic cell trafficking. Nat Immunol 2004;5(10):1045–51.

[58] Zhou D, Medoff BD, Chen L, Li L, Zhang XF, Praskova M, et al. The Nore1B/Mst1 complex restrains antigen receptor-induced proliferation of naive T cells. Proc Natl Acad Sci USA 2008;105(51):20321–6.

[59] Ishiguro K, Avruch J, Landry A, Qin S, Ando T, Goto H, et al. Nore1B regulates TCR signaling via Ras and Carma1. Cell Signal 2006;18(10):1647–54.

[60] Miertzschke M, Stanley P, Bunney TD, Rodrigues-Lima F, Hogg N, Katan M. Characterization of interactions of adapter protein RAPL/Nore1B with RAP GTPases and their role in T cell migration. J Biol Chem 2007;282(42):30629–42.

[61] Macheiner D, Heller G, Kappel S, Bichler C, Stattner S, Ziegler B, et al. NORE1B, a candidate tumor suppressor, is epigenetically silenced in human hepatocellular carcinoma. J Hepatol 2006;45(1):81–9.

[62] Macheiner D, Gauglhofer C, Rodgarkia-Dara C, Grusch M, Brachner A, Bichler C, et al. NORE1B is a putative tumor suppressor in hepatocarcinogenesis and may act via RASSF1A. Cancer Res 2009;69(1):235–42.

[63] Vos MD, Ellis CA, Elam C, Ulku AS, Taylor BJ, Clark GJ. RASSF2 is a novel K-Ras-specific effector and potential tumor suppressor. J Biol Chem 2003;278(30):28045–51.

[64] Akino K, Toyota M, Suzuki H, Mita H, Sasaki Y, Ohe-Toyota M, et al. The Ras effector RASSF2 is a novel tumor-suppressor gene in human colorectal cancer. Gastroenterology 2005;129(1):156–69.

[65] Clark J, Freeman J, Donninger H. Loss of RASSF2 enhances tumorigencity of lung cancer cells and confers resistance to chemotherapy. Mol Biol Int 2012;2012:705948.

[66] Yi Y, Nandana S, Case T, Nelson C, Radmilovic T, Matusik RJ, et al. Candidate metastasis suppressor genes uncovered by array comparative genomic hybridization in a mouse allograft model of prostate cancer. Mol Cytogenet 2009;2:18.

[67] Donninger H, Hesson L, Vos M, Beebe K, Gordon L, Sidransky D, et al. The Ras effector RASSF2 controls the PAR-4 tumor suppressor. Mol Cell Biol 2010;30(11):2608–20.

[68] Payne SR, Serth J, Schostak M, Kamradt J, Strauss A, Thelen P, et al. DNA methylation biomarkers of prostate cancer: confirmation of candidates and evidence urine is the most sensitive body fluid for non-invasive detection. Prostate 2009;69(12):1257–69.

[69] Perez-Janices N, Blanco-Lugin I, Tunon MT, Barba-Ramos E, Ibanez B, Zazpe-Cenoz I, et al. EPB41L3, TSP-1 and RASSF2 as new clinically relevant prognostic biomarkers in diffuse gliomas. Oncotarget 2015;6(1):368–80.

[70] Song H, Oh S, Oh HJ, Lim DS. Role of the tumor suppressor RASSF2 in regulation of MST1 kinase activity. Biochem Biophys Res Commun 2010;391(1):969–73.

[71] Cooper WN, Hesson LB, Matallanas D, Dallol A, von Kriegsheim A, Ward R, et al. RASSF2 associates with and stabilizes the proapoptotic kinase MST2. Oncogene 2009;28(33):2988–98.

[72] Imai T, Toyota M, Suzuki H, Akino K, Ogi K, Sogabe Y, et al. Epigenetic inactivation of RASSF2 in oral squamous cell carcinoma. Cancer Sci 2008;99(5):958–66.

[73] Burghel GJ, Lin WY, Whitehouse H, Brock I, Hammond D, Bury J, et al. Identification of candidate driver genes in common focal chromosomal aberrations of microsatellite stable colorectal cancer. PLoS One 2013;8(12):e83859.

[74] Safavi S, Hansson M, Karlsson K, Bilogalv A, Johansson B, Paulsson K. Novel gene targets detected by genomic profiling in a consecutive series of 126 adults with acute lymphoblastic leukemia. Haematologica 2015;100(1):55–61.

[75] Guo H, Liu H, Wei J, Li Y, Yu H, Huan X, et al. Functional single nucleotide polymorphisms of the RASSF3 gene and susceptibility to squamous cell carcinoma of the head and neck. Eur J Cancer 2014;50(3):582–92.

[76] Peng H, Liu H, Zhao S, Wu J, Fan J, Liao J. Silencing of RASSF3 by DNA hypermethylation is associated with tumorigenesis in somatotroph adenomas. PLoS One 2013;8(3):e59024.

[77] Fukatsu A, Ishiguro F, Tanaka I, Kudo T, Nakagawa K, Shinjo K, et al. RASSF3 downregulation increases malignant phenotypes of non-small cell lung cancer. Lung Cancer 2014;83(1):23–9.

[78] Kudo T, Ikeda M, Nishikawa M, Yang Z, Ohno K, Nakagawa K, et al. The RASSF3 candidate tumor suppressor induces apoptosis and G1-S cell-cycle arrest via p53. Cancer Res 2012;72(11):2901–11.

[79] Eckfeld K, Hesson L, Vos MD, Bieche I, Latif F, Clark GJ. RASSF4/AD037 is a potential ras effector/tumor suppressor of the RASSF family. Cancer Res 2004;64(23):8688–93.

[80] Chow LS, Lo KW, Kwong J, Wong AY, Huang DP. Aberrant methylation of RASSF4/AD037 in nasopharyngeal carcinoma. Oncol Rep 2004;12(4):781–7.

[81] Michifuri Y, Hirohashi Y, Torigoe T, Miyazaki A, Fujino J, Tamura Y, et al. Small proline-rich protein-1B is overexpressed in human oral squamous cell cancer stem-like cells and is related to their growth through activation of MAP kinase signal. Biochem Biophys Res Commun 2013;439(1):96–102.

[82] Crose LE, Galindo KA, Kephart JG, Chen C, Fitamant J, Bardeesy N, et al. Alveolar rhabdomyosarcoma-associated PAX3-FOXO1 promotes tumorigenesis via Hippo pathway suppression. J Clin Invest 2014;124(1):285–96.

[83] Allen NP, Donninger H, Vos MD, Eckfeld K, Hesson L, Gordon L, et al. RASSF6 is a novel member of the RASSF family of tumor suppressors. Oncogene 2007;26(42):6203–11.

[84] Ikeda M, Hirabayashi S, Fujiwara N, Mori H, Kawata A, Iida J, et al. Ras-association domain family protein 6 induces apoptosis via both caspase-dependent and caspase-independent pathways. Exp Cell Res 2007;313(7):1484–95.

[85] Djos A, Martinsson T, Kogner P, Caren H. The RASSF gene family members RASSF5, RASSF6 and RASSF7 show frequent DNA methylation in neuroblastoma. Mol Cancer 2012;11:40.

[86] Hesson LB, Dunwell TL, Cooper WN, Catchpoole D, Brini AT, Chiaramonte R, et al. The novel RASSF6 and RASSF10 candidate tumour suppressor genes are frequently epigenetically inactivated in childhood leukaemias. Mol Cancer 2009;8:42.

[87] Mezzanotte JJ, Hill VC, Schmidt ML, Shinawi T, Tommasi S, Krex D, et al. RASSF6 exhibits promoter hypermethylation in metastatic melanoma and inhibits invasion in melanoma cells. Epigenetics 2014;9(11):1496–503.

[88] Ikeda M, Kawata A, Nishikawa M, Tateishi Y, Yamaguchi M, Nakagawa K, et al. Hippo pathway-dependent and -independent roles of RASSF6. Sci Signal 2009;2(90):ra59.

[89] Iwasa H, Kudo T, Maimaiti S, Ikeda M, Maruyama J, Nakagawa K, et al. The RASSF6 tumor suppressor protein regulates apoptosis and the cell cycle via MDM2 protein and p53 protein. J Biol Chem 2013;288(42):30320–9.

[90] O'Neill E, Rushworth L, Baccarini M, Kolch W. Role of the kinase MST2 in suppression of apoptosis by the proto-oncogene product Raf-1. Science 2004;306(5705):2267–70.

[91] Thaler S, Hahnel PS, Schad A, Dammann R, Schuler M. RASSF1A mediates p21Cip1/Waf1-dependent cell cycle arrest and senescence through modulation of the Raf-MEK-ERK pathway and inhibition of Akt. Cancer Res 2009;69(5):1748–57.

[92] Hesson LB, Cooper WN, Latif F. The role of RASSF1A methylation in cancer. Dis Markers 2007;23(1–2):73–87.

[93] Pfeifer GP, Dammann R. Methylation of the tumor suppressor gene RASSF1A in human tumors. Biochemistry (Mosc) 2005;70(5):576–83.

[94] Agathanggelou A, Cooper WN, Latif F. Role of the Ras-association domain family 1 tumor suppressor gene in human cancers. Cancer Res 2005;65(9):3497–508.

[95] Kashuba VI, Pavlova TV, Grigorieva EV, Kutsenko A, Yenamandra SP, Li J, et al. High mutability of the tumor suppressor genes RASSF1 and RBSP3 (CTDSPL) in cancer. PLoS One 2009;4(5):e5231.

[96] Basu TN, Gutmann DH, Fletcher JA, Glover TW, Collins FS, Downward J. Aberrant regulation of ras proteins in malignant tumour cells from type 1 neurofibromatosis patients. Nature 1992;356(6371):713–5.

[97] Walker SA, Kupzig S, Bouyoucef D, Davies LC, Tsuboi T, Bivona TG, et al. Identification of a Ras GTPase-activating protein regulated by receptor-mediated Ca^{2+} oscillations. EMBO J 2004;23(8):1749–60.

[98] Wang Z, Tseng CP, Pong RC, Chen H, McConnell JD, Navone N, et al. The mechanism of growth-inhibitory effect of DOC-2/DAB2 in prostate cancer. Characterization of a novel GTPase-activating protein associated with N-terminal domain of DOC-2/DAB2. J Biol Chem 2002;277(15):12622–31.

[99] Kim DH, Kim JS, Park JH, Lee SK, Ji YI, Kwon YM, et al. Relationship of Ras association domain family 1 methylation and K-ras mutation in primary non-small cell lung cancer. Cancer Res 2003;63(19):6206–11.

[100] Cox AD, Fesik SW, Kimmelman AC, Luo J, Der CJ. Drugging the undruggable RAS: mission possible? Nat Rev Drug Discov 2014;13(11):828–51.

[101] Lyko F, Brown R. DNA methyltransferase inhibitors and the development of epigenetic cancer therapies. J Natl Cancer Inst 2005;97(20):1498–506.

[102] Hatzimichael E, Crook T. Cancer epigenetics: new therapies and new challenges. J Drug Deliv 2013;2013:529312.

[103] Palakurthy RK, Wajapeyee N, Santra MK, Gazin C, Lin L, Gobeil S, et al. Epigenetic silencing of the RASSF1A tumor suppressor gene through HOXB3-mediated induction of DNMT3B expression. Mol Cell 2009;36(2):219–30.

[104] Kwon O, Jeong SJ, Kim SO, He L, Lee HG, Jang KL, et al. Modulation of E-cadherin expression by K-Ras; involvement of DNA methyltransferase-3b. Carcinogenesis 2010;31(7):1194–201.

[105] Kuck D, Caulfield T, Lyko F, Medina-Franco JL. Nanaomycin A selectively inhibits DNMT3B and reactivates silenced tumor suppressor genes in human cancer cells. Mol Cancer Ther 2010;9(11):3015–23.

[106] Sztiller-Sikorska M, Koprowska K, Majchrzak K, Harman M, Czyz M. Natural compounds' activity against cancer stem-like or fast-cycling melanoma cells. PLoS One 2014;9(3):e90783.

CHAPTER 2

Ras and the Hippo Pathway in Cancer

E. O'Neill
University of Oxford, Oxford, United Kingdom

CONTENTS

INTRODUCTION

The Ras signal transduction pathway is the most prevalent oncogenic pathway across all tumor types, with activating mutations or amplifications directly responsible for constitutive growth and survival signals [1]. Mutations in all RAS isoforms, K-Ras, H-Ras, and N-Ras, have been found to increase the GTP-bound "active" form of the protein. In addition, membrane tyrosine kinase receptors that normally signal to wild-type Ras can be hyper-activated either by direct mutation, gene amplification, or elevated levels of the receptor ligand, eg, epidermal growth factor (EGF), stimulating supra-physiological pathway activation [2]. Mitogens and chemicals, such as lysophosphatidic acid (LPA) and sphingosine-1-phosphate, also promote growth and can signal to RAS directly via G-protein–coupled receptors (GPCRs) that activate heterotrimeric G-proteins, in particular $G\alpha_i$ [3], and can also trigger RAS activity via G-protein–independent activation of β-arrestin and SRC [4]. Mutation or activation of GPCRs in cancer can also result in hyper-activation of cellular proliferation and survival pathways that include Ras.

Enhanced wild-type Ras can occur via alteration of endogenous control mechanisms, eg, via up-regulation of GTPase exchange factors (Ras GEFs) (Ras GAPs) that load Ras with GTP or down-regulation of GTPase activating proteins that switch Ras off by promotion hydrolysis of GTP to GDP such as occurs with deletion of neurofibromatosis protein 1 (NF1) [5]. Once activated, Ras signals to a range of downstream effectors by direct interaction with RAS-binding domains, classically RAF kinases, A-Raf, B-Raf, or C-Raf (Raf-1), Pi3K, Tiam1, and RGS12/14 (themselves GAPs for heterotrimeric G-proteins) or through a structurally distinct Ras association (RA) domain similar to that found in RalGDS, Afadin-6, Rin1, Phospholipase C epsilon, and the RASSF proteins (RASSF1–10) [6,7]. Many of these downstream effectors are oncogenes or tumor suppressor genes in their own right, being directly mutated or silenced in cancers independent of Ras, eg, B-Raf, Pi3K, RalGDS, and RASSF1. The main downstream pathways activated

Conquering RAS. http://dx.doi.org/10.1016/B978-0-12-803505-4.00002-3

in these cases are mitogen-activated protein kinase (MAPK), Pi3K-AKT, and, of interest to this chapter, RASSF1-MST (Hippo), especially given the emerging cross talk of epidermal growth factor receptor (EGFR)-RAS signaling with the Hippo tumor suppressor pathway. The MAPK cascade involves a number of related parallel pathways that activate a set of effector kinases extracellular regulated kinase (ERK)1/2, JNK1/2, $p38_{\alpha/\beta/\gamma/\delta}$, or ERK5 [8]. Although developmental, inflammatory or stress cues can activate these to different extents, Ras has been linked to activation of JNK, p38 but most notably ERK1 and ERK2. RAS activation of RAF kinases promotes activation of MEK1/2, which in turn activates ERK1/2 kinases to directly stimulate proliferation by promoting entry into the cell cycle. The process is assisted by a number of scaffolding proteins that bring constituent units in the cascade in close proximity within a cell, resulting in enhanced control of the overall signal output of the pathway [9]. Scaffolds may in this way be necessary for increasing the overall level of signal propagation, may direct the localization of a signal within a cell, or may temper the amplitude of signal output so that an appropriate cellular response can be achieved. Not surprisingly, many scaffolding proteins are also disrupted in cancer cells and lead to an inappropriate level of Ras signaling, such as the RAS-RAF scaffold CNK, the RAF-MEK scaffolds KSR and IQGAP [10,11], MP1, which supports and localizes MEK-ERK [12], and the RASSF1 scaffold, which limits MEK activation [13] (Fig. 2.1).

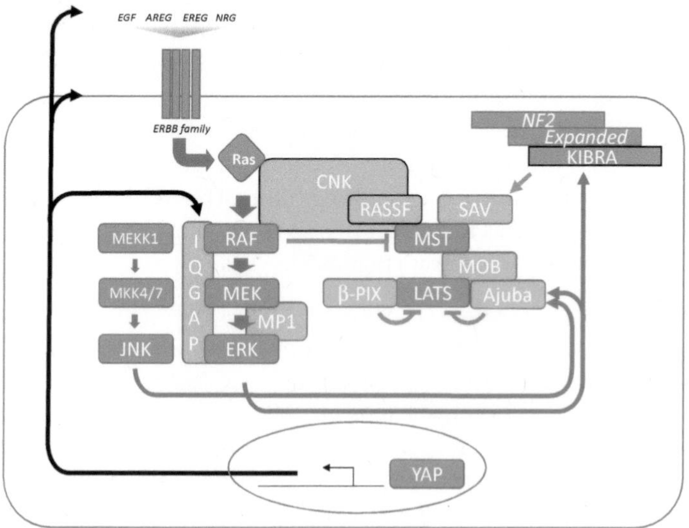

FIGURE 2.1

Scaffolding proteins RASSF, SAV, and MOB are central to the activation of the Hippo pathway. Conversely, additional scaffolds (b-PIX, Ajuba) block the ability of LATS to phosphorylate YAP (*red lines*) and are positively responsive to JNK or ERK phosphorylation (*green arrows*). ERK can also phosphorylate KIBRA, an upstream activator of SAV, to promote growth. Ras-MAPK scaffolds amplify MAPK signaling and therefore contribute to repression of Hippo signaling. YAP transcriptional activation promotes IQGAP, EGFR, and growth factors presenting a positive feedback loop.

RAS is without doubt a potent oncogene, and the obvious reciprocity with additional mutations within the pathway demonstrates a clear requirement for pathway hyperactivity in developing neoplasms, the best example being N-Ras and B-Raf mutation in melanoma [14]. Despite this, however, K-Ras, B-Raf, or Pi3K mutations can exist in normal human or mouse tissue without developing tumor, implying that RAS pathway mutations require signal amplification and/or tumor suppressor pathways limit manifestation until lost [15–18]. This latter fact is attributed to p53 that responds to RAS activation by promoting a permanent cell cycle arrest known as senescence. P53 is commonly mutated in cancer; however, while germline mutations or loss of *TP53* clearly precipitate tumorigenesis independently of Ras, somatic alterations appear be a late event in tumor progression, often correlating more with metastatic spread [19–21]. This implies that additional *TP53*-related family members, such as p63 and p73, or additional tumor suppressor pathways must play a role in early tumorigenesis and an understanding of these events will undoubtedly help in finding the Achilles' heel of the Ras pathway in cancer.

THE HIPPO TUMOR SUPPRESSOR PATHWAY

The Hippo tumor suppressor pathway has emerged as a key fundamental developmentally conserved pathway that limits cell growth to regulate tissue growth and organ size [22]. The pathway is activated in response to signals that require cell cycle arrest such as cell polarity, mechanotransduction, and DNA damage and is inhibited by growth factors or mitogens related to EGF and LPA [23,24]. The core pathway consists of a central kinase related to the *Saccharomyces cerevisiae* Ste20 MAPK, *Drosophila melanogaster* Hpo, or mammalian MST kinases, which activates the effector AGC kinase *dm*Wts or LATS in mammals (collectively referred to as MST and LATS, respectively) [25]. The main output of the pathway currently is LATS-mediated phosphorylation of the transcription co-factor yes-associated protein (YAP) (*dm*Yki), which functions in a diverse, sometimes opposing, set of biological responses via association with different transcription factors [26]. Although a number of inputs from apico-basal and planar cell polarity regulators (eg, Crumbs, Fat, Par6/aPkc) and FERM domain proteins transduce upstream signals, the only direct binding modulators of Hippo kinase activity are the scaffolds Sav1 (h*WW45*, *dm*Salvador), Rassf1-6 (*dm*Rassf), and Mob1 (*dm*Mats), which support either dimerization or association with LATS [27–29]. Similarly to the role in development, advances have pointed to a role for the pathway in limiting the emergence and expansion of cancer cells in both model organisms and human tumors [23]. Genetic alterations in pathway components lead to increased tumor risk [30–32]; however, somatic mutations in the central players are notably rare and often below the general mutational background implying that constitutive activation of the pathway may be selected against in tumors. This may in part be due to the diverse roles that YAP can play in addition to proliferation, including differentiation [33,34], apoptosis, and cell cycle arrest [35], that may limit progression of establishing tumor lesions in humans. There is increasing evidence for inactivation of Hippo pathway signaling in sporadic cancer by amplification of YAP [26,36], deletion (eg, MST2) [37], or epigenetic gene silencing (eg, MST, LATS, and RASSF1) [38–40]. Interestingly, evidence now exists for the methylation-associated gene silencing of the RAS and Hippo pathway scaffold, RASSF1A, correlating with increased YAP activation in human breast, bladder, and glioma cohorts [41]. Loss of this Hippo activator has been shown to correlate with adverse clinical outcomes reminiscent of constitutive Hippo pathway activity in mouse models [42–44]. Moreover, Ras signaling has become an appealing candidate for Hippo inactivation as increasing reports outline how EGFR-RAS-RAF-MEK-ERK–mediated cross talk controls Hippo signaling [45–48]. Therefore, as Hippo pathway loss may have important implications for personalized approaches to targeting RAS-RAF mutations in cancer [48], we aim now to discuss the pathway cross talk in detail.

MST AND LATS AS THE CENTRAL MAMMALIAN KINASES UNDER THE INFLUENCE OF HIPPO SIGNALING

As mentioned earlier, the mammalian pathway consists of two central kinases: (1) MST; the Mammalian STe20-like kinases 1 and 2 (MST1 and MST2, respectively), homologs of the *Saccharomyces cerevisiae* Ste20 kinase, and (2) LATS; the warts related Large Tumor Suppressors LATS1 and LATS2, homologs of the yeast Dbf2 mitotic exit kinase [49]. Activation of MST1/2 occurs via homo-dimerization or hetero-dimerization and trans-phosphorylation of the T183/T180 residue in the ATP-binding pocket, essential for coordinating ATP during hydrolysis [25]. Dimerization of MST kinases occurs via a C-terminal leucine [22] zipper motif referred to as the SARAH domain as it is found in SAv, RAssf1-6, and Hippo kinases [49]. The kinase domains of MST1 and MST2 are located in the N-termini and are constitutively active when released by caspase cleavage during apoptosis, implying that kinase activity is restricted by sequences in the C-terminal. Located between the kinase domain and the SARAH domains, there is an auto-inhibitory a-helix that restricts kinase activity of the monomer [25].

SAV or RASSF proteins can bind monomeric MST and either scaffold kinase activity to LATS, eg, SAV [22], or remove this innate repressive activity via dimerization of RASSF-MST heterodimers, essentially a quaternary RASSF1-MST:MST-RASSF1 complex, that allows auto-transphosphorylation of T183/T180 [28,50]. Importantly, RASSF1, 2, and 5 (NORE1/RAPL) appear to hold MST kinases in an inhibitory state until activation induces dimerization, whereas RASSF3, 4, and 6 are more similar to dRASSF in that they appear to repress activity [51,52], potentially through a failure to dimerize or possibly because the activation signal for these molecules has not yet been determined. For RASSF1 and RASSF5, the association of RAS with the RA domains may influence dimerization capacity and/or MST activation [53], in keeping with observation that wild-type RAS activation may play an inhibitory role on Hippo pathway activation, whereas mutant RAS may trigger activity [37]. Importantly, RAF, itself a RAS effector, binds directly to the MST1/2 SARAH domain to prevent dimerization and association with RASSF1, and limits Hippo pathway activity in mice and *Drosophila* [50,54]. Much like tissue overgrowth in cancer, RAF promotes hypertrophy in the heart by stimulating Yap activity, which suggests RAF-mediated inhibition of MST kinases is a conserved regulatory nexus, and in keeping with the role for Sav1 and RASSF1A in restricting tissue growth during development [55]and in response to injury, respectively [43]. Moreover, the regulation between RAF and MST appears to be reciprocal as activation of MST kinase promotes association with RAF and restricts the ability of RAF to activate MEK [13] (Fig. 2.2). This is suggestive of a mechanism by which inhibitor bound RAF or MEK molecules, known to increase B-Raf/Raf-1 dimerization [56], may cause collateral inhibition of MST by supporting RAF-MST inactivation and could thus facilitate the YAP transcription observed under these conditions.

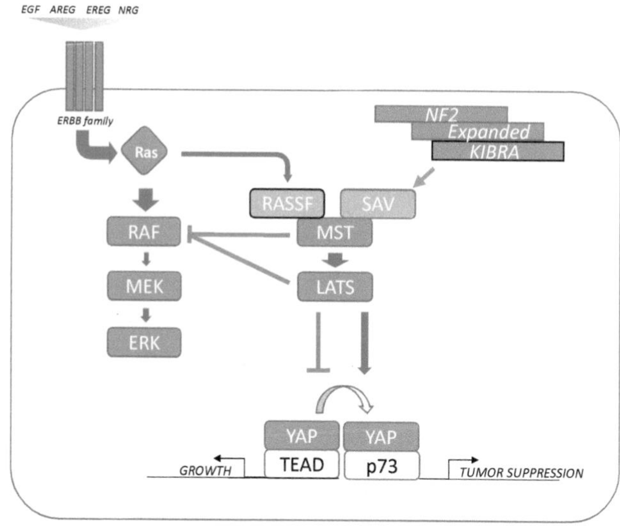

FIGURE 2.2

Hippo pathway centers on activation of MST. This is promoted via signals from the NF2/Expanded/Kibra complex to SAV or via the RAS interacting scaffold RASSF. Activation of MST leads to LATS kinase activity and phosphorylation of YAP. This prevents association with oncogenic TEAD growth-promoting transcription factors. Phosphorylation of YAP not only promotes increased cytoplasmic levels but also switches transcription factor association under conditions of tumor suppression and differentiation. Interestingly, increased MST activity prevents MEK activation in an RASSF-dependent manner. LATS also represses RAF activity directly by phosphorylating S259 and switching RAF off.

Once activated, MST kinase associates with and phosphorylates the Mob1 scaffold causing a conformation change that allows activation of LATS kinase activity [57]. Interestingly, RAS-GTPases are responsible for directing the localization of the activated LATS-Mob1 heterodimer to spindles where the role in cell division and mitosis also appears evolutionarily conserved from early eukaryotes [58]. The MST activation of LATS results in phosphorylation of the transcriptional co-activator and YAP (homolog of *Drosophila* Yorkie, *Yki*) [23]. Although MST and LATS have dramatic effects in vivo (although redundancy between individual isoforms is observed) [59–63], they are only partially epistatic with YAP [64], suggesting the possibility of further substrates for LATS that have not yet been described. The scaffold protein Ajuba is a conserved scaffold in the Hippo pathway that interacts with LATS kinases and Sav to limit signaling at the centrosome and reduce inhibitory phosphorylation of YAP [65]. EGFR-RAS signaling leads to activation of both ERK and JNK MAPKs, which have both been demonstrated to phosphorylate Ajuba proteins, enhancing inhibitory binding to LATS and ultimately allowing the activation of YAP transcription [66,67]. Mammalian MST kinases were originally implicated in the activation of JNK via MKK7 and MEKK1 [68,69] and consistent with these results, MST1 could not activate JNK in cells deleted for MKK7 [70]. MST kinases were also demonstrated to respond to disruption of actin stress fibers and stimulate JNK activity, indicating a physiologically relevant pathway downstream of the Ras-like GTPase Rho [71]. Moreover, JNK-mediated modulation of Ajuba, a Lim domain protein, enhances f-actin, which would sequester MST1/2, reducing Hippo pathway activation and allowing YAP nuclear transit [72]. In line with these more recent studies showing that JNK prevents YAP proliferative functions, YAP has been previously shown to be directly phosphorylated by JNK and lead to p73-mediated apoptosis [73,74].

The neurofibromatosis 2 tumor suppressor, NF2 (dmMerlin)/Kibra/Expanded complex lies upstream of the core Hippo cassette and is reported to activate Sav1 and MST [75]. ERK-mediated phosphorylation of Kibra has also been reported to be a MAPK signaling event that is required for cell proliferation implying this is potentially another means of the RAS pathway–mediated repression of Hippo activity [76]. Interestingly, Kibra itself may play a role in collagen-induced ERK signaling, again indicating another level of reciprocal regulatory activity between the pathways [77]. Alternatively, ARFGEF7 (or beta-Pix) is another Hippo pathway scaffold, which, although does not require GEF activity, promotes LATS associated with YAP and Taz to restrict Hippo pathway output and YAP/Taz-mediated transcription [78]. That this latter scaffold is linked more to RAS-like GTPases Rac1 and Cdc42 demonstrates the potential of pathway cross talk under different conditions where RAS, RHO, RAC, or CD42 may be active, such as in cancer.

As discussed earlier, up-regulation of KRAS activity is now being widely observed to have multiple points, eg, RAF-MST, ERK-Kibra or ERK-Ajuba, and JNK-Ajuba, through which it can exert a repressive effect on Yap and Taz transcriptional activity (Fig. 2.1). Just as with Hippo pathway scaffolding proteins, KRAS pathway scaffolds, such as IQGAP, which enhance RAS-RAF-MEK-ERK signaling (Fig. 2.1), have also been described to suppress the Hippo pathway possibly by driving some or all of these same inhibitory events [79]. Upstream receptor-driven activation of RAS similarly leads to repression of the Hippo pathway output. Notably in mammals, activators of the ERBB family of EGF-related receptors, EGF, Epiregulin (EREG), Amphiregulin (AREG), heparin-binding EGF (hbEGF), and Neuregulin-1 and -2 (NRG1, NRG2) have all been documented to lead to inactivation of the Hippo pathway [42,45,80] (Fig. 2.3). As mentioned earlier, LPA can also inhibit the pathway via activation of heterotrimeric G-proteins [81]; although this occurred via Rho/Rock mediated in direct repression of LATS [82,83], the effects of LPA on RAS [3], PKA on Ras/Raf-1 [84], and Rho/Rock on MST [71] could be confounding factors not taken into account (Fig. 2.3). Surprisingly, all EGF-regulated ligands were ruled out and LPA activation of Gα was found to be the sole serum-mediated signal repressing the Hippo pathway, independent of RAS [81], but given the

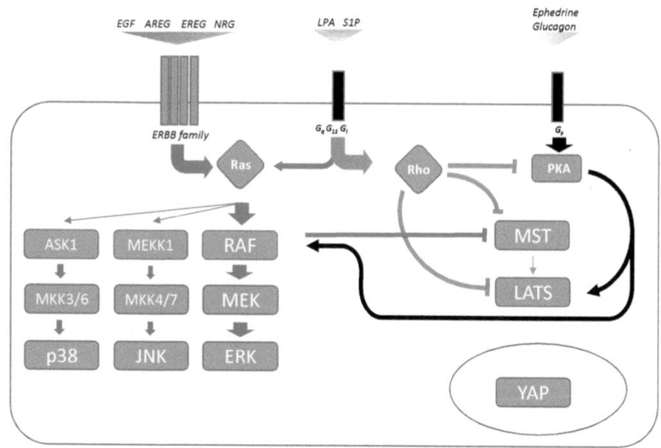

FIGURE 2.3

Activation of the ERBB family of receptors by growth factors activates Ras MAPK signaling (*green arrows*). The Ras effector Raf inactivates MST1 and MST2 by preventing dimerization. Ras can also be activated by G-protein–coupled receptors receptor. GPCR activation of the heterotrimeric G-proteins G_a, G_{11}, and G_q have a strong stimulation of Rho activity, which inhibits MST and LATS via modulating actin dynamics. In addition, Rho mediates inhibition of cAMP-dependent protein kinase A (PKA). In contrast, GPCR activation of G_S results in PKA activation and repression of LATS, thereby facilitating YAP nuclear localization. Notably, PKA mediates a wide range of effects through multiple substrates including RAF, which has been linked to MST repression, and as with LATS would facilitate YAP activity.

accumulating data it may be interesting to determine whether these are contextual differences where EGF ligands or LPA plays independent roles. Irrespective of the specific mechanisms, there are clearly multiple levels at which the inhibition can occur in Ras-activated cancers, with direct RAF-MST inhibition and LATS repression via ERK-mediated phosphorylation of Ajuba being the clearest examples to date.

YAP SIGNALING AND ITS REGULATION BY HIPPO

YAP is a modular protein that has functions in both the cytoplasm, primarily at cell junctions, and in the nucleus where it serves as a co-factor for a wide variety of transcription factors involved in proliferation (eg, TEA domain family, TEAD1-4), stem cells (eg, the core pluripotency factor Oct4 and TGFb effector SMADs), differentiation (eg, RUNX2), and cell cycle control/apoptosis (eg, the p53 family members, p63 and p73) [85]. The pro-proliferative activity is mostly observed in growing tissue during development and in replicating cancer cells where YAP drives TEA-domain containing transcription factors (TEAD1-4) transcriptional activity. Yap and its close homolog WWTR1 (Taz) are also responsible for nuclear localization and transcriptional activity of SMADs in response to TGFβ signaling [85]. YAP and Taz also influence differentiation via modulating RUNX and PPARγ, which define differentiation into osteoclasts or adipocytes, respectively [85]. In addition, there is a well-described role for YAP in p73-mediated tumor suppression [24], suggesting that YAP transcription activity must be coordinated to switch between proliferation, cell cycle arrest, and differentiation. Indeed, direct phosphorylation by LATS was first shown to be a pro-p73 association signal [35], whereas the same signal disrupted association with TEAD [86], ultimately resulting in the same outcome—a switch from proliferation to tumor suppression. Under "Hippo on" conditions, YAP both binds to p73 and accumulates in the cytoplasm via association with the catenin complex at cell junctions, integrating into the polarity monitoring pathway. Free cytoplasmic YAP is degraded by the βTrCP-SCF ubiquitin ligase complex; however, in the presence of RAS, the ligase targeting components are down-regulated, resulting in increased YAP stability [87]. Thus active RAS can drive increased levels of YAP, which can achieve nuclear localization, independently of "Hippo on" conditions, and the increased levels of YAP are required for RAS-driven tumorigenesis.

Invariably, the loss of Hippo pathway activity disables the tumor suppressor activity of p73 and supports tumor growth through TEAD transcription. Surprisingly, however, despite numerous mouse models to implicate TEAD activation in tumorigenesis, as mentioned earlier, there are few cases of somatic alteration leading to constitutive YAP-TEAD activity in common cancers [36].

Having appreciated the increasing evidence that the RAS-RAF-MEK-ERK cascade intersects with the Hippo tumor suppressor pathway to facilitate tumorigenesis. We can consider further reports of crossover between RAS signaling and the Hippo pathway via feedback loops, whereby inactivation of the Hippo signaling leads to increased YAP/TEAD-mediated transcription of target genes that increase RAS activation. As mentioned earlier, YAP has been shown to be an important activator of EGFR by promoting transcription of the ligands EGF and AREG, thereby increasingly the MAPK signal presenting a feedback that further represses Hippo activity. This is believed to additionally contribute to therapeutic resistance by increasing survival signals such as AKT in response to stimulation of the EGFR family [37], as earlier, and increasing levels of insulin growth factor (IGF) [88], the receptor for which, IGFR, is a more classical activator of Pi3K-AKT signaling. Viral oncogenesis accelerates these effects as high levels of YAP are maintained by the viral oncoprotein E6, through suppression of the proteasome [89]. Importantly, inactivation of the tumor suppressor NF2 (*dm*Merlin) promotes neurofibromatosis, a nerve sheath tumor that is also linked to RAS activation via inactivation of the RAS-GAP, NF1. Loss of NF2 is associated with reduced Hippo pathway activation and Yap, in this context at least, is observed to be a transcription factor for H-Ras, K-Ras, and N-Ras amplifying mRNA and therein increasing K-RAS pro-oncogenic activity [90]. Taken together, activation of Yap directly promotes increased transcription of receptor ligands for the EGFR family, EGFR and RAS, elevating mRNA levels in a TEAD-dependent manner and resulting in a positive feedback loop to stabilize repression of Hippo and amplify MAPK activity (Fig. 2.1). Notably, up-regulation of EGFR itself is a major route of Ras pathway activation in tumors, either by gene amplification or by increased transcription, and is of great clinical interest given the association with resistance to anti-EGFR targeted therapies [91].

CONCLUSIONS AND FUTURE DIRECTIONS

In light of all the evidence, it is clear that we need an understanding of exactly how the Hippo pathway is inactivated in cancer. Although we are seeing clear repression of the Hippo at the signaling level by RAS-MAPK, the fact that the Hippo pathway increases RAS-MAPK activity leaves us with a "which comes first the chicken or the egg?" scenario. If Hippo loss can activate EGFR-RAS-MAPK and conversely EGFR-RAS-MAPK represses Hippo signaling, we cannot be completely confident what the primary lesion is in tumors, especially where activation of the RAS pathway is present in the absence of mutations. Genetic polymorphisms that weaken the Hippo pathway do exist and carriers are prone to not only neurofibromatosis but also sporadic cancers such as soft-tissue sarcomas and hepatocellular, lung, or breast cancers [30–32]. Moreover, somatic alterations have been observed in cancers such as uveal melanoma [92,93], but these are rare and do not explain the causative factor behind increasing clinical observations of Hippo loss in sporadic malignancies. Clues are arising from development, where in endocrine pancreatic β-cells, RASSF1 responds to KRAS and overrides MAPK activation, arresting cells in line with Hippo pathway activation [94,95]. This is similar to colorectal cancer (CRC) where K-Ras activation promotes RASSF1-mediated Hippo tumor suppression, but where loss of RASSF1 or MST kinases is required for progression [37]. Intriguingly, RASSF1 expression is lost in CRC, not by mutation, although genomic deletions of ch3p21 have been known for some time to be associated with lung, breast, and ovarian cancers [96], but by epigenetic silencing of the RASSF1A promoter [40]. This not only occurs frequently but also a clinically confirmed prognostic factor for tumor onset and for overall survival, suggesting that YAP activation in pancreas, CRC, and gastrointestinal tumors may be a result of RASSF1-mediated Hippo pathway loss [45–47]. In line with this, RASSF1 methylation has also been linked to pre-cancerous dysplasia in the colon [43]. This also suggests that inactivation of RASSF1 is likely to be linked to reduced Hippo pathway and YAP activation in further sporadic malignancies where RASSF1 is known to also be a prognostic factor [41]. Taken together, this suggests that, epigenetic inactivation, also observed for MST and LATS, may be a key mark of Hippo pathway loss and an important consideration for susceptible patient cohorts where EGFR-RAS-RAF-MEK genetic lesions are being targeted.

References

[1] Pylayeva-Gupta Y, Grabocka E, Bar-Sagi D. RAS oncogenes: weaving a tumorigenic web. Nat Rev Cancer 2011;11:761–74.

[2] Bronte G, Silvestris N, Castiglia M, Galvano A, Passiglia F, Sortino G, et al. New findings on primary and acquired resistance to anti-EGFR therapy in metastatic colorectal cancer: do all roads lead to RAS? Oncotarget 2015;6:24780–96.

[3] Kranenburg O, Moolenaar WH. Ras-MAP kinase signaling by lysophosphatidic acid and other G protein-coupled receptor agonists. Oncogene 2001;20:1540–6.

[4] Magalhaes AC, Dunn H, Ferguson SS. Regulation of GPCR activity, trafficking and localization by GPCR-interacting proteins. Br J Pharmacol 2012;165:1717–36.

[5] Ratner N, Miller SJ. A RASopathy gene commonly mutated in cancer: the neurofibromatosis type 1 tumour suppressor. Nat Rev Cancer 2015;15:290–301.

[6] Raaijmakers JH, Bos JL. Specificity in Ras and Rap signaling. J Biol Chem 2009;284:10995–9.

[7] Chan JJ, Katan M. PLCvarepsilon and the RASSF family in tumour suppression and other functions. Adv Biol Regul 2013;53:258–79.

[8] Raman M, Chen W, Cobb MH. Differential regulation and properties of MAPKs. Oncogene 2007;26:3100–12.

[9] Brown MD, Sacks DB. Protein scaffolds in MAP kinase signalling. Cell Signal 2009;21:462–9.

[10] Claperon A, Therrien M. KSR and CNK: two scaffolds regulating RAS-mediated RAF activation. Oncogene 2007;26:3143–58.

[11] White CD, Brown MD, Sacks DB. IQGAPs in cancer: a family of scaffold proteins underlying tumorigenesis. FEBS Lett 2009;583:1817–24.

[12] Teis D, Taub N, Kurzbauer R, Hilber D, de Araujo ME, Erlacher M, et al. p14-MP1-MEK1 signaling regulates endosomal traffic and cellular proliferation during tissue homeostasis. J Cell Biol 2006;175:861–8.

[13] Romano D, Nguyen LK, Matallanas D, Halasz M, Doherty C, Kholodenko BN, et al. Protein interaction switches coordinate Raf-1 and MST2/Hippo signalling. Nat Cell Biol 2014;16:673–84.

[14] Dhomen N, Marais R. New insight into BRAF mutations in cancer. Curr Opin Genet Dev 2007;17:31–9.

[15] Hafner C, Toll A, Fernandez-Casado A, Earl J, Marques M, Acquadro F, et al. Multiple oncogenic mutations and clonal relationship in spatially distinct benign human epidermal tumors. Proc Natl Acad Sci USA 2010;107:20780–5.

[16] Michaloglou C, Vredeveld LC, Soengas MS, Denoyelle C, Kuilman T, van der Horst CM, et al. BRAFE600-associated senescence-like cell cycle arrest of human naevi. Nature 2005;436:720–4.

[17] Junttila MR, Karnezis AN, Garcia D, Madriles F, Kortlever RM, Rostker F, et al. Selective activation of p53-mediated tumour suppression in high-grade tumours. Nature 2010;468:567–71.

[18] Feldser DM, Kostova KK, Winslow MM, Taylor SE, Cashman C, Whittaker CA, et al. Stage-specific sensitivity to p53 restoration during lung cancer progression. Nature 2010;468:572–5.

[19] Tan EH, Morton JP, Timpson P, Tucci P, Melino G, Flores ER, et al. Functions of TAp63 and p53 in restraining the development of metastatic cancer. Oncogene 2014;33:3325–33.

[20] Morton JP, Timpson P, Karim SA, Ridgway RA, Athineos D, Doyle B, et al. Mutant p53 drives metastasis and overcomes growth arrest/senescence in pancreatic cancer. Proc Natl Acad Sci USA 2010;107:246–51.

[21] Oren M, Rotter V. Mutant p53 gain-of-function in cancer. Cold Spring Harb Perspect Biol 2010;2:a001107.

[22] Irvine KD, Harvey KF. Control of organ growth by patterning and hippo signaling in *Drosophila*. Cold Spring Harb Perspect Biol 2015;7.

[23] Yu FX, Zhao B, Guan KL. Hippo pathway in organ size control, tissue homeostasis, and cancer. Cell 2015;163:811–28.

[24] O'Neill E, Hamilton G. Hippo pathway and apoptosis. In: Oren M, Aylon Y, editors. The hippo signaling pathway and cancer. Springer; 2013. p. 117–45.

[25] Rawat SJ, Chernoff J. Regulation of mammalian Ste20 (Mst) kinases. Trends Biochem Sci 2015;40:149–56.

[26] Harvey K, Tapon N. The Salvador-Warts-Hippo pathway – an emerging tumour-suppressor network. Nat Rev Cancer 2007;7:182–91.

[27] Couzens AL, Knight JD, Kean MJ, Teo G, Weiss A, Dunham WH, et al. Protein interaction network of the mammalian Hippo pathway reveals mechanisms of kinase-phosphatase interactions. Sci Signal 2013;6:rs15.

[28] Hamilton G, Yee KS, Scrace S, O'Neill E. ATM regulates a RASSF1A-dependent DNA damage response. Curr Biol 2009;19:2020–5.

[29] Avruch J, Xavier R, Bardeesy N, Zhang XF, Praskova M, Zhou D, et al. Rassf family of tumor suppressor polypeptides. J Biol Chem 2009;284:11001–5.

[30] Bayram S. Association between RASSF1A Ala133Ser polymorphism and cancer susceptibility: a meta-analysis involving 8,892 subjects. Asian Pac J Cancer Prev 2014;15:3691–8.

[31] Wu C, Xu B, Yuan P, Miao X, Liu Y, Guan Y, et al. Genome-wide interrogation identifies YAP1 variants associated with survival of small-cell lung cancer patients. Cancer Res 2010;70:9721–9.

[32] Yuan H, Liu H, Liu Z, Zhu D, Amos CI, Fang S, et al. Genetic variants in Hippo pathway genes YAP1, TEAD1 and TEAD4 are associated with melanoma-specific survival. Int J Cancer 2015;137:638–45.

[33] Hong JH, Hwang ES, McManus MT, Amsterdam A, Tian Y, Kalmukova R, et al. TAZ, a transcriptional modulator of mesenchymal stem cell differentiation. Science 2005;309:1074–8.

[34] van der Weyden L, Papaspyropoulos A, Poulogiannis G, Rust AG, Rashid M, Adams DJ, et al. Loss of RASSF1A synergizes with deregulated RUNX2 signaling in tumorigenesis. Cancer Res 2012;72:3817–27.

[35] Matallanas D, Romano D, Yee K, Meissl K, Kucerova L, Piazzolla D, et al. RASSF1A elicits apoptosis through an MST2 pathway directing proapoptotic transcription by the p73 tumor suppressor protein. Mol Cell 2007;27:962–75.

[36] Harvey KF, Zhang X, Thomas DM. The Hippo pathway and human cancer. Nat Rev Cancer 2013;13:246–57.

[37] Matallanas D, Romano D, Al-Mulla F, O'Neill E, Al-Ali W, Crespo P, et al. Mutant K-Ras activation of the proapoptotic MST2 pathway is antagonized by wild-type K-Ras. Mol Cell 2011;44:893–906.

[38] Takahashi Y, Miyoshi Y, Takahata C, Irahara N, Taguchi T, Tamaki Y, et al. Down-regulation of LATS1 and LATS2 mRNA expression by promoter hypermethylation and its association with biologically aggressive phenotype in human breast cancers. Clin Cancer Res 2005;11:1380–5.

[39] Seidel C, Schagdarsurengin U, Blumke K, Wurl P, Pfeifer GP, Hauptmann S, et al. Frequent hypermethylation of MST1 and MST2 in soft tissue sarcoma. Mol Carcinog 2007;46:865–71.

[40] Grawenda AM, O'Neill E. Clinical utility of RASSF1A methylation in human malignancies. Br J Cancer 2015;113:372–81.

[41] Vlahov N, Scrace S, Soto MS, Grawenda AM, Bradley L, Pankova D, et al. Alternate RASSF1 transcripts control SRC activity, E-cadherin contacts, and YAP-mediated invasion. Curr Biol 2015;25:3019–34.

[42] Gregorieff A, Liu Y, Inanlou MR, Khomchuk Y, Wrana JL. Yap-dependent reprogramming of Lgr5(+) stem cells drives intestinal regeneration and cancer. Nature 2015;526:715–8.

[43] Gordon M, El-Kalla M, Zhao Y, Fiteih Y, Law J, Volodko N, et al. The tumor suppressor gene, RASSF1A, is essential for protection against inflammation -induced injury. PLoS One 2013;8:e75483.

[44] Halder G, Johnson RL. Hippo signaling: growth control and beyond. Development 2011;138:9–22.

[45] Zhang W, Nandakumar N, Shi Y, Manzano M, Smith A, Graham G, et al. Downstream of mutant KRAS, the transcription regulator YAP is essential for neoplastic progression to pancreatic ductal adenocarcinoma. Sci Signal 2014;7:ra42.

[46] Shao DD, Xue W, Krall EB, Bhutkar A, Piccioni F, Wang X, et al. KRAS and YAP1 converge to regulate EMT and tumor survival. Cell 2014;158:171–84.

[47] Kapoor A, Yao W, Ying H, Hua S, Liewen A, Wang Q, et al. Yap1 activation enables bypass of oncogenic Kras addiction in pancreatic cancer. Cell 2014;158:185–97.

[48] Lin L, Sabnis AJ, Chan E, Olivas V, Cade L, Pazarentzos E, et al. The Hippo effector YAP promotes resistance to RAF- and MEK-targeted cancer therapies. Nat Genet 2015;47:250–6.

[49] Avruch J, Zhou D, Fitamant J, Bardeesy N, Mou F, Barrufet LR. Protein kinases of the Hippo pathway: regulation and substrates. Semin Cell Dev Biol 2012;23:770–84.

[50] O'Neill E, Rushworth L, Baccarini M, Kolch W. Role of the kinase MST2 in suppression of apoptosis by the proto-oncogene product Raf-1. Science 2004;306:2267–70.

[51] Volodko N, Gordon M, Salla M, Ghazaleh HA, Baksh S. RASSF tumor suppressor gene family: biological functions and regulation. FEBS Lett 2014;588:2671–84.

[52] Polesello C, Huelsmann S, Brown NH, Tapon N. The *Drosophila* RASSF homolog antagonizes the hippo pathway. Curr Biol 2006;16:2459–65.

[53] Avruch J, Praskova M, Ortiz-Vega S, Liu M, Zhang XF. Nore1 and RASSF1 regulation of cell proliferation and of the MST1/2 kinases. Methods Enzymol 2006;407:290–310.

[54] Yu L, Daniels JP, Wu H, Wolf MJ. Cardiac hypertrophy induced by active Raf depends on Yorkie-mediated transcription. Sci Signal 2015;8:ra13.

[55] Tapon N, Harvey KF, Bell DW, Wahrer DC, Schiripo TA, Haber D, et al. salvador Promotes both cell cycle exit and apoptosis in *Drosophila* and is mutated in human cancer cell lines. Cell 2002;110:467–78.

[56] Rushworth LK, Hindley AD, O'Neill E, Kolch W. Regulation and role of Raf-1/B-Raf heterodimerization. Mol Cell Biol 2006;26:2262–72.

[57] Hergovich A, Hemmings BA. Mammalian NDR/LATS protein kinases in hippo tumor suppressor signaling. Biofactors 2009;35:338–45.

[58] Muller-Taubenberger A, Kastner PM, Schleicher M, Bolourani P, Weeks G. Regulation of a LATS-homolog by Ras GTPases is important for the control of cell division. BMC Cell Biol 2014;15:25.

[59] Anguera MC, Liu M, Avruch J, Lee JT. Characterization of two Mst1-deficient mouse models. Dev Dyn 2008;237:3424–34.

[60] Oh S, Lee D, Kim T, Kim TS, Oh HJ, Hwang CY, et al. Crucial role for Mst1 and Mst2 kinases in early embryonic development of the mouse. Mol Cell Biol 2009;29:6309–20.

[61] Du X, Dong Y, Shi H, Li J, Kong S, Shi D, et al. Mst1 and mst2 are essential regulators of trophoblast differentiation and placenta morphogenesis. PLoS One 2014;9:e90701.

[62] St John MA, Tao W, Fei X, Fukumoto R, Carcangiu ML, Brownstein DG, et al. Mice deficient of Lats1 develop soft-tissue sarcomas, ovarian tumours and pituitary dysfunction. Nat Genet 1999;21:182–6.

[63] McPherson JP, Tamblyn L, Elia A, Migon E, Shehabeldin A, Matysiak-Zablocki E, et al. Lats2/Kpm is required for embryonic development, proliferation control and genomic integrity. EMBO J 2004;23:3677–88.

[64] Morin-Kensicki EM, Boone BN, Howell M, Stonebraker JR, Teed J, Alb JG, et al. Defects in yolk sac vasculogenesis, chorioallantoic fusion, and embryonic axis elongation in mice with targeted disruption of Yap65. Mol Cell Biol 2006;26:77–87.

[65] Das Thakur M, Feng Y, Jagannathan R, Seppa MJ, Skeath JB, Longmore GD. Ajuba LIM proteins are negative regulators of the Hippo signaling pathway. Curr Biol 2010;20:657–62.

[66] Sun G, Irvine KD. Ajuba family proteins link JNK to Hippo signaling. Sci Signal 2013;6:ra81.

[67] Reddy BV, Irvine KD. Regulation of Hippo signaling by EGFR-MAPK signaling through Ajuba family proteins. Dev Cell 2013;24:459–71.

[68] Graves JD, Gotoh Y, Draves KE, Ambrose D, Han DK, Wright M, et al. Caspase-mediated activation and induction of apoptosis by the mammalian Ste20-like kinase Mst1. EMBO J 1998;17:2224–34.

[69] Graves JD, Draves KE, Gotoh Y, Krebs EG, Clark EA. Both phosphorylation and caspase-mediated cleavage contribute to regulation of the Ste20-like protein kinase Mst1 during CD95/Fas-induced apoptosis. J Biol Chem 2001;276:14909–15.

[70] Ura S, Nishina H, Gotoh Y, Katada T. Activation of the c-Jun N-terminal kinase pathway by MST1 is essential and sufficient for the induction of chromatin condensation during apoptosis. Mol Cell Biol 2007;27:5514–22.

[71] Densham RM, O'Neill E, Munro J, Konig I, Anderson K, Kolch W, et al. MST kinases monitor actin cytoskeletal integrity and signal via c-Jun N-terminal kinase stress-activated kinase to regulate p21Waf1/Cip1 stability. Mol Cell Biol 2009;29:6380–90.

[72] Enomoto M, Kizawa D, Ohsawa S, Igaki T. JNK signaling is converted from anti- to pro-tumor pathway by Ras-mediated switch of Warts activity. Dev Biol 2015;403:162–71.

[73] Danovi SA, Rossi M, Gudmundsdottir K, Yuan M, Melino G, Basu S. Yes-associated protein (YAP) is a critical mediator of c-Jun-dependent apoptosis. Cell Death Differ 2008;15:217–9.

[74] Tomlinson V, Gudmundsdottir K, Luong P, Leung KY, Knebel A, Basu S. JNK phosphorylates Yes-associated protein (YAP) to regulate apoptosis. Cell Death Dis 2010;1:e29.

[75] Genevet A, Tapon N. The Hippo pathway and apico-basal cell polarity. Biochem J 2011;436:213–24.

[76] Yang S, Ji M, Zhang L, Chen Y, Wennmann DO, Kremerskothen J, et al. Phosphorylation of KIBRA by the extracellular signal-regulated kinase (ERK)-ribosomal S6 kinase (RSK) cascade modulates cell proliferation and migration. Cell Signal 2014;26:343–51.

[77] Hilton HN, Stanford PM, Harris J, Oakes SR, Kaplan W, Daly RJ, et al. KIBRA interacts with discoidin domain receptor 1 to modulate collagen-induced signalling. Biochim Biophys Acta 2008;1783:383–93.

[78] Heidary Arash E, Song KM, Song S, Shiban A, Attisano L. Arhgef7 promotes activation of the Hippo pathway core kinase Lats. EMBO J 2014;33:2997–3011.

[79] Anakk S, Bhosale M, Schmidt VA, Johnson RL, Finegold MJ, Moore DD. Bile acids activate YAP to promote liver carcinogenesis. Cell Rep 2013;5:1060–9.

[80] Zhang J, Ji JY, Yu M, Overholtzer M, Smolen GA, Wang R, et al. YAP-dependent induction of amphiregulin identifies a non-cell-autonomous component of the Hippo pathway. Nat Cell Biol 2009;11:1444–50.

[81] Yu FX, Zhao B, Panupinthu N, Jewell JL, Lian I, Wang LH, et al. Regulation of the Hippo-YAP pathway by G-protein-coupled receptor signaling. Cell 2012;150:780–91.

[82] Kim M, Kim M, Lee S, Kuninaka S, Saya H, Lee H, et al. cAMP/PKA signalling reinforces the LATS-YAP pathway to fully suppress YAP in response to actin cytoskeletal changes. EMBO J 2013;32:1543–55.

[83] Yu FX, Zhang Y, Park HW, Jewell JL, Chen Q, Deng Y, et al. Protein kinase A activates the Hippo pathway to modulate cell proliferation and differentiation. Genes Dev 2013;27:1223–32.

[84] Dumaz N, Marais R. Integrating signals between cAMP and the RAS/RAF/MEK/ERK signalling pathways. Based on the anniversary prize of the Gesellschaft fur Biochemie und Molekularbiologie Lecture delivered on 5 July 2003 at the Special FEBS Meeting in Brussels. FEBS J 2005;272:3491–504.

[85] Mauviel A, Nallet-Staub F, Varelas X. Integrating developmental signals: a Hippo in the (path) way. Oncogene 2012;31:1743–56.

[86] Dong J, Feldmann G, Huang J, Wu S, Zhang N, Comerford SA, et al. Elucidation of a universal size-control mechanism in *Drosophila* and mammals. Cell 2007;130:1120–33.

[87] Hong X, Nguyen HT, Chen Q, Zhang R, Hagman Z, Voorhoeve PM, et al. Opposing activities of the Ras and Hippo pathways converge on regulation of YAP protein turnover. EMBO J 2014;33:2447–57.

[88] Xin M, Kim Y, Sutherland LB, Qi X, McAnally J, Schwartz RJ, et al. Regulation of insulin-like growth factor signaling by Yap governs cardiomyocyte proliferation and embryonic heart size. Sci Signal 2011;4:ra70.

[89] He C, Mao D, Hua G, Lv X, Chen X, Angeletti PC, et al. The Hippo/YAP pathway interacts with EGFR signaling and HPV oncoproteins to regulate cervical cancer progression. EMBO Mol Med 2015;7:1426–49.

[90] Garcia-Rendueles ME, Ricarte-Filho JC, Untch BR, Landa I, Knauf JA, Voza F, et al. NF2 loss promotes oncogenic RAS-induced thyroid cancers via YAP-dependent transactivation of RAS proteins and sensitizes them to MEK inhibition. Cancer Discov 2015;5:1178–93.

[91] Harris TJ, McCormick F. The molecular pathology of cancer. Nat Rev Clin Oncol 2010;7:251–65.

[92] Feng X, Degese MS, Iglesias-Bartolome R, Vaque JP, Molinolo AA, Rodrigues M, et al. Hippo-independent activation of YAP by the GNAQ uveal melanoma oncogene through a trio-regulated rho GTPase signaling circuitry. Cancer Cell 2014;25:831–45.

[93] Yu FX, Luo J, Mo JS, Liu G, Kim YC, Meng Z, et al. Mutant Gq/11 promote uveal melanoma tumorigenesis by activating YAP. Cancer Cell 2014;25:822–30.

[94] Garcia-Ocana A, Stewart AF. "RAS"ling beta cells to proliferate for diabetes: why do we need MEN? J Clin Invest 2014;124:3698–700.

[95] Chamberlain CE, Scheel DW, McGlynn K, Kim H, Miyatsuka T, Wang J, et al. Menin determines K-RAS proliferative outputs in endocrine cells. J Clin Invest 2014;124:4093–101.

[96] Kok K, Naylor SL, Buys CH. Deletions of the short arm of chromosome 3 in solid tumors and the search for suppressor genes. Adv Cancer Res 1997;71:27–92.

CHAPTER 3

The Many Roles of Ral GTPases in Ras-Driven Cancer

D.F. Kashatus
The University of Virginia School of Medicine, Charlottesville, VA, United States

INTRODUCTION

Ras mutations are found in up to a third of human malignancies and finding a way to inhibit Ras signaling has been one of the most intensive quests over the past several decades [1]. The inability, thus far, to directly target Ras has driven a massive effort to characterize and understand the myriad effector pathways that become engaged following activation of Ras, in the hopes that targeting these pathways may be an effective strategy to combat Ras-driven tumors. Because of these efforts, we now have a wealth of research on the importance of mitogen-activated protein kinase (MAPK) [2] signaling and phosphoinositide 3-kinase (PI3K) [3] signaling and a deep toolbox of compounds that allow us to disrupt these critical signaling nodes. Despite this success, directly targeting these pathways in tumors harboring Ras mutations has led to only marginal success, underscoring the need to better understand the complete array of physiological changes elicited by activation of Ras and how these changes collaborate to drive the tumorigenic phenotype.

Among the effectors engaged by GTP-bound Ras molecules is a family of guanine nucleotide exchange factors that promote the activity of two small Ras-related GTPases, RalA and RalB. First identified in 1986 [4], these proteins garnered a lot of attention when it was shown that inhibition of RalA could block transformation downstream of Ras in immortalized epithelial cells and block tumor growth in a subcutaneous xenograft model [5,6]. Since that time, we have learned much about the biological activities elicited by Ral signaling and have uncovered its important role in a number of different types of tumors. Further, excitement has grown following the discovery of small molecules that target this pathway [7]. As a number of excellent reviews have been written about this pathway in the past several years [8,9], this chapter will focus on recent discoveries related to the role of Ral signaling in Ras-driven cancers and the therapeutic potential of inhibiting Ral to combat tumor growth.

CONTENTS

Conquering RAS. http://dx.doi.org/10.1016/B978-0-12-803505-4.00003-5

RAL REGULATION

RalA and RalB are two highly related small GTPases that were first identified in the late 1980s in a screen designed to find new GTPases related to Ras [4,10]. The two proteins share nearly 100% identity in both the GTPase domain and the effector binding region, but diverge significantly in their hyper-variable C-terminal domain [10]. As members of the Ras family of small GTPases, Ral proteins are activated by guanine nucleotide exchange factors (RalGEFs) that catalyze the exchange of GDP for GTP to promote binding to downstream effectors and then inactivated following hydrolysis of GTP back to GDP. This GTPase activity is enhanced several fold through the activity of GTPase-activating proteins (RalGAPs).

RalGEFs

Several guanine nucleotide exchange factors have been identified that regulate GDP/GTP exchange and promote activation of both RalA and RalB. Four of these GEFs, Ral guanine nucleotide dissociation stimulator (RalGDS) [11,12], Ral guanine nucleotide dissociation stimulator-like 1 (RGL) [13], RGL2 [14], and RGL3 [15], interact directly with the effector binding region of activated Ras and thus place Ral signaling directly downstream of Ras activation. Two others, Ral GEF with PH domain and SH3 binding motif 1 (RalGPS1) [16,17] and RalGPS2 [18], do not bind Ras proteins and thus have the potential to mediate Ras-independent Ral signaling. The protein telophase disk protein of 60 KDa (TD-60, also known as RCC2, regulator of chromosome condensation 2) was shown to have GEF activity for RalA. Activation of RalA by TD-60 was shown to contribute to the regulation of the interactions between microtubules and kinetochores in mitosis [19]. With seven identified proteins capable of activating RalA and RalB, we still have very little understanding of the circumstances under which each of these proteins promotes activation of Ral and whether and how the spatiotemporal regulation of RalA activation by the myriad RalGEFs contributes to the specificity of Ral signaling under varying conditions. Individual depletion of the RalGEFs indicates non-overlapping functions of the different GEFs and suggests that the diversity in RalGEF activity indeed contributes to specificity in signaling at least for certain biological functions. For example, knockdown of RalGDS and RalGPS2 phenocopies RalA depletion, whereas knockdown of RGL and RalGPS1 phenocopies depletion of RalB in their respective effects on cytokinesis in HeLa cells [20]. In further support of this notion, non-overlapping functions, and potential Ral-independent functions, have been identified for the different RalGEFs in pancreatic cancer cells [21], highlighting the need for additional studies to tease apart this important mechanism driving the diversity of Ral signaling.

RalGAPs

Although the existence of RalGAP activity was identified more than two decades ago [22], the molecular identity of the RalGAP complex was only discovered in 2009 with the identification and molecular characterization of Ral GTPase activating protein, alpha subunits 1 and 2 (RalGAPα1 and RalGAPα2) [23]. These catalytic subunits require heterodimerization with a common RalGAPβ subunit to become active and are currently the only known GTPase activating proteins (GAPs) for Ral proteins [24]. In part due to their recent discovery, we still know very little about the regulation of this complex and how it integrates upstream signals to modulate Ral activity. The Serine/Threonine kinase Akt phosphorylates both RalGAPα1 and RalGAPα2 downstream of insulin stimulation, leading to activation of Ral and increased Glut4 translocation and glucose uptake [24–26]. The inhibitory effect of the phosphorylation is due to induced binding between the RalGAP complex and 14-3-3 [25,26]. Furthermore, the RalGAP complex was shown to directly interact with the Ras family GTPase κB-Ras and this interaction was shown to inhibit Ral activity. Loss of κB-Ras in cells was further shown to increase tumorigenicity in a xenograft model through increases in both Ral and nuclear factor κB (NF-κB) signaling κ[27]. It is clear from these studies that modulation of GAP activity can have profound effects on Ral activity and that a better understanding of the regulation of the RalGAP complex will add considerably to our understanding of how the regulation of Ral signaling contributes to tumor growth.

Localization

RalA and RalB are both ubiquitously expressed and localized to the plasma membrane, in addition to several other membranes within the cell, including Golgi [28], endosomes [29,30], and mitochondria [31]. Understanding how Ral localization is regulated is very important as specific subcellular localization can have a profound impact on which effectors Ral encounters as well as on the consequence of effector activation. Both Ral proteins end in a C-terminal CAAX motif (C = cysteine, A = aliphatic amino acid, X = any amino acid) that is post-translationally modified to regulate localization to specific membranes within the endomembrane system [32,33]. The presence of leucine in the X position (RalA = CCIL, RalB = CCLL) dictates that both proteins are modified by transfer of a geranylgeranyl moiety on the first cysteine by geranylgeranyltransferase I (GGTase-I) and subsequent cleavage of the AAX tripeptide by Ras converting endopeptidase 1 (RCE1). The remaining lipid-modified cysteine is then methylated by isoprenylcysteine carboxymethyltransferase (ICMT) [34]. An alternative pathway also exists in which RCE1 does not remove the AAX and the cysteine occupying the first A position is palmitoylated [35]. The precise role of these two alternative modifications is not fully understood. However, recent evidence suggests that, although both RalA and RalB require RCE1 for targeting to the plasma membrane, only RalB requires ICMT for this localization, whereas RalA requires ICMT to localize to recycling endosomes. Furthermore, palmitoylation is required for RalB, but not RalA, to localize to the plasma membrane [36].

In addition to these modifications, several other post-translational modifications of the hyper-variable region have been shown to influence the subcellular localization of RalA and RalB. For example, differential serine phosphorylation in this region has been shown to mediate translocation of both Ral proteins to different membranes within the cytoplasm [31,37,38] and mono-ubiquitination has been demonstrated to promote enrichment of RalA in lipid rafts within the plasma membrane [39]. Whether these modifications directly or indirectly affect RalA and RalB activity in addition to affecting their subcellular distribution is unclear, although phosphorylation of RalA has been proposed to promote its GTP binding [40] and ubiquitination of RalB has been shown to promote its interaction with Sec5 over Exo84 [41]. Notably, several of these modifications appear to be important for tumor growth. In particular, phosphorylation of RalA on Serine 194 by Aurora A has been shown to re-localize RalA to internal membranes, including the mitochondria, and inhibition of this phosphorylation was shown to block tumor growth of pancreatic cancer cell lines in a xenograft model [31,38,40]. Furthermore, loss of RalA S183 and S194 de-phosphorylation by the tumor suppressor protein phosphatase 2A Aβ (PP2A Aβ) was shown to promote a transformed phenotype in vitro [42]. In addition, RalB is phosphorylated on Serine 198 by protein kinase C alpha (PKCα) and potentially protein kinase A (PKA), which promotes a number of RalB-dependent downstream effects, including actin cytoskeletal rearrangements and vesicular trafficking [37,43]. Importantly, inhibition of this phosphorylation was shown to inhibit both tumor growth and metastatic potential in a xenograft model of bladder cancer [37]. It is not well-understood how phosphorylation of the hyper-variable C-terminus alters the localization of RalA and RalB, although it has been proposed that neutralization of the charge on the poly-basic region by the negatively charged phosphate group is sufficient to alter specific membrane affinity, similar to what has been proposed for the proteins myristoylated alanine-rich protein kinase C substrate (MARCKS) and KRas [44,45]. It is also unclear how these changes in subcellular localization impact interaction with both upstream regulators and downstream effectors. Given how important this dynamic intra-cellular trafficking of RalA and RalB seem to be for tumor growth, understanding the impact of this post-translational regulation will prove to be an important area of research in the next several years.

RAL EFFECTORS

A number of effector proteins have been identified that bind to Ral proteins, but it has been a challenge to determine how RalA and RalB integrate upstream signaling pathways to engage the appropriate effectors and how each of these effector pathways contributes to the changes in cellular physiology elicited by activation of Ral (Table 3.1). In particular, we still lack a complete understanding of the differences between RalA and RalB in effector engagement and how those differences are regulated. Furthermore, we do not have a complete picture of which effectors preferentially engage Ral proteins under different physiological conditions and how these dynamics are regulated.

Table 3.1 Effectors of RalA and RalB

Effector	Functions
RalBP1	Cdc42 and Rac1 GAP [46–48] Endocytosis [50–54] Mitochondrial fission [31] Transport of glutathione conjugates [56–58]
Sec5	Exocytosis [64–66] TBK1 activation [41,76]
Exo84	Exocytosis [64–66] Autophagy regulation [41,75]
Filamin	Actin dynamics [85]
ZONAB	Transcription [84]
MLK3 (putative)	Regulation of JNK signaling [92]
Nucleotide-independent Effectors	***Functions***
Phospholipase D1	Vesicular trafficking [86–88]
Phospholipase C	Calcium signaling [91]

GAP, *GTPase activating protein*; JNK, *C-Jun N-Terminal kinase 1*; MLK3, *Mixed lineage kinase 3*; RalBP1, *Ral binding protein 1*; TBK1, *TANK-binding kinase 1*; ZONAB, *ZO-1-associated nucleic acid-binding protein*

RalBP1

Ral binding protein 1 (RalBP1, also known as RLIP76 and Rip1) is a large protein that was first identified as a Ral-interacting protein with GTPase activating activity toward the Rho family GTPases CDC42 and Rac1 [46–48]. Whether this GAP activity, which has been demonstrated in vitro [46], is physiologically relevant and contributes significantly to RalBP1 function remains to be determined. Indeed, RalBP1 has been shown to be a positive regulator of Rac and to be required for adhesion-induced Rac activity [49]. Notably, the RhoGAP domain, but not its function, was shown to be critical for this activity [49].

Subsequent studies have identified several additional functional activities for RalBP1, but its exact role downstream of activated Ral remains poorly understood. One fairly well-characterized activity associated with RalBP1 is its regulation of receptor-mediated endocytosis. RalBP1 interacts with partner Of RalBP1 (POB1), which links it to the endocytic machinery through an interaction with the protein Eps15 [50,51]. It can also bind directly to the endocytic adapter AP2 [52]. Inhibition of these interactions have been demonstrated to block the ligand -dependent internalization of a number of growth factor receptors, including the epidermal growth factor receptor (EGFR), insulin receptor, and the α-amino-3-hydroxy-5-methyl-4-isoxazolepropionic acid receptor (AMPAR) [52–54].

The function of RalBP1 in endocytosis appears to be mediated not through any particular enzymatic function, but by acting as a molecular scaffold to bridge the endocytic machinery with its regulators. Consistent with this, it was shown to act as a bridge between Epsin and the mitotic kinase cyclinB/Cdk1, enabling the inhibition of endocytosis during mitosis [55]. This scaffold function may extend beyond the endocytic machinery as RalBP1 can also promote mitochondrial fission during mitosis by facilitating the phosphorylation of the fission GTPase dynamin related protein 1 (Drp1) [31].

RalBP1 also functions as an ATP-dependent transporter in a manner depending on two ATP-binding sites in its N-terminus [56]. This activity has been proposed to be important for chemo-resistance in tumor cells and may also contribute to the endocytic function of RalBP1 [57,58]. Clearly, a more detailed analysis of RalBP1 function will be required to fully understand the role it plays downstream of Ral signaling. This is especially important given that RalBP1 appears to be one of the key effectors that mediates Ral tumorigenic potential. RalBP1 over-expression has been observed in colorectal cancer and breast cancer, where it is associated with poor prognosis [59,60]. It is also highly expressed in human glioblastoma and was shown to promote proliferation and survival through activation of Ras-related C3 botulinum toxin substrate 1 (Rac1) and C-Jun N-Terminal kinase 1 (JNK) signaling [61]. Mechanistically, RalBP1 has been proposed to promote the tumorigenic phenotype through effects on invasion and cell motility, as it was shown to promote invadopodia formation in a GAP-independent, but ATPase-dependent manner [62]. This activity may contribute to the enhanced metastatic activity associated with RalBP1 observed in a panel of human cancer cell lines [63].

Exocyst

The other most well-validated effectors of RalA and RalB, and the ones most associated with Ral tumorigenic function, are Sec5 and exocyst complex 84 KDa subunit (Exo84). These two proteins are components of the octameric exocyst complex, whose activity regulates the delivery of secretory components to specific membrane compartments [64–66]. Ral-dependent exocyst function has been implicated in regulated exocytosis and polarized membrane delivery in a number of different systems. The RalA-exocyst interaction is required for the delivery of E-cadherin to the basolateral membrane in MDCK cells [30], and knockdown of RalA and RalB inhibits GTP-dependent exocytosis in PC12 cells [67]. Ral was also shown to participate in platelet-dense granule secretion [68], regulated exocytosis in Weibel–Palade bodies [69], induced delivery of Glut4 to the plasma membrane [70], and insulin secretion in pancreatic β-cells [71].

Notably, the Ral–exocyst interaction also contributes to additional cellular processes. For example, the RalA–exocyst interaction is required for filopodia formation induced by tumor necrosis factor alpha (TNFα) and interleukin-1 beta (IL-1β) [65]. RalA tethers the exocyst to the cytokinetic furrow during exocytosis and RalB recruits it to the mid-body during cytokinesis [20]. As a consequence, disruption of RalA–exocyst function results in a failure to complete cytokinesis [29]. Furthermore, RalA and RalB have opposing, but exocyst-dependent, functions in the development of tight junctions [72].

Although originally proposed to promote exocyst assembly by tethering Sec5 and Exo84 [66], structural analysis suggests an overlap in the binding sites such that Ral is unable to interact with both proteins simultaneously [73,74]. These data are consistent with a number of subsequent studies that detail non-overlapping functions for both the Ral–Sec5 and Ral–Exo84 interactions. RalB promotes autophagosome formation during nutrient starvation through its interaction with Exo84 and consequent activation of Unc-51 like autophagy activating kinase 1 (ULK1) [75]. This activity is independent of its interaction with Sec5. The RalB–Sec5 interaction, on the other hand, is critical to recruit and activate the atypical IκB kinase TANK-binding kinase 1 (TBK1) [76]. Activation of this kinase, normally involved in innate immune signaling, protects cells against apoptosis in the presence of oncogenic stress. The differential regulation of these two independent roles for RalB is controlled by ubiquitination at Lysine 47, which inhibits the RalB–Exo84 interaction while promoting the RalB–Sec5 interaction [41]. The result of this switch is a decrease in Exo84-Ulk1-driven autophagy and an increase in Sec5-TBK1-dependent cell survival (Fig. 3.1). This phenotype is further enhanced through the activation of mammalian target of Rapamycin complex 1 (mTORC1) [77], which depends on

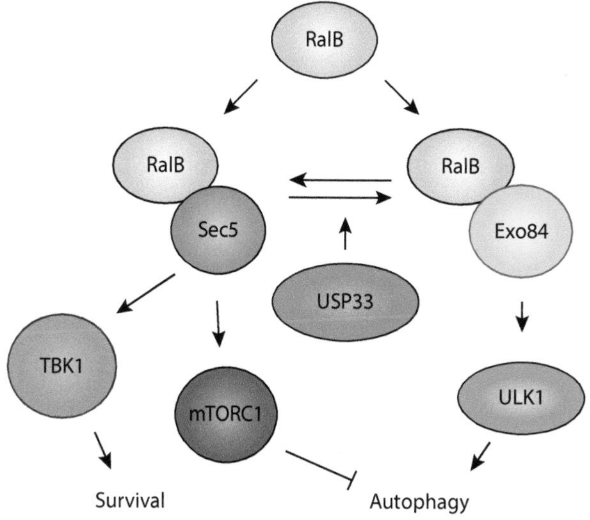

FIGURE 3.1
RalB regulates both autophagy and survival through its interactions with Sec5 and Exo84. The coordinated reciprocal regulation of autophagy and survival in response to changing nutrient conditions represents a potential mechanism through which RalB contributes to tumor growth when it is activated downstream of oncogenic Ras.

the Sec5–Ral interaction and requires the exocyst complex [78]. Activation of mTORC1 further inhibits autophagy and promotes cell survival.

The myriad biological processes that Exo84 and Sec5 control suggest a number of potential mechanisms through which their interactions with Ral proteins may contribute to oncogenesis [79]. Indeed, disrupting the RalA–exocyst interaction was shown to cause defects in the migration of prostate tumor cells [80]. Furthermore, RalB and the exocyst were shown to be required for RhoA activation and dissemination in an in vitro model of invasion in A549 lung cancer cells [81]. Despite these findings, there are few studies that directly link the Ral–exocyst interaction to tumor growth in vivo. Pharmacological inhibition of TBK1 has been shown to inhibit tumor growth in a genetically engineered model of Ras-driven lung cancer [82]. Furthermore, the RalB-TBK1 axis was shown to be critical for tumor initiation, anchorage independence, self-renewal, and erlotinib resistance in both in vitro and in vivo models of both lung and pancreatic cancer [83]. Despite these elegant studies, further in vivo studies will clearly be required to solidify the hypothesis that the interactions between Ral and Sec5 and Exo84 are broadly playing a role in the many tumor types shown to be associated with the activation of Ral.

Additional Effectors

Although much of the research focus has been on RalBP1 and the exocyst, a number of other Ral effectors have been identified that could potentially contribute to its tumorigenic function. The RalA interaction with the Y-box transcription factor, ZO-1-associated nucleic acid-binding protein (ZONAB), leads to a loss of inhibition at ZONAB-dependent promoters and activation of genes that may contribute to proliferation [84]. GTP-bound RalA also binds to the protein filamin to induce filopodia formation [85].

Several other proteins bind to Ral to either modulate its function or to propagate Ral-dependent signaling. Phospholipase D1 (PLD1) binds to Ral GTPases in a nucleotide-independent manner [86]. PLD, which catalyzes the generation of phosphatidic acid, is activated downstream of tyrosine kinases in a RalA-dependent fashion [87] and EGF-induced transformation of EGFR-expressing rat fibroblasts was shown to depend on RalA and PLD [88]. Calmodulin, a calcium-dependent messenger protein, binds to RalA and contributes to calcium-dependent, and Ras-independent, activation of Ral signaling [89,90]. In addition, RalA and calmodulin are both involved in the activation of the enzyme phospholipase C delta 1 (PLCD1) [91].

It is probable that novel effectors of RalA and RalB will continue to be found as more attention is paid to these proteins. It was demonstrated that GTP-bound RalA regulates activation of the transcription factor forkhead box O4 (FOXO4) downstream of reactive oxygen species [92]. Active RalA directly promotes the assembly of a JNK-interacting protein 1 (JIP1) scaffold complex consisting of mixed lineage kinase 3 (MLK3), MAP kinase kinase 4 (MKK4), and JNK, leading to JNK activity and its subsequent phosphorylation of FOXO4. GTP-bound, but not GDP-bound, RalA directly binds to and activates MLK3 within this complex, suggesting MLK3 may be a novel effector of RalA. The importance of these less-well-characterized Ral effectors is underscored by the fact that certain tumorigenic functions ascribed to RalA and RalB, such as radio-resistance, are independent of both RalBP1 and the exocyst [93].

RAL IN MODEL SYSTEMS

Studying Ral signaling in various model systems has provided great insight into the regulation of these pathways and has facilitated the growth in our understanding of how Ral may impact human tumorigenesis downstream of oncogenic Ras. Work in the nematode *Caenorhabditis elegans* has demonstrated that Ras-Ral signaling antagonizes Ras-Raf signaling to promote fate determination during vulva development [94]. In addition, Ral-mediated regulation of FOXO4 activation has been shown to be conserved in *C. elegans*, reinforcing the utility of worm genetics in interrogating novel mechanisms of Ral signaling [92]. The fruit fly *Drosophila melanogaster* has also been proved to be a useful organism to understand the complexity of Ral signaling. For example, it was shown that RalA regulates furrow formation in early *Drosophila* embryos through the recruitment of Sec5 and the exocyst to the plasma membrane to allow the tethering of Rab8 vesicular compartments and ultimately, the ingression of incipient furrows [95]. Studies in *Drosophila* have also characterized an antagonistic relationship between Ral-GTPases and JNK in development and cell fate determination [96,97].

In addition to these lower eukaryotes, our understanding of the role of Ral signaling in cancer has been enhanced greatly by the generation of knockout mice, both for RalA and RalB, as well as for many of the regulators of Ral and their downstream effectors. Somewhat surprisingly, single knockout of neither RalA nor RalB had an effect on KRas-driven lung tumor growth although deletion of both blocked tumor formation, confirming the importance of Ral signaling for tumor growth downstream of Ras, but suggesting a level of functional redundancy that was not apparent in the various tumor models performed with human cells [98]. Notably, however, knockout of RalA leads to embryonic lethality, whereas RalB knockout mice are viable, confirming independent functions for the two proteins, at least during development. Knockout of RalGAPα2 further implicates Ral in tumorigenesis as these mice exhibit increased Ral activity and increased tumor growth in a chemically induced model of bladder cancer [99]. In addition, although it is difficult to tease apart the roles of the RalGEF proteins due to their functional redundancy, knockout of RalGDS was shown to protect against tumor initiation in a chemically induced model of HRas-driven skin carcinoma [100]. In terms of the Ral effectors, RalBP1 knockout mice are viable, but exhibit impaired tumor growth in a xenograft model due to impaired angiogenesis [101]. In addition to knockout models, transgenic over-expression of Ral-related proteins has provided some insight into the tumorigenic potential of this pathway. For example, transgenic tissue-specific over-expression of the RalGEF Rgr leads to tumor growth in a number of different tissues tested [102].

RAL IN TUMORIGENESIS

Since the time RalA was initially cloned, much of the focus on Ral research has been on its potential role in mediating the pro-tumorigenic effects of activated Ras. Indeed, much of what we know about Ral biology was discovered in the context of studies aimed at understanding this role. The sequence homology to Ras [4], as well as the discovery that four of the RalGEFs bind directly to GTP-bound Ras [12–15], fueled much of the early interest in the role of Ral in cancer. Subsequent studies, however, performed mostly in murine cells, argued for a mostly complementary role for Rals in Ras-driven cancer and suggested a more dominant role for the MAPK and PI3K pathways [103,104]. However, with the development of genetically defined, immortalized human cell lines in the late 1990s [105], the focus shifted back to Ral when it was demonstrated that expression of RalGEF-specific effector domain mutants of oncogenic Ras ($HRas^{G12V,E37G}$) could promote anchorage-independent growth in a number of immortalized human cell lines [5]. In addition, targeting the RalGEF Rlf to the plasma membrane by fusing it in frame to the C-terminus of Ras (Rlf-CAAX) was shown to promote transformation, and both $HRas^{G12V}$- and Rlf-CAAX-mediated transformation was blocked by the expression of the dominant negative $RalA^{S28N}$ [5]. Follow-up studies suggested that RalA, rather than RalB, was playing a dominant role in Ras-mediated transformation, as knockdown of RalA inhibited anchorage-independent growth and expression of an activated $RalA^{G23V}$ mutant by itself showed transformation potential [6,106].

These early studies made it clear that despite their high degree of sequence similarity, RalA and RalB had distinct biological roles at least in the context of Ras-driven oncogenic signaling. A wealth of subsequent studies have begun to clarify these roles, mostly through the expression of RalA- and RalB-specific small interfering RNAs, as well as over-expression of activated mutants of each protein, but our understanding is still very incomplete. Knockdown of RalB in transformed cells was shown to promote apoptosis [106] and inhibit migration [107], leading to the paradigm that, although RalA may be important for tumor initiation, RalB plays more of a role in invasion and metastasis. This was supported by the finding that RalA knockdown inhibited tumor initiation in a xenograft model using a panel of pancreatic cell lines, whereas knockdown of RalB led to a loss of metastasis in a tail vein injection assay [108]. Furthermore, RalB inhibition blocks migration in a multiple myeloma cell line [109], inhibits migration metastatic growth in a xenograft model of bladder cancer [37], and decreases migration and invasion in glioma cells [110].

Experimental support for a dominant role for RalA in tumor initiation and progression comes from a number of different model systems. Knockdown of RalA, but not RalB, inhibits anchorage-independent growth in a colorectal cancer cell line [111,112] and knockdown of RalA blocks tumor growth to a greater extent than RalB in a xenograft model of melanoma [113]. In addition, RalA activity is increased to a greater extent than RalB activity in non–small-cell lung cancer (NSCLC) [114], and knockdown of RalA in NSCLC cell lines blocked both proliferation and invasiveness [115]. RalA expression is elevated in human ovarian cancer and hepatocellular carcinoma tissues and cell lines, and its genetic inhibition leads to decreased proliferation and invasiveness in both [116,117].

As we accumulate further evidence about the different roles RalA and RalB play in the tumorigenic process, the challenge will be to understand the mechanistic underpinnings of these differences. Several studies have suggested that understanding the RalB-TBK1 axis may be key to understanding many of the RalB-specific roles [41,76,82,83] (Fig. 3.1). However, a link between RalA-specific functions, such as the regulation of exocyst function in the polarized delivery of membrane proteins in epithelial cells [30] and the regulation of mitochondrial fission [31], and its tumorigenic activities, remains to be uncovered.

Many, although not all, of these studies have directly addressed the differential contributions of RalA and RalB under controlled experimental conditions, although the redundancy between RalA and RalB apparent in a genetically engineered mouse model of lung cancer [98] underscores the need to continue to explore these differences using more physiological models of the human disease, such as patient-derived xenografts. Further mechanistic studies that employ both murine and human cell lines will provide additional insight into whether the distinct tumorigenic roles uncovered in human xenograft studies represent true physiological differences or artifacts of the techniques available to study human tumorigenesis.

INHIBITION OF RAL

The myriad studies over the past several decades linking the activities of both RalA and RalB to tumorigenic growth downstream of Ras has fueled a great deal of excitement that inhibition of these signaling pathways may represent a powerful therapeutic approach to cancer treatment [8,9,118–121] (Table 3.2). Despite this, the search for inhibitors of Ral signaling has encountered many of the same issues that have plagued the search for Ras inhibitors over the past 30 years. Inhibition of

Table 3.2 Therapeutic Strategies Targeting Ral Signaling

Target	Strategy	References
Membrane localization	Geranylgeranyltransferase inhibitors	[33,110,115,117,122,123]
Effector binding	Small molecule inhibitors	[7]
Downstream signaling	TANK-binding kinase 1 (TBK1) inhibitors	[82]

Ral membrane targeting, through the use of inhibitors against GGTase-I, has been the most widely used pharmacological approach to inhibit Ral to date. Knockdown of GGTase-I was demonstrated to decrease migration and invasion in glioma cells [110], and GGTase inhibitors (GGTIs) were shown to have anti-tumorigenic effects in a xenograft model of pancreatic cancer [122] and to block cellular proliferation in vitro and tumor growth in vivo in a model of hepatocellular carcinoma [117]. Furthermore, GGTIs have been shown to inhibit a number of the pro-tumorigenic effects of Ral signaling in both pancreatic and oral squamous cell carcinoma cells [33,123] and GGTI treatment inhibited tumor growth in a xenograft model of non–small-cell lung carcinoma [115]. Although these and other studies highlight the promise of GGTIs as a therapeutic option, the number of other important cellular proteins modified by the geranylgeranyl moiety raises issues of specificity and suggests that toxicity may hamper their efficacy when used clinically. Similarly, inhibitors of both ICMT and RCE1 have been shown to affect Ral localization and activity [36], but a lack of specificity for Ral GTPases likely indicates that their use in the clinic will be limited.

Given the specificity issues inherent to inhibition of Ral localization, the identification of small molecule inhibitors that directly target Ral activity has been met with a great deal of excitement [7]. A structure-based in silico screening approach identified several molecules that specifically disrupt the interaction between Ral proteins and their effectors and these molecules were shown to have efficacy in both in vitro growth assays and in a xenograft assay of lung cancer. Beyond the discovery of these molecules themselves, these studies validate the possibility of specific inhibitors of Ral GTPases and suggest that future molecules may be found with specificity for RalA and RalB, which will not only represent promising therapeutic leads but also help tremendously in our quest to experimentally unravel the roles of these GTPases in a number of different tumor types.

Finally, like Ras, successful inhibition of Ral signaling may be ultimately achieved through inhibition of the key molecules and pathways that become engaged downstream of its activation. Consistent with this idea, inhibition of TBK1 by the compound CYT387 was shown to block tumor growth in a genetically engineered model of KRas-driven lung cancer [82]. As we continue to unravel the specific role each effector pathway is playing downstream of RalA and RalB, a number of additional therapeutic targets should reveal themselves, allowing us to increase our arsenal against Ral signaling in cancer.

CONCLUSIONS AND FUTURE DIRECTIONS

Despite all that we have learned since RalA and RalB were discovered nearly 30 years ago, we still have a long way to go in fully understanding how these two small GTPases contribute to tumor growth downstream of oncogenic Ras. The rapid development of new tools to study these proteins, including novel pharmacological inhibitors and advanced genetic manipulations, should finally allow us to gain more insights into the key remaining questions such as: How does differential localization affect effector usage? What are the roles of different RalGEFs? What are the key pathways that underlie the different tumorigenic activities of RalA and RalB and can these be explored using murine models of tumorigenesis? Hopefully, as these questions are answered, we will emerge with a host of new strategies in the fight against Ras-driven malignancies.

References

[1] Pylayeva-Gupta Y, Grabocka E, Bar-Sagi D. RAS oncogenes: weaving a tumorigenic web. Nat Rev Cancer November 2011;11(11):761–74.

[2] Santarpia L, Lippman SM, El-Naggar AK. Targeting the MAPK-RAS-RAF signaling pathway in cancer therapy. Expert Opin Ther Targets January 2012;16(1):103–19.

[3] Fruman DA, Rommel C. PI3K and cancer: lessons, challenges and opportunities. Nat Rev Drug Discov February 2014;13(2):140–56.

[4] Chardin P, Tavitian A. The ral gene: a new ras related gene isolated by the use of a synthetic probe. EMBO J September 1986;5(9):2203–8.

[5] Hamad NM, Elconin JH, Karnoub AE, Bai W, Rich JN, Abraham RT, et al. Distinct requirements for Ras oncogenesis in human versus mouse cells. Gene Dev August 15, 2002;16(16):2045–57.

[6] Lim K-H, Baines AT, Fiordalisi JJ, Shipitsin M, Feig LA, Cox AD, et al. Activation of RalA is critical for Ras-induced tumorigenesis of human cells. Cancer Cell June 2005;7(6):533–45.

[7] Yan C, Liu D, Li L, Wempe MF, Guin S, Khanna M, et al. Discovery and characterization of small molecules that target the GTPase Ral. Nature November 20, 2014;515(7527):443–7.

[8] Shirakawa R, Horiuchi H. Ral GTPases: crucial mediators of exocytosis and tumourigenesis. J Biochem May 2015;157(5):285–99.

[9] Gentry LR, Martin TD, Reiner DJ, Der CJ. Ral small GTPase signaling and oncogenesis: more than just 15 minutes of fame. Biochim Biophys Acta December 2014;1843(12):2976–88.

[10] Chardin P, Tavitian A. Coding sequences of human ralA and ralB cDNAs. Nucleic Acids Res June 12, 1989;17(11):4380.

[11] Spaargaren M, Bischoff JR. Identification of the guanine nucleotide dissociation stimulator for Ral as a putative effector molecule of R-ras, H-ras, K-ras, and Rap. Proc Natl Acad Sci USA December 20, 1994;91(26):12609–13.

[12] Hofer F, Fields S, Schneider C, Martin GS. Activated Ras interacts with the Ral guanine nucleotide dissociation stimulator. Proc Natl Acad Sci USA November 8, 1994;91(23):11089–93.

[13] Kikuchi A, Demo SD, Ye ZH, Chen YW, Williams LT. ralGDS family members interact with the effector loop of ras p21. Mol Cell Biol November 1994;14(11):7483–91.

[14] Peterson SN, Trabalzini L, Brtva TR, Fischer T, Altschuler DL, Martelli P, et al. Identification of a novel RalGDS-related protein as a candidate effector for Ras and Rap1. J Biol Chem November 22, 1996;271(47):29903–8.

[15] Shao H, Andres DA. A novel RalGEF-like protein, RGL3, as a candidate effector for rit and Ras. J Biol Chem September 1, 2000;275(35):26914–24.

[16] de Bruyn KM, de Rooij J, Wolthuis RM, Rehmann H, Wesenbeek J, Cool RH, et al. RalGEF2, a pleckstrin homology domain containing guanine nucleotide exchange factor for Ral. J Biol Chem September 22, 2000;275(38):29761–6.

[17] Rebhun JF, Chen H, Quilliam LA. Identification and characterization of a new family of guanine nucleotide exchange factors for the ras-related GTPase Ral. J Biol Chem May 5, 2000;275(18):13406–10.

[18] Martegani E, Ceriani M, Tisi R, Berruti G. Cloning and characterization of a new Ral-GEF expressed in mouse testis. Ann NY Acad Sci November 2002;973:135–7.

[19] Papini D, Langemeyer L, Abad MA, Kerr A, Samejima I, Eyers PA, et al. TD-60 links RalA GTPase function to the CPC in mitosis. Nat Commun 2015;6:7678.

[20] Cascone I, Selimoglu R, Ozdemir C, Del Nery E, Yeaman C, White M, et al. Distinct roles of RalA and RalB in the progression of cytokinesis are supported by distinct RalGEFs. EMBO J September 17, 2008;27(18):2375–87.

[21] Vigil D, Martin RD, Williams F, Yeh JJ, Campbell SL, Der CJ. Aberrant overexpression of the Rgl2 Ral small GTPase-specific guanine nucleotide exchange factor promotes pancreatic cancer growth through Ral-dependent and Ral-independent mechanisms. J Biol Chem November 5, 2010;285(45):34729–40.

[22] Emkey R, Freedman S, Feig LA. Characterization of a GTPase-activating protein for the Ras-related Ral protein. J Biol Chem May 25, 1991;266(15):9703–6.

[23] Shirakawa R, Fukai S, Kawato M, Higashi T, Kondo H, Ikeda T, et al. Tuberous sclerosis tumor suppressor complex-like complexes act as GTPase-activating proteins for Ral GTPases. J Biol Chem August 7, 2009;284(32):21580–8.

[24] Chen X-W, Leto D, Xiong T, Yu G, Cheng A, Decker S, et al. A Ral GAP complex links PI 3-kinase/Akt signaling to RalA activation in insulin action. Mol Biol Cell January 1, 2011;22(1):141–52.

[25] Leto D, Uhm M, Williams A, Chen XW, Saltiel AR. Negative regulation of the RalGAP complex by 14-3-3. J Biol Chem February 5, 2013;288(13):9272–83.

[26] Chen Q, Quan C, Xie B, Chen L, Zhou S, Toth R, et al. GARNL1, a major RalGAP α subunit in skeletal muscle, regulates insulin-stimulated RalA activation and GLUT4 trafficking via interaction with 14-3-3 proteins. Cell Signal August 2014;26(8):1636–48.

[27] Oeckinghaus A, Postler TS, Rao P, Schmitt H, Schmitt V, Grinberg-Bleyer Y, et al. κB-Ras proteins regulate both NF-κB-dependent inflammation and Ral-dependent proliferation. Cell Rep September 25, 2014;8(6):1793–807.

[28] Fernández RMH, Ruiz-Miro M, Dolcet X, Aldea M, Gari E. Cyclin D1 interacts and collaborates with Ral GTPases enhancing cell detachment and motility. Oncogene April 21, 2011;30(16):1936–46.

[29] Chen X-W, Inoue M, Hsu SC, Saltiel AR. RalA-exocyst-dependent recycling endosome trafficking is required for the completion of cytokinesis. J Biol Chem December 15, 2006;281(50):38609–16.

[30] Shipitsin M, Feig LA. RalA but not RalB enhances polarized delivery of membrane proteins to the basolateral surface of epithelial cells. Mol Cell Biol July 2004;24(13):5746–56.

[31] Kashatus DF, Lim KH, Brady DC, Pershing NL, Cox AD, Counter CM. RALA and RALBP1 regulate mitochondrial fission at mitosis. Nat Cell Biol September 2011;13(9):1108–15.

[32] Kinsella BT, Erdman RA, Maltese WA. Carboxyl-terminal isoprenylation of ras-related GTP-binding proteins encoded by rac1, rac2, and ralA. J Biol Chem May 25, 1991;266(15):9786–94.

[33] Falsetti SC, Wang DA, Peng H, Carrico D, Cox AD, Der CJ, et al. Geranylgeranyltransferase I inhibitors target RalB to inhibit anchorage-dependent growth and induce apoptosis and RalA to inhibit anchorage-independent growth. Mol Cell Biol November 2007;27(22):8003–14.

[34] Michaelson D, Ali W, Chiu VK, Bergo M, Silletti J, Wright L, et al. Postprenylation CAAX processing is required for proper localization of Ras but not Rho GTPases. Mol Biol Cell April 2005;16(4):1606–16.

[35] Nishimura A, Linder ME. Identification of a novel prenyl, palmitoyl CaaX modification of Cdc42 that regulates RhoGDI binding. Mol Cell Biol January 28, 2013;33(7):1417–29.

[36] Gentry LR, Nishimura A, Cox AD, Martin TD, Tsygankov D, Nishida M, et al. Divergent roles of CAAX motif-signaled posttranslational modifications in the regulation and subcellular localization of Ral GTPases. J Biol Chem September 11, 2015;290(37):22851–61.

[37] Wang H, Owens C, Chandra N, Conaway MR, Brautigan DL, Theodorescu D. Phosphorylation of RalB is important for bladder cancer cell growth and metastasis. Cancer Res November 1, 2010;70(21):8760–9.

[38] Lim K-H, Brady DC, Kashatus DF, Ancrille BB, Der CJ, Cox AD, et al. Aurora-A phosphorylates, activates, and relocalizes the small GTPase RalA. Mol Cell Biol January 2010;30(2):508–23.

[39] Neyraud V, Aushev VN, Hatzoglou A, Meunier B, Cascone I, Camonis J. RalA and RalB proteins are ubiquitinated GTPases, and ubiquitinated RalA increases lipid raft exposure at the plasma membrane. J Biol Chem August 24, 2012;287(35):29397–405.

[40] Wu J-C, Chen TY, Yu CT, Tsai SJ, Hsu JM, Tang MJ, et al. Identification of V23RalA-Ser194 as a critical mediator for Aurora-A-induced cellular motility and transformation by small pool expression screening. J Biol Chem March 11, 2005;280(10):9013–22.

[41] Simicek M, Lievens S, Laga M, Guzenko D, Aushev VN, Kalev P, et al. The deubiquitylase USP33 discriminates between RALB functions in autophagy and innate immune response. Nat Cell Biol September 22, 2013;15(10):1220–30.

[42] Sablina AA, Chen W, Arroyo JD, Corral L, Hector M, Bulmer SE, et al. The tumor suppressor PP2A Abeta regulates the RalA GTPase. Cell June 1, 2007;129(5):969–82.

[43] Martin TD, Mitin N, Cox AD, Yeh JJ, Der CJ. Phosphorylation by protein kinase Cα regulates RalB small GTPase protein activation, subcellular localization, and effector utilization. J Biol Chem April 27, 2012;287(18):14827–36.

[44] Bivona TG, Quatela SE, Bodemann BO, Ahearn IM, Soskis MJ, Mor A, et al. PKC regulates a farnesyl-electrostatic switch on K-Ras that promotes its association with Bcl-XL on mitochondria and induces apoptosis. Mol Cell February 17, 2006;21(4):481–93.

[45] McLaughlin S, Aderem A. The myristoyl-electrostatic switch: a modulator of reversible protein-membrane interactions. Trends Biochem Sci July 1995;20(7):272–6.

[46] Jullien-Flores V, Dorseuli O, Romero F, Letourneur F, Saragosti S, Berger R, et al. Bridging Ral GTPase to Rho pathways. RLIP76, a Ral effector with CDC42/Rac GTPase-activating protein activity. J Biol Chem September 22, 1995;270(38):22473–7.

[47] Cantor SB, Urano T, Feig LA. Identification and characterization of Ral-binding protein 1, a potential downstream target of Ral GTPases. Mol Cell Biol August 1995;15(8):4578–84.

[48] Park SH, Weinberg RA. A putative effector of Ral has homology to Rho/Rac GTPase activating proteins. Oncogene December 7, 1995;11(11):2349–55.

[49] Goldfinger LE, Ptak C, Jeffery ED, Shabanowitz J, Hunt DF, Ginsberg MH. RLIP76 (RalBP1) is an R-Ras effector that mediates adhesion-dependent Rac activation and cell migration. J Cell Biol September 11, 2006;174(6):877–88.

[50] Yamaguchi A, Urano T, Goi T, Feig LA. An Eps homology (EH) domain protein that binds to the Ral-GTPase target, RalBP1. J Biol Chem December 12, 1997;272(50):31230–4.

[51] Ikeda M, Ishida O, Hinoi T, Kishida S, Kikuchi A. Identification and characterization of a novel protein interacting with Ral-binding protein 1, a putative effector protein of Ral. J Biol Chem January 9, 1998;273(2):814–21.

[52] Jullien-Flores V, Mahe Y, Mirey G, Leprince C, Meunier-Bisceuil B, Sorkin A, et al. RLIP76, an effector of the GTPase Ral, interacts with the AP2 complex: involvement of the Ral pathway in receptor endocytosis. J Cell Sci August 2000;113(Pt 16):2837–44.

[53] Han K, Kim MH, Seeburg D, Seo J, Verpelli C, Han S, et al. Ehlers MD, editor. Regulated RalBP1 Binding to RalA and PSD-95 Controls AMPA Receptor Endocytosis and LTD. PLoS Biol September 2009;7(9):e1000187.

[54] Nakashima S, Morinaka K, Koyama S, Ikeda M, Kishida M, Okawa K, et al. Small G protein Ral and its downstream molecules regulate endocytosis of EGF and insulin receptors. EMBO J July 1, 1999;18(13):3629–42.

[55] Rosse C, L'Hoste S, Offner N, Picard A, Camonis J. RLIP, an effector of the Ral GTPases, is a platform for Cdk1 to phosphorylate epsin during the switch off of endocytosis in mitosis. J Biol Chem August 15, 2003;278(33):30597–604.

[56] Awasthi S, Cheng J, Singhal SS, Saini MK, Pandya U, Pikula S, et al. Novel function of human RLIP76: ATP-dependent transport of glutathione conjugates and doxorubicin. Biochemistry August 8, 2000;39(31):9327–34.

[57] Singhal SS, Yadav S, Singhal J, Zajac E, Awasthi YC, Awasthi S. Depletion of RLIP76 sensitizes lung cancer cells to doxorubicin. Biochem Pharmacol August 1, 2005;70(3):481–8.

[58] Singhal SS, Wickramarachchi D, Yadav S, Singhal J, Leake K, Vatsyayan R, et al. Glutathione-conjugate transport by RLIP76 is required for clathrin-dependent endocytosis and chemical carcinogenesis. Mol Cancer Ther January 2011;10(1):16–28.

[59] Mollberg NM, Steinert G, Aigner M, Hamm A, Lin FJ, Elbers H, et al. Overexpression of RalBP1 in colorectal cancer is an independent predictor of poor survival and early tumor relapse. Cancer Biol Ther June 2012;13(8):694–700.

[60] Wang C-Z, Yuan P, Xu B, Yuan L, Yang HZ, Liu X. RLIP76 expression as a prognostic marker of breast cancer. Eur Rev Med Pharmacol Sci June 2015;19(11):2105–11.

[61] Wang Q, Wang JY, Zhang XP, Lv ZW, Fu D, Lu YC, et al. RLIP76 is overexpressed in human glioblastomas and is required for proliferation, tumorigenesis and suppression of apoptosis. Carcinogenesis January 15, 2013;34(4):916–26.

[62] Neel NF, Rossman KL, Martin TD, Kayes TK, Yeh JJ, Der CJ. The RalB small GTPase mediates formation of invadopodia through a GTPase-activating protein-independent function of the RalBP1/RLIP76 effector. Mol Cell Biol April 2012;32(8):1374–86.

[63] Wu Z, Owens C, Chandra N, Popovic K, Conaway M, Theodorescu D. RalBP1 is necessary for metastasis of human cancer cell lines. Neoplasia December 2010;12(12):1003–12.

[64] Moskalenko S, Henry DO, Rosse C, Mirey G, Camonis JH, White MA. The exocyst is a Ral effector complex. Nat Cell Biol January 2002;4(1):66–72.

[65] Sugihara K, Asano S, Tanaka K, Iwamatsu A, Okawa K, Ohta Y. The exocyst complex binds the small GTPase RalA to mediate filopodia formation. Nat Cell Biol January 2002;4(1):73–8.

[66] Moskalenko S, Tong C, Tosse C, Mirey G, Formstecher E, Daviet L, et al. Ral GTPases regulate exocyst assembly through dual subunit interactions. J Biol Chem December 19, 2003;278(51):51743–8.

[67] Li G, Han L, Chou TC, Fujita Y, Arunachalam L, Xu A, et al. RalA and RalB function as the critical GTP sensors for GTP-dependent exocytosis. J Neurosci January 3, 2007;27(1):190–202.

[68] Kawato M, Shirakawa R, Kondo H, Higashi T, Ikeda T, Okawa K, et al. Regulation of platelet dense granule secretion by the Ral GTPase-exocyst pathway. J Biol Chem November 12, 2007;283(1):166–74.

[69] de Leeuw HP, Wijers-Koster PM, van Mourik JA, Voorberg J. Small GTP-binding protein RalA associates with Weibel-Palade bodies in endothelial cells. Thromb Haemost September 1999;82(3):1177–81.

[70] Chen X-W, Leto D, Chiang SH, Wang Q, Saltiel AR. Activation of RalA is required for insulin-stimulated Glut4 trafficking to the plasma membrane via the exocyst and the motor protein Myo1c. Dev Cell September 2007;13(3):391–404.

[71] Ljubicic S, Bezzi P, Vitale N, Regazzi R. Maedler K, editor. The GTPase RalA Regulates Different Steps of the Secretory Process in Pancreatic Beta-Cells. PLoS One 2009;4(11):e7770.

[72] Hazelett CC, Sheff D, Yeaman C. RalA and RalB differentially regulate development of epithelial tight junctions. Mol Biol Cell December 2011;22(24):4787–800.

[73] Fukai S, Matern HT, Jagath JR, Scheller RH, Brunger AT. Structural basis of the interaction between RalA and Sec5, a subunit of the sec6/8 complex. EMBO J July 1, 2003;22(13):3267–78.

[74] Jin R, Junutula JR, Matern HT, Ervin KE, Scheller RH, Brunger AT. Exo84 and Sec5 are competitive regulatory Sec6/8 effectors to the RalA GTPase. EMBO J May 26, 2005;24(12):2064–74.

[75] Bodemann BO, Orvedahl A, Cheng T, Ram RR, Ou YH, Formstecher E, et al. RalB and the exocyst mediate the cellular starvation response by direct activation of autophagosome assembly. Cell January 21, 2011;144(2):253–67.

[76] Chien Y, Kim S, Bumeister R, Loo YM, Kwon SW, Johnson CL, et al. RalB GTPase-mediated activation of the IkappaB family kinase TBK1 couples innate immune signaling to tumor cell survival. Cell October 6, 2006;127(1):157–70.

[77] Maehama T, Tanaka M, Nishina H, Murakami M, Kanaho Y, Hanada K. RalA functions as an indispensable signal mediator for the nutrient-sensing system. J Biol Chem December 12, 2008;283(50):35053–9.

[78] Martin TD, Chen XW, Kaplan RE, Saltiel AR, Walker CL, Reiner DJ, et al. Ral and Rheb GTPase activating proteins integrate mTOR and GTPase signaling in aging, autophagy, and tumor cell invasion. Mol Cell January 23, 2014;53(2):209–20.

[79] Camonis JH, White MA. Ral GTPases: corrupting the exocyst in cancer cells. Trends Cell Biol June 2005;15(6):327–32.

[80] Hazelett CC, Yeaman C. Sec5 and Exo84 mediate distinct aspects of RalA-dependent cell polarization. PLoS One 2012;7(6):e39602.

[81] Biondini M, Duclos G, Meyer-Schaller N, Silberzan P, Camonis J, Parrini MC. RalB regulates contractility-driven cancer dissemination upon TGFβ stimulation via the RhoGEF GEF-H1. Sci Rep 2015;5:11759.

[82] Zhu Z, Aref AR, Cohoon TJ, Barbie TU, Imamura Y, Yang S, et al. Inhibition of KRAS-driven tumorigenicity by interruption of an autocrine cytokine circuit. Cancer Discov April 2014;4(4):452–65.

[83] Seguin L, Kato S, Franovic A, Camargo MF, Lesperance J, Elliott KC, et al. An integrin β_3-KRAS-RalB complex drives tumour stemness and resistance to EGFR inhibition. Nat Cell Biol May 2014;16(5):457–68.

[84] Frankel P, Aronheim A, Kavanagh E, Balda MS, Matter K, Bunney TD, et al. RalA interacts with ZONAB in a cell density-dependent manner and regulates its transcriptional activity. EMBO J January 12, 2005;24(1):54–62.

[85] Ohta Y, Suzuki N, Nakamura S, Hartwig JH, Stossel TP. The small GTPase RalA targets filamin to induce filopodia. Proc Natl Acad Sci USA March 2, 1999;96(5):2122–8.

[86] Jiang H, Luo JQ, Urano T, Frankel P, Lu Z, Foster DA, et al. Involvement of Ral GTPase in v-Src-induced phospholipase D activation. Nature November 23, 1995;378(6555):409–12.

[87] Voss M, Weernink PA, Haupenthal S, Moller U, Cool RH, Bauer B, et al. Phospholipase D stimulation by receptor tyrosine kinases mediated by protein kinase C and a Ras/Ral signaling cascade. J Biol Chem December 3, 1999;274(49):34691–8.

[88] Lu Z, Hornia A, Joseph T, Sukezane T, Frankel P, Zhong M, et al. Phospholipase D and RalA cooperate with the epidermal growth factor receptor to transform 3Y1 rat fibroblasts. Mol Cell Biol January 2000;20(2):462–7.

[89] Wang KL, Roufogalis BD. Ca^{2+}/calmodulin stimulates GTP binding to the ras-related protein ral-A. J Biol Chem May 21, 1999;274(21):14525–8.

[90] Wang KL, Kahn MT, Roufogalis BD. Identification and characterization of a calmodulin-binding domain in Ral-A, a Ras-related GTP-binding protein purified from human erythrocyte membrane. J Biol Chem June 20, 1997;272(25):16002–9.

[91] Sidhu RS, Clough RR, Bhullar RP. Regulation of phospholipase C-delta1 through direct interactions with the small GTPase Ral and calmodulin. J Biol Chem June 10, 2005;280(23):21933–41.

[92] van den Berg MCW, van Gogh IJ, Smits AM, van Triest M, Dansen TB, Visscher M, et al. The small GTPase RALA controls c-Jun N-terminal kinase-mediated FOXO activation by regulation of a JIP1 scaffold complex. J Biol Chem July 26, 2013;288(30):21729–41.

[93] Kidd AR, Snider JL, Martin TD, Graboski SF, Der CJ, Cox AD. Ras-related small GTPases RalA and RalB regulate cellular survival after ionizing radiation. Int J Radiat Oncol Biol Phys September 1, 2010;78(1):205–12.

[94] Zand TP, Reiner DJ, Der CJ. Ras effector switching promotes divergent cell fates in *C. elegans* vulval patterning. Dev Cell January 18, 2011;20(1):84–96.

[95] Holly RM, Mavor LM, Zuo Z, Blankenship JT. A rapid, membrane-dependent pathway directs furrow formation through RalA in the early *Drosophila* embryo. Development July 1, 2015;142(13):2316–28.

[96] Sawamoto K, Winge P, Koyama S, Hirota Y, Yamada C, Miyao S, et al. The *Drosophila* Ral GTPase regulates developmental cell shape changes through the Jun NH(2)-terminal kinase pathway. J Cell Biol July 26, 1999;146(2):361–72.

[97] Balakireva M, Rosse C, Langevin J, Chien YC, Gho M, Gonzy-Treboul G, et al. The Ral/exocyst effector complex counters c-Jun N-terminal kinase-dependent apoptosis in *Drosophila melanogaster*. Mol Cell Biol December 2006;26(23):8953–63.

[98] Peschard P, McCarthy A, Leblanc-Dominguez V, Yeo M, Guichard S, Stamp G, et al. Genetic deletion of RALA and RALB small GTPases reveals redundant functions in development and tumorigenesis. Curr Biol November 6, 2012;22(21):2063–8.

[99] Saito R, Shirakawa R, Nishiyama H, Kobayashi T, Kawato M, Kanno T, et al. Downregulation of Ral GTPase-activating protein promotes tumor invasion and metastasis of bladder cancer. Oncogene February 14, 2013;32(7):894–902.

[100] González-García A, Pritchard CA, Paterson HF, Mavria G, Stamp G, Marshall CJ. RalGDS is required for tumor formation in a model of skin carcinogenesis. Cancer Cell 2005;7(3):219–26.

[101] Lee S, Wurtzel JG, Singhal SS, Awasthi S, Goldfinger LE. RALBP1/RLIP76 depletion in mice suppresses tumor growth by inhibiting tumor neovascularization. Cancer Res October 15, 2012;72(20):5165–73.

[102] Jiménez M, Perez de Castro I, Benet M, Garcia JF, Inghirami G, Pellicer A. The Rgr oncogene induces tumorigenesis in transgenic mice. Cancer Res September 1, 2004;64(17):6041–9.

[103] Urano T, Emkey R, Feig LA. Ral-GTPases mediate a distinct downstream signaling pathway from Ras that facilitates cellular transformation. EMBO J February 15, 1996;15(4):810–6.

[104] White MA, Vale T, Camonis JH, Schafer E, Wigler MH. A role for the Ral guanine nucleotide dissociation stimulator in mediating Ras-induced transformation. J Biol Chem July 12, 1996;271(28):16439–42.

[105] Hahn WC, Counter CM, Lundberg AS, Beijersbergen RL, Brooks MW, Weinberg RA. Creation of human tumour cells with defined genetic elements. Nature July 29, 1999;400(6743):464–8.

[106] Chien Y, White MA. RAL GTPases are linchpin modulators of human tumour-cell proliferation and survival. EMBO Rep August 2003;4(8):800–6.

[107] Oxford G, Owens CR, Titus BJ, Foreman TL, Herlevsen MC, Smith SC, et al. RalA and RalB: antagonistic relatives in cancer cell migration. Cancer Res August 15, 2005;65(16):7111–20.

[108] Lim K-H, O'Hayer K, Adam SJ, Kendall SD, Campbell PM, Der CJ, et al. Divergent roles for RalA and RalB in malignant growth of human pancreatic carcinoma cells. Curr Biol December 19, 2006;16(24):2385–94.

[109] de Gorter DJJ, Reijmers RM, Beuling EA, Naber HP, Kuil A, Kersten MJ, et al. The small GTPase Ral mediates SDF-1-induced migration of B cells and multiple myeloma cells. Blood April 1, 2008;111(7):3364–72.

[110] Song X, Hua L, Xu Y, Fang Z, Wang Y, Gao J, et al. Involvement of RalB in the effect of geranylgeranyltransferase I on glioma cell migration and invasion. Clin Transl Oncol June 2015;17(6):477–85.

[111] Martin TD, Der CJ. Differential involvement of RalA and RalB in colorectal cancer. Small GTPases April 2012;3(2):126–30.

[112] Győrffy B, Stelniec-Klotz I, Sigler C, Kasack K, Redmer T, Qian Y, et al. Effects of RAL signal transduction in KRAS- and BRAF-mutated cells and prognostic potential of the RAL signature in colorectal cancer. Oncotarget May 30, 2015;6(15):13334–46.

[113] Zipfel PA, Brady DC, Kashatus DF, Ancrile BD, Tyler DS, Counter CM. Ral activation promotes melanomagenesis. Oncogene August 26, 2010;29(34):4859–64.

[114] Guin S, Ru Y, Wynes MW, Mishra R, Lu X, Owens C, et al. Contributions of KRAS and RAL in non-small-cell lung cancer growth and progression. J Thorac Oncol December 2013;8(12):1492–501.

[115] Male H, Patel V, Jacob MA, Borrego-Diaz E, Wang K, Young DA, et al. Inhibition of RalA signaling pathway in treatment of non-small cell lung cancer. Lung Cancer August 2012;77(2):252–9.

[116] Wang K, Terai K, Peng W, Rouyanian A, Liu J, Roby KF, et al. The role of RalA in biology and therapy of ovarian cancer. Oncotarget December 10, 2013;5:1–14.

[117] Ezzeldin M, Borrego-Diaz E, Taha M, Esfandyari T, Wise AL, Peng W, et al. RalA signaling pathway as a therapeutic target in hepatocellular carcinoma (HCC). Mol Oncol July 2014;8(5):1043–53.

[118] Feig LA. Ral-GTPases: approaching their 15 minutes of fame. Trends Cell Biol August 2003;13(8):419–25.

[119] Bodemann BO, White MA. Ral GTPases and cancer: linchpin support of the tumorigenic platform. Nat Rev Cancer February 2008;8(2):133–40.

[120] Kashatus DF. Ral GTPases in tumorigenesis: emerging from the shadows. Exp Cell Res July 25, 2013;319(15):1–6.

[121] Neel NF, Martin TD, Stratford JK, Zand TP, Reiner DJ, Der CJ. The RalGEF-ral effector signaling network: the road less traveled for anti-ras drug discovery. Genes Cancer March 2011;2(3):275–87.

[122] Lu J, Chan L, Fiji HD, Dahl R, Kwon O, Tamanoi F. In vivo antitumor effect of a novel inhibitor of protein geranylgeranyltransferase-I. Mol Cancer Ther May 2009;8(5):1218–26.

[123] Hamada M, Miki T, Iwai S, Shimizu H, Yura Y. Involvement of RhoA and RalB in geranylgeranyltransferase I inhibitor-mediated inhibition of proliferation and migration of human oral squamous cell carcinoma cells. Cancer Chemother Pharmacol September 2011;68(3):559–69.

CHAPTER 4

The Biology, Prognostic Relevance, and Targeted Treatment of Ras Pathway–Positive Childhood Acute Lymphoblastic Leukemia

J.A.E. Irving

Newcastle University, Newcastle upon Tyne, United Kingdom

CONTENTS

INTRODUCTION

This chapter describes Ras/RAF/MEK/Erk (Ras) pathway activation in the most common childhood cancer, namely acute lymphoblastic leukaemia. It includes the diverse mechanisms that activate the pathway, the related prognostic significance, role in disease progression and the prospects for delivering new therapies into the clinic for relapsed and high risk disease.

CHILDHOOD ALL

ALL is the most prevalent cancer in children, representing 25% of all childhood cancers, and has a peak incidence between 2 and 5 years of age. It is a clonal disorder of developing lymphocytes and is therefore classified immunophenotypically as B or T lineage. ALL is characterized by chromosomal alterations, which can be either numerical or structural [1]. In B lineage ALL, numerical abnormalities include both hyper-diploidy and hypo-diploidy and structural aberrations are usually chromosomal translocations that create fusion genes such as the t(12;21)(p13;q22) that encodes the *ETV6-RUNX1* fusion gene, the t(9;22)(q34;q11) that encodes *BCR-ABL1*, as well as rearrangements involving the *MLL* gene, sited at 11q23, which has various gene fusions partners. In T ALL, dysregulation of *TAL1*, *TLX1*, *TLX3*, and *LYL1* genes is caused by translocations with T-cell antigen receptor loci. Translocations often disrupt genes involved in lymphoid development (eg, *RUNX1* and *ETV6*) or activate proto-oncogenes (eg, *ABL1*). Several of these alterations are significantly associated with outcome: high hyper-diploidy (>50 chromosomes) and *ETV6-RUNX1* are associated with a very favorable outcome, whereas low hypo-diploidy (<40 chromosomes) and *MLL* rearrangements are associated with a dismal prognosis [2]. Other prognostic factors at diagnosis include age, with children less than 1 year or greater than 10 years of age faring less well, and also levels of

Conquering RAS. http://dx.doi.org/10.1016/B978-0-12-803505-4.00004-7

peripheral leukemia as monitored by the white blood cell count, with values more than $50 \times 10^9/l$ associated with a poorer prognosis. The clinical response to induction chemotherapy is a powerful prognostic factor and has played an increasing role in risk stratification. This is monitored by morphology but more importantly by highly sensitive assays that measure low-level minimal residual disease.

Although contemporary clinical trials for newly diagnosed ALL report 5-year survival rates approaching 90% [3,4], for those children who have a relapse the outlook is much poorer and relapsed ALL remains a frequent cause of death in children with cancer [5–7]. For those children classified as high risk at relapse, including those with early relapse, T-cell disease, and marrow involvement, survival rates are only 25%. Clearly, new therapeutic strategies are needed for relapsed ALL and the Ras pathway may play an important role in achieving this goal.

RAS PATHWAY ACTIVATION

Ras pathway activation is found in over a third of ALL and is brought about by several different routes, highlighting its importance in leukemogenesis. A diagram of the pathway, with genes identified to be mutated in ALL is shown in Fig. 4.1. They include *NRAS* (neuroblastoma RAS viral oncogene homolog),

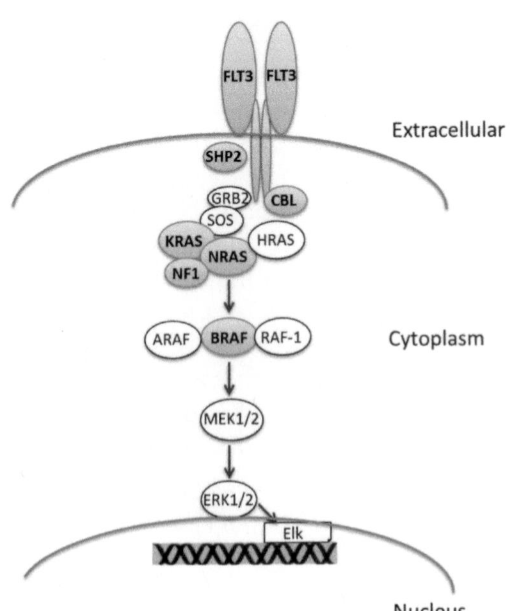

FIGURE 4.1
The Ras/RAF/MEK/Erk pathway, with mutated genes in acute lymphoblastic leukemia, shown in red.

KRAS (Kirsten rat sarcoma viral oncogene homolog), *FLT3* (FMS-related tyrosine kinase 3), *PTPN11* (protein tyrosine phosphatase, non-receptor type 11), *CBL* (casitas B-lineage lymphoma), *BRAF* (v-raf murine sarcoma viral oncogene homolog B), and *NF1* (neurofibromin). As in other cancers, the most common mechanism of activation is somatic mutation in the GTPases, *NRAS*, and *KRAS* [8–24]. Mutations cluster in the classic codons 12, 13, and 61 and dramatically reduce the rate of GTP hydrolysis by inhibiting interaction with GTPase-activating proteins (GAPs), leading to constitutive activation. The incidence of *NRAS*/*KRAS* mutations in ALL is reported as 15–30% and they are more common in B lineage than in T lineage ALL. There are also differences in the incidence between cytogenetic subgroups, with a preponderance in hyper-diploidy but they are seldom seen in the *ETV6–RUNX1* subgroup [9,14,15,18].

Flt3 is a receptor tyrosine kinase (RTK) that is expressed on the surface of hematopoietic progenitors. *FLT3* mutations are reported in around 2–9% of ALL cases and similar to *NRAS* and *KRAS* are often associated with hyper-diploidy [14,18,25–31]. Mutations are localized to either the tyrosine kinase or the juxta-membrane region domain and result in constitutive activation of Flt3, hyper-signaling of the Ras pathway, and ligand-independent proliferation in vitro [32,33]. In *MLL*-rearranged ALL, Flt3 is often constitutively activated because of high-level expression rather than mutation [34,35].

Shp2 is encoded by the *PTPN11* gene and is a protein tyrosine phosphatase that is highly expressed in hematopoietic cells. It has a positive influence on cell signaling pathways, including Ras and Janus kinase-signal transducers and activators of transcription. Although the mechanism is not clear, postulated targets of Shp2 include Ras GAP-binding sites on RTKs and/or Sprouty proteins and they are also implicated in increasing the production of reactive oxygen species, which hyper-sensitizes cytokine signaling [36]. Mutations in *PTPN11* are usually heterozygote, missense, gain-of-function mutations sited in the SH2 or phosphatase regions of the protein and are found in 2–10% of ALL, again often associated with hyper-diploidy [9,13,14,18,37].

The Cbl proteins are a highly conserved family of RING (really interesting new gene) finger ubiquitin E3 ligases that target a variety of RTKs for degradation and consequently can impact on the Ras pathway. Somatic mutations of *c-CBL* have been identified in 1–2% of ALL cases and are often associated with acquired uni-parental disomy at the c-*CBL* gene locus, in which there is duplication of a mutated allele and loss of the wild-type allele, resulting in a homozygous mutant state [38–41]. Functionally, inactivating mutations of Cbl proteins lead to stabilization of RTK receptors in an active state, with associated constitutive activation of the Ras pathway.

Microdeletions and mutations have also been reported in the GAP, *NF1* [42,43] as well as mutations in BRAF [11,23,24,44,45]. In a subtype of T ALL, known as early thymocyte precursor (ETP), focal amplification of *BRAF* was found in one patient, thus amplification of *BRAF* may be an additional mechanism of activation [24]. In addition, poor-risk chromosomal translocations, including *BCR/ABL* and those involving the *MLL* locus, are associated with constitutive activation of the Ras pathway [46–48]. A study that assessed p-ERK (extracellular signal–regulated kinase) levels and mutation status in primary ALL cells showed good concordance between the presence of Ras pathway mutations and pathway activation although the levels of p-ERK were found to be quite variable [49]. The biological significance of this variability is not known. Activation of the pathway has also been demonstrated in the absence of known canonical Ras pathway mutations, suggesting that additional mechanisms of pathway activation remain to be identified [22,49]. One possible mechanism is the de-regulated expression of a microRNA, MiR335, that targets and regulates levels of ERK2 [50].

ARE RAS PATHWAY MUTATIONS AN INITIATING OR SECONDARY EVENT?

One of the key questions in Ras pathway ALL is whether mutations act as initiation events in leukemia development or are secondary, cooperating genetic events that follow on from another initiator. In fact there is evidence to support both scenarios. For example, germline mutations in components of the Ras signaling pathway, including *KRAS* and *PTPN11*, cause inherited developmental disorders and sufferers have an increased risk of hematological malignancies, including ALL, suggesting they are the first "hit" in the multi-step leukemogenic process [51]. Studies in mouse models suggest that *NRAS/KRAS*, *NF1*, and *PTPN11* are indeed initiating events in the development of ALL and can give rise to both B and T lineage ALL but require cooperating genetic events for full-blown leukemia [52–60]. Hematopoietic stem/progenitor cell populations from *KRASG12D* mice show similar basal levels of phosphorylated ERK relative to wild type but are hyper-sensitive and show an exaggerated response to growth factor stimulation [61]. However, a mouse study showed that the *PTPN11 E76K* mutation was clearly sufficient to induce acute leukemia alone and this residue is recurrently mutated in ALL [58]. Interestingly, the *E76K*-mutated mice leukemias were associated with centrosome amplification and aneuploidy, which is highly relevant given the predominance of *PTPN11* mutations in hyper-diploid ALL cases. Elegant experiments reported by Li et al. revealed that *NRAS* mutation in hematopoietic stem cells have a bimodal effect, generating not only an expanding, rapidly dividing cell population but also a more quiescent cell population, which has the capacity for long-term self-renewal [62]. This may allow mutated cells to have long-term dominance over wild-type cells.

On the other hand, there are several lines of evidence in primary samples suggesting that Ras pathway mutations occur as a second, cooperative genetic "hit," at least in some patients. For example, *PTPN11* and *RAS* mutations have been shown to be present at diagnosis of ALL but can be lost at relapse, indicating a secondary role [37,63]. In some patients, it is clear that mutations are present in only a minor population of leukemia cells at diagnosis [9,14]. In fact, one study used a highly sensitive, quantitative polymerase chain reaction assay for common *KRAS* mutations and identified a very low level of mutated cells (<1%) in a significant number of patients with diagnostic ALL who had been classified as Ras pathway wild type by standard mutation screening [49]. These data suggest that *RAS* mutations are a relatively common event in leukemogenesis but that these mutated clones do not necessarily become the dominant clone at diagnosis. In MLL-rearranged ALL, *KRAS* itself acts as a cooperative lesion and the introduction of *KRAS* mutations into *MLL/AF4* transgenic mouse model results in a more aggressive leukemia that resembles the human equivalent [64,65]. Taken together, the evidence suggests that similar to acute myeloid leukemia (AML), Ras pathway–activating mutations may act as either initiating or cooperating events.

DISEASE EVOLUTION

A surprising finding from a study of Ras pathway mutations in relapsed ALL was that analyses of the matched diagnostic sample from the same patient frequently showed no evidence for the mutation [14,49,66]. However, highly sensitive, mutation-specific assays often revealed a minority subclone of cells with the same mutation, suggesting that they conferred a degree of resistance to therapy and enabled expansion of a pre-existing mutated *RAS* subclone, present at diagnosis, to predominate at relapse [14,49]. This hypothesis is supported by in vitro studies in which activation of the Ras pathway in hematopoietic cells is associated with resistance to glucocorticoids and anthracyclines, key drugs used in ALL therapy [45,67,68]. Resistance is mediated by transcriptional influences of ERK target proteins as well as those regulating the apoptotic regulatory machinery, such as the pro-apoptotic protein Bim. In one anecdotal case, *KRAS* mutations found at low level at diagnosis were enriched to the extent that the leukemia persisting after several weeks of combination chemotherapy was all *KRAS* mutated [49].

PROGNOSTIC SIGNIFICANCE

Several studies have investigated the prognostic significance of Ras pathway mutations. Although one early study demonstrated a higher relapse rate and reduced remission rate in children with *NRAS*-mutated ALL, these associations have not been replicated in recent studies [8,11,17,69]. In the largest investigation to date, which included more than 800 children enrolled in a US study, there was no relationship between *NRAS/KRAS* mutation status and high-risk clinical features and no effect on outcome [8]. A smaller UK study showed a significant association with the presence of Ras pathway mutation and some high-risk features including a higher presenting white cell count but saw no effect on early disease clearance [14]. A confounding factor in these studies relates to the high prevalence of Ras pathway mutations in very-good-risk and very-poor-risk subgroups, which may mask any prognostic association when analyzed together. For example, Ras pathway mutations are consistently associated with high hyper-diploidy, which represents a third of all newly diagnosed children and confers a very favorable prognosis [12,14,15,18,29,66,70]. Reported incidences of Ras pathway mutations in this subgroup range from 24% to almost 60%. Yet, they also occur at high frequency in very-poor-risk groups including hypo-diploid ALL, ETP, and another group classified as "high risk," based principally on the absence of good risk cytogenetic features [22–24].

Thus, more recent investigations have questioned the impact of mutation status on prognosis within the context of specific groups. For example, in infant ALL (<12 months of age), which is often associated with translocations of the *MLL* gene, the presence of *NRAS/KRAS* mutations was associated with an extremely poor outcome [45,71]. The comparative 5-year event-free survival rates for *RAS* mutated was 0.0% compared with 32.7% for *RAS* wild type [45]. Patients with mutated ALL presented with a higher peripheral white blood cell count and their leukemia cells were more resistant to glucocorticoids in vitro. Synthetic glucocorticoids, such as dexamethasone, are pivotal agents in the treatment of ALL because of their ability to specifically induce apoptosis in developing lymphocytes and thus these in vitro observations may explain the poor clinical response [45]. Conversely, in high hyper-diploid ALL, there was no effect of the presence of Ras pathway mutation on prognosis, although this was a relatively small study of <80 patients with few relapses, given the good prognosis associated with this cytogenetic group [18]. Mutation screening studies of large current trials for newly diagnosed ALL will define whether Ras pathway status has prognostic relevance and if it can be used to enhance current risk stratification.

At relapse of ALL, two large clinical trials have demonstrated prognostic significance of *NRAS* and *KRAS* mutations, with an association seen for poor outcome. Mutations in the Ras pathway are found in approximately 25–39% of B lineage cases at relapse [14,49,63,72]. In a study of over 200 children with B lineage ALL treated in the *ALL-REZ BFM 2002* trial, the frequency of Ras pathway mutations was 35% and these data are summarized in Fig. 4.2 [49]. The presence of any Ras pathway mutations was associated with high-risk features such as early relapse and specifically for *NRAS/KRAS* mutations, there was a greater proportion of patients with on-treatment relapse, central nervous system (CNS) involvement, and reduced remission rates following re-induction

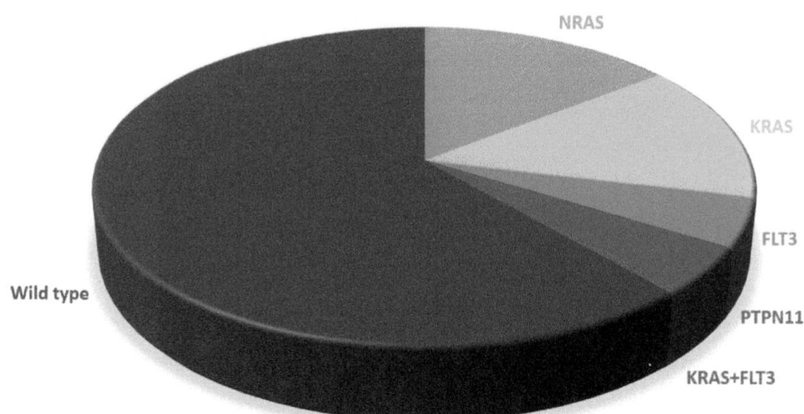

FIGURE 4.2
Percentage of Ras pathway mutations in children with relapsed acute lymphoblastic leukemia entered into the *ALL-REZ BFM 2002* clinical trial. *Adapted from Irving J, Matheson E, Minto L, Blair H, Case M, Halsey C, et al. Ras pathway mutations are prevalent in relapsed childhood acute lymphoblastic leukemia and confer sensitivity to MEK inhibition. Blood November 27, 2014;124(23):3420–30. PubMed PMID:25253770. Pubmed Central PMCID: 4246039.*

chemotherapy. For *KRAS* mutations, there was a reduced overall survival, with overall survival rates of 36% for mutated compared with 55% for wild type. In another international trial for relapsed ALL (UKR3), *NRAS* mutations were associated with an inferior survival in children with "good risk" cytogenetics [72]. *NRAS* mutation status was integrated with other genetic abnormalities along with clinical and cytogenetic parameters to generate a new risk index that was significantly better at predicting outcome than the current standard. The risk index needs confirmation in an independent patient cohort but may allow improved risk stratification and thus better outcome for relapsed ALL.

TARGETED THERAPIES

De-regulation of the Ras pathway is common across numerous cancer types, including ALL, and is thus an attractive therapeutic target. Small molecule inhibitors targeting different components of the pathway are currently being evaluated in clinical trials and may benefit children with ALL.

Initial efforts to target de-regulated Ras focused on disrupting Ras post-translational processing using inhibitors of farnesyl transferase (FTase). One of these, tipifarnib (R-115777), was shown to have in vitro activity against both B and T lineage primary ALL cells although the sample set was too small to evaluate whether *RAS* mutation status correlated with enhanced drug response [73]. Phase I clinical trials of tipifarnib were initially done in adults with ALL and

AML. Clinical responses were seen in approximately one-third of patients and p-ERK inhibition was demonstrated in some leukemias which were p-ERK positive before drug administration. However, the few patients with ALL in the trial, all demonstrated disease progression within 7–21 days [74]. A more recent phase I study of tipifarnib in children, principally AML and ALL, showed clear inhibition of FTase activity in leukemic cells but no clinical responses were observed [75]. Targeting FTase is less than ideal as when FTase is inhibited, Nras and Kras can undergo alternate post-translational modification by geranylgeranyltransferase, allowing the crucial cellular trafficking of Ras proteins to be maintained [76]. In addition, FTase inhibitors are not specific for Ras and will inhibit farnesylation of other proteins containing CAAX motifs and may explain the spectrum of toxicities seen with this drug.

Because the sole substrate of MEK1/2 is ERK1/2, this restricted substrate specificity has prompted the development of inhibitors of MEK, because one would expect these drugs to be associated with less "off-target" activity than FTase, described earlier, and may inhibit the pathway regardless of the mechanism of upstream activation. There are a number of MEK inhibitors (MEKi) in the advanced stages of clinical trial including trametinib (GSK1120212), pimasertib (MSC1936369B), PD0325901, and selumetinib (AZD6244, ARRY-142886) [77–81]. In general, sensitivity to MEKi is enhanced in tumor cells harboring activating Ras pathway mutations, including ALL cells, although there are exceptions [14,38,60,80,82,83]. However, a prospective in vitro study using leukemia cells from children with ALL clearly demonstrated differential sensitivity to MEKi, with 50% lethal dose values being significantly lower in Ras pathway–positive cells (mean 250 nM) than in cells that were negative (mean 68 μM) [49]. These primary ALL cells included *NRAS*, *KRAS*, as well as *FLT3/CBL* mutants. These in vitro data were replicated in vivo using an orthotopic xenograft mouse model in which primary ALL cells are engrafted in the bone marrow of immune-compromised, nonobese diabetic severe combined immunodeficiency gamma null mice, extravasate into the peripheral blood, spread to the spleen and extra-medullary sites, including the CNS [84]. The model is a robust and reliable representation of human disease that recapitulates the aggressiveness of the leukemia seen in the patient. Using this model, oral dosing with MEKi was found to have activity against *NRAS* and *KRAS* mutant ALL primagrafts, significantly reducing peripheral blood leukemia and reducing spleen size. Accompanying pharmacodynamic assessments showed inhibition of p-ERK and induction of apoptosis in ALL cells. Histological analysis of post-mortem brains found extensive meningeal leukemic infiltration in control vehicle treated, but not in MEKi mice, suggesting that this drug may eradicate CNS ALL, which is important given the association of *RAS* mutations and CNS disease identified in the *ALL-REZ BFM 2002* trial. There was no activity of MEKi against wild-type ALL primagrafts. The observed high specificity against RAS-positive ALL may translate into the clinic an effective drug with little off-target toxicity. The MEKi used in these studies was selumetinib, a potent, selective, allosteric inhibitor of MEK1/2, which has a favorable toxicity profile and demonstrated anti-tumor activity and has reached phase III clinical trial for *KRAS* mutant advanced non–small-cell lung cancer [79,80,85]. It has undergone phase I clinical testing in children with BRAF-driven recurrent/refractory pediatric low-grade glioma that defined a maximum tolerated dose and partial responses were observed in some patients [86].

Some activity of the MEKi, PD0325901, has also been demonstrated in vivo for *KRAS*-mutated T ALL [60]. MEK inhibition has also been studied in a mouse model in which bi-allelic inactivation of *NF1* causes a myeloproliferative disorder that can be progressed to AML by inducing secondary genetic aberrations using retro-viral mutagenesis [57]. Although the initial disorder was shown to be relatively resistant to MEKi, the *NF1*-deficient leukemias developed greater sensitivity, indicating that the cooperating mutations that progressed them to the leukemia made them highly dependent on Ras signaling [57]. However, in vivo testing of MEKi in *RAS*-mutated hypo-diploid ALL showed no activity of MEKi but instead there was responsiveness to PI3K inhibitors, highlighting dependency on the PI3K/Akt/mTOR effector pathway [22]. Linked to this, an interesting study has teased apart the relative importance of Raf/MEK/ERK and PI3K/Akt/mTOR pathways in a mouse model of *KRAS G12D* mutated T ALL and shown that both effector pathways are drivers of aberrant growth [87]. Thus, combining inhibitors of both pathways may be an effective therapeutic strategy and indeed significant synergy has been shown in solid tumors, and clinical trials of this drug combination are underway [88,89].

Synergism with MEKi has also been demonstrated with Nutli-3a, which induces the MDM2-p53 axis, the Bcl-2 and Bcl-x(L) inhibitor, navitoclax (ABT263), docetaxel, as well as with the key agent in all ALL treatment regimens, dexamethasone [50,85,90–92]. Dexamethasone induces apoptosis in ALL cells via induction of the pro-apoptotic Bim and because Bim is inactivated by ERK phosphorylation, there is a biological rationale for potential synergism with MEKi. In fact, benchmark MEKi have demonstrated synergy with dexamethasone in ALL cell line models in vitro [91]. Using Ras pathway–activated ALL primagraft cells, very strong synergy was observed with selumetinib and dexamethasone dosing and was associated with an enhancement of BIM induction, suggesting co-exposure may be highly effective for the treatment of Ras pathway–positive ALL (J. Irving, unpublished observations). An international clinical trial of this combination is planned for multiple refractory/relapsed Ras pathway–mutated ALL.

Other potential drugs include small molecule inhibitors of Flt3 and there are several multi-targeted agents undergoing clinical trial, which inhibit this

RTK, including lestaurtinib, sorafenib, and AT9283. Infant leukemia, defined as leukemia diagnosed at less than 1 year of age, is often associated with *MLL* rearrangements with Flt3 being over-expressed rather than mutated [35,93]. Following promising pre-clinical data with Flt3 inhibitors, a phase III trial is now underway in which newly diagnosed infants are randomized to receive combination chemotherapy plus or minus lestaurtinib. Evidence also suggests that direct targeting of mutant *PTPN11* may be possible. A screen of naturally occurring compounds identified cryptotanshinone as a direct inhibitor of Shp2, preventing both its signaling and cellular function [94]. Pre-clinical studies found that mouse myeloid progenitors and primary leukemia cells bearing the activating *PTPN11 E76K* mutation were sensitive to this inhibitor. This drug is an active ingredient of the traditional medicinal herbal plant *Salvia miltiorrhiza Bunge* and is commonly used in Asian countries to treat a number of conditions including coronary heart disease, hyper-lipidemia, acute ischemic stroke, hepatitis, chronic renal failure, and Alzheimer disease. In addition, *PTPN11* gain-of-function mutations have been shown to cause cytokine hyper-sensitivity in hematopoietic cells, in part by enhancing the production of reactive oxygen species [36]. This hyper-sensitivity could be partially corrected with antioxidants; thus this may be an additional therapeutic approach. Finally, Raf inhibitors, specifically Braf inhibitors, given the identification of *BRAF* mutations in some patients with ALL, may also be useful.

Targeting mutant Ras protein itself is an obvious strategy but has proved to be very challenging. However, Ostrem et al. reported on the development of such an inhibitor. They discovered a new binding pocket on the Kras protein, which they exploited to create a small molecule that irreversibly binds to Kras (G12C), shifting its affinity to favor GDP over GTP and thus impairing binding to Raf. Dosing of lung cancer cell lines bearing *KRAS G12C* caused a reduction in cell viability and induced apoptosis but had no effect on wild-type cells or those bearing other *KRAS* mutations [95]. This is a key drug development milestone that may pave the way for the development of other specific mutant Ras inhibitors.

Intriguingly, there is evidence that cells with mutated Ras may be more sensitive to some traditional chemotherapeutic agents, particularly the nucleoside analogue, cytarabine. These in vitro observations have been translated in the clinic, with patients with Ras mutant–positive AML responding significantly better to high-dose cytarabine, with 10-year cumulative relapse rates of 45% compared with 68% for patients with wild-type RAS. In the low-dose cytarabine arm, there was no differential effect, with relapse rates of 100% versus 80%, respectively [96]. The mechanism behind this preferential sensitivity appears to be synergism of mutant Ras with cytarabine in activating DNA damage checkpoints, resulting in increased differentiation and an associated reduction in clonogenic potential [97]. Similar investigations in ALL have not been reported to date.

CONCLUSIONS

As in other malignancies, de-regulation of the Ras pathway in childhood ALL is common in leukemogenesis and also contributes to the emergence of drug-resistant clones at relapse. The affected genes and mechanism of activation are diverse, highlighting its importance in leukemogenesis. The presence of *NRAS/KRAS* mutations at relapse is associated with a significantly poorer overall survival and may be used to enhance risk stratification in future trials. Genomic analyses of large diagnostic cohorts will test whether Ras pathway mutations are useful prognostic biomarkers at this stage of disease too. Importantly, activation of the pathway may direct new therapies needed for relapsed ALL and possibly those children at high risk of relapse. To this end, a phase I/II clinical trial of MEK inhibitors is planned for multiple relapsed/refractory Ras pathway mutant ALL.

List of Acronyms and Abbreviations

ALL Acute lymphoblastic leukemia
AML Acute myeloid leukemia
BRAF v-raf murine sarcoma viral oncogene homolog B
CBL Casitas B-lineage lymphoma
CNS Central nervous system
FLT3 FMS-Related Tyrosine Kinase 3
FTase Farnesyl transferase
GAP GTPase-activating protein
KRAS Kirsten rat sarcoma viral oncogene homolog
MEKi MEK inhibitors
NF1 Neurofibromin
NRAS Neuroblastoma RAS viral oncogene homolog
PTPN11 Protein Tyrosine Phosphatase Non-Receptor Type 11

References

[1] Harrison CJ. Acute lymphoblastic leukemia. Clin Laboratory Med December 2011;31(4): 631–47. ix. PMID:22118741.

[2] Moorman AV, Harrison CJ, Buck GA, Richards SM, Secker-Walker LM, Martineau M, et al. Karyotype is an independent prognostic factor in adult acute lymphoblastic leukemia (ALL): analysis of cytogenetic data from patients treated on the Medical Research Council (MRC) UKALLXII/Eastern Cooperative Oncology Group (ECOG) 2993 trial. Blood April 15, 2007;109(8):3189–97. PMID:17170120.

[3] Vora A, Goulden N, Mitchell C, Hancock J, Hough R, Rowntree C, et al. Augmented post-remission therapy for a minimal residual disease-defined high-risk subgroup of children and young people with clinical standard-risk and intermediate-risk acute lymphoblastic leukaemia (UKALL 2003): a randomised controlled trial. Lancet Oncol July 2014;15(8):809–18. PMID:24924991.

[4] Vora A, Goulden N, Wade R, Mitchell C, Hancock J, Hough R, et al. Treatment reduction for children and young adults with low-risk acute lymphoblastic leukaemia defined by minimal residual disease (UKALL 2003): a randomised controlled trial. Lancet Oncol March 2013;14(3):199–209. PMID:23395119.

[5] Parker C, Waters R, Leighton C, Hancock J, Sutton R, Moorman AV, et al. Effect of mitoxantrone on outcome of children with first relapse of acute lymphoblastic leukaemia (ALL R3): an open-label randomised trial. Lancet December 11, 2010;376(9757):2009–17. PMID:21131038. Pubmed Central PMCID: 3010035.

[6] Hof J, Krentz S, van Schewick C, Korner G, Shalapour S, Rhein P, et al. Mutations and deletions of the TP53 gene predict nonresponse to treatment and poor outcome in first relapse of childhood acute lymphoblastic leukemia. J Clin Oncol August 10, 2011;29(23):3185–93. PMID:21747090.

[7] Malempati S, Gaynon PS, Sather H, La MK, Stork LC, Children's Oncology G. Outcome after relapse among children with standard-risk acute lymphoblastic leukemia: children's Oncology Group study CCG-1952. J Clin Oncol December 20, 2007;25(36):5800–7. PMID:18089878.

[8] Perentesis JP, Bhatia S, Boyle E, Shao Y, Shu XO, Steinbuch M, et al. RAS oncogene mutations and outcome of therapy for childhood acute lymphoblastic leukemia. Leukemia April 2004;18(4):685–92. PMID:14990973. Epub 2004/03/03. eng.

[9] Tartaglia M, Martinelli S, Cazzaniga G, Cordeddu V, Iavarone I, Spinelli M, et al. Genetic evidence for lineage-related and differentiation stage-related contribution of somatic PTPN11 mutations to leukemogenesis in childhood acute leukemia. Blood July 15, 2004;104(2): 307–13. PMID:14982869. Epub 2004/02/26. eng.

[10] Shu XO, Perentesis JP, Wen W, Buckley JD, Boyle E, Ross JA, et al. Parental exposure to medications and hydrocarbons and ras mutations in children with acute lymphoblastic leukemia: a report from the Children's Oncology Group. Cancer Epidemiol Biomarkers Prev July 2004;13(7):1230–5. PMID:15247135. Epub 2004/07/13. eng.

[11] Gustafsson B, Angelini S, Sander B, Christensson B, Hemminki K, Kumar R. Mutations in the BRAF and N-ras genes in childhood acute lymphoblastic leukaemia. Leukemia February 2005;19(2):310–2. PMID:15538400. Epub 2004/11/13. eng.

[12] Wiemels JL, Zhang Y, Chang J, Zheng S, Metayer C, Zhang L, et al. RAS mutation is associated with hyperdiploidy and parental characteristics in pediatric acute lymphoblastic leukemia. Leukemia March 2005;19(3):415–9. PMID:15674422. Epub 2005/01/28. eng.

[13] Yamamoto T, Isomura M, Xu Y, Liang J, Yagasaki H, Kamachi Y, et al. PTPN11, RAS and FLT3 mutations in childhood acute lymphoblastic leukemia. Leuk Res 2006;30(9):1085–9.

[14] Case M, Matheson E, Minto L, Hassan R, Harrison CJ, Bown N, et al. Mutation of genes affecting the RAS pathway is common in childhood acute lymphoblastic leukemia. Cancer Res August 15, 2008;68(16):6803–9. PMID:18701506.

[15] Wiemels JL, Kang M, Chang JS, Zheng L, Kouyoumji C, Zhang L, et al. Backtracking RAS mutations in high hyperdiploid childhood acute lymphoblastic leukemia. Blood Cells Mol Dis October 15, 2010;45(3):186–91. PMID:20688547. Pubmed Central PMCID: 2943008. Epub 2010/08/07. eng.

[16] Ahuja HG, Foti A, Bar-Eli M, Cline MJ. The pattern of mutational involvement of RAS genes in human hematologic malignancies determined by DNA amplification and direct sequencing. Blood April 15, 1990;75(8):1684–90. PMID:2183888. Epub 1990/04/15. eng.

[17] Lubbert M, Mirro Jr J, Miller CW, Kahan J, Isaac G, Kitchingman G, et al. N-ras gene point mutations in childhood acute lymphocytic leukemia correlate with a poor prognosis. Blood March 1, 1990;75(5):1163–9. PMID:2407301.

[18] Paulsson K, Horvat A, Strombeck B, Nilsson F, Heldrup J, Behrendtz M, et al. Mutations of FLT3, NRAS, KRAS, and PTPN11 are frequent and possibly mutually exclusive in high hyperdiploid childhood acute lymphoblastic leukemia. Genes Chromosomes Cancer January 2008;47(1):26–33. PubMed PMID:17910045.

[19] Kawamura M, Ohnishi H, Guo SX, Sheng XM, Minegishi M, Hanada R, et al. Alterations of the p53, p21, p16, p15 and RAS genes in childhood T-cell acute lymphoblastic leukemia. Leuk Res February 1999;23(2):115–26. PMID:10071127.

[20] Kawamura M, Kikuchi A, Kobayashi S, Hanada R, Yamamoto K, Horibe K, et al. Mutations of the p53 and ras genes in childhood t(1;19)-acute lymphoblastic leukemia. Blood May 1, 1995;85(9):2546–52. PMID:7727782.

[21] Terada N, Miyoshi J, Kawa-Ha K, Sasai H, Orita S, Yumura-Yagi K, et al. Alteration of N-ras gene mutation after relapse in acute lymphoblastic leukemia. Blood January 15, 1990;75(2): 453–7. PMID:1967219.

[22] Holmfeldt L, Wei L, Diaz-Flores E, Walsh M, Zhang J, Ding L, et al. The genomic landscape of hypodiploid acute lymphoblastic leukemia. Nat Genet March 2013;45(3):242–52. PMID:23334668. Epub 2013/01/22. eng.

[23] Zhang J, Mullighan CG, Harvey RC, Wu G, Chen X, Edmonson M, et al. Key pathways are frequently mutated in high-risk childhood acute lymphoblastic leukemia: a report from the Children's Oncology Group. Blood September 15, 2011;118(11):3080–7. PMID:21680795. Pubmed Central PMCID: 3175785. Epub 2011/06/18. eng.

[24] Zhang J, Ding L, Holmfeldt L, Wu G, Heatley SL, Payne-Turner D, et al. The genetic basis of early T-cell precursor acute lymphoblastic leukaemia. Nature January 12, 2012;481(7380):157–63. PMID:22237106. Pubmed Central PMCID: 3267575.

[25] Taketani T, Taki T, Sugita K, Furuichi Y, Ishii E, Hanada R, et al. FLT3 mutations in the activation loop of tyrosine kinase domain are frequently found in infant ALL with MLL rearrangements and pediatric ALL with hyperdiploidy. Blood February 1, 2004;103(3):1085–8. PMID:14504097. Epub 2003/09/25. eng.

[26] Chang P, Kang M, Xiao A, Chang J, Feusner J, Buffler P, et al. FLT3 mutation incidence and timing of origin in a population case series of pediatric leukemia. BMC Cancer 2010;10:513. PMID:20875128. Pubmed Central PMCID: 2955609.

[27] Armstrong SA, Mabon ME, Silverman LB, Li A, Gribben JG, Fox EA, et al. FLT3 mutations in childhood acute lymphoblastic leukemia. Blood May 1, 2004;103(9):3544–6. PMID:14670924.

[28] Leow S, Kham SK, Ariffin H, Quah TC, Yeoh AE. FLT3 mutation and expression did not adversely affect clinical outcome of childhood acute leukaemia: a study of 531 Southeast Asian children by the Ma-Spore study group. Hematol Oncol December 2011;29(4):211–9. PMID:21387358.

[29] Braoudaki M, Karpusas M, Katsibardi K, Papathanassiou C, Karamolegou K, Tzortzatou-Stathopoulou F. Frequency of FLT3 mutations in childhood acute lymphoblastic leukemia. Med Oncol December 2009;26(4):460–2. PMID:19085113. Epub 2008/12/17. eng.

[30] Andersson A, Paulsson K, Lilljebjorn H, Lassen C, Strombeck B, Heldrup J, et al. FLT3 mutations in a 10year consecutive series of 177 childhood acute leukemias and their impact on global gene expression patterns. Genes Chromosomes Cancer January 2008;47(1):64–70. PMID:17943971. Epub 2007/10/19. eng.

[31] Karabacak BH, Erbey F, Bayram I, Yilmaz S, Acipayam C, Kilinc Y, et al. Fms-like tyrosine kinase 3 mutations in childhood acute leukemias and their association with prognosis. Asian Pac J Cancer Prev 2010;11(4):923–7. PMID:21133602.

[32] Yamamoto Y, Kiyoi H, Nakano Y, Suzuki R, Kodera Y, Miyawaki S, et al. Activating mutation of D835 within the activation loop of FLT3 in human hematologic malignancies. Blood April 15, 2001;97(8):2434–9. PMID:11290608. Epub 2001/04/06. eng.

[33] Choudhary C, Schwable J, Brandts C, Tickenbrock L, Sargin B, Kindler T, et al. AML-associated Flt3 kinase domain mutations show signal transduction differences compared with Flt3 ITD mutations. Blood July 1, 2005;106(1):265–73. PMID:15769897. Epub 2005/03/17. eng.

[34] Armstrong SA, Kung AL, Mabon ME, Silverman LB, Stam RW, Den Boer ML, et al. Inhibition of FLT3 in MLL. Validation of a therapeutic target identified by gene expression based classification. Cancer Cell February 2003;3(2):173–83. PMID:12620411.

[35] Stam RW, den Boer ML, Schneider P, Nollau P, Horstmann M, Beverloo HB, et al. Targeting FLT3 in primary MLL-gene-rearranged infant acute lymphoblastic leukemia. Blood October 1, 2005;106(7):2484–90. PMID:15956279.

[36] Xu D, Zheng H, Yu WM, Qu CK. Activating mutations in protein tyrosine phosphatase Ptpn11 (Shp2) enhance reactive oxygen species production that contributes to myeloproliferative disorder. PLoS One 2013;8(5):e63152. PMID:23675459. Pubmed Central PMCID: 3651249.

[37] Molteni CG, Te Kronnie G, Bicciato S, Villa T, Tartaglia M, Basso G, et al. PTPN11 mutations in childhood acute lymphoblastic leukemia occur as a secondary event associated with high hyperdiploidy. Leukemia January 2010;24(1):232–5. PMID:19776760. Epub 2009/09/25. eng.

[38] Nicholson L, Knight T, Matheson E, Minto L, Case M, Sanichar M, et al. Casitas B lymphoma mutations in childhood acute lymphoblastic leukemia. Genes Chromosomes Cancer March 2012;51(3):250–6. PMID:22072526.

[39] Martinelli S, Checquolo S, Consoli F, Stellacci E, Rossi C, Silvano M, et al. Loss of CBL E3-ligase activity in B-lineage childhood acute lymphoblastic leukaemia. Br J Haematol October 2012;159(1):115–9. PMID:22834886.

[40] Saito Y, Aoki Y, Muramatsu H, Makishima H, Maciejewski JP, Imaizumi M, et al. Casitas B-cell lymphoma mutation in childhood T-cell acute lymphoblastic leukemia. Leuk Res August 2012;36(8):1009–15. PMID:22591685. Pubmed Central PMCID: PMC3693942. Epub 2012/05/18. eng.

[41] Shiba N, Park MJ, Taki T, Takita J, Hiwatari M, Kanazawa T, et al. CBL mutations in infant acute lymphoblastic leukaemia. Br J Haematol March 2012;156(5):672–4. PMID:21988239.

[42] Balgobind BV, Van Vlierberghe P, van den Ouweland AM, Beverloo HB, Terlouw-Kromosoeto JN, van Wering ER, et al. Leukemia-associated NF1 inactivation in patients with pediatric T-ALL and AML lacking evidence for neurofibromatosis. Blood April 15, 2008;111(8):4322–8. PMID:18172006. Epub 2008/01/04. eng.

[43] Mullighan CG, Goorha S, Radtke I, Miller CB, Coustan-Smith E, Dalton JD, et al. Genome-wide analysis of genetic alterations in acute lymphoblastic leukaemia. Nature April 12, 2007;446(7137):758–64. PMID:17344859.

[44] Davidsson J, Lilljebjorn H, Panagopoulos I, Fioretos T, Johansson B. BRAF mutations are very rare in B- and T-cell pediatric acute lymphoblastic leukemias. Leukemia August 2008;22(8):1619–21. PMID:18273045.

[45] Driessen EM, van Roon EH, Spijkers-Hagelstein JA, Schneider P, de Lorenzo P, Valsecchi MG, et al. Frequencies and prognostic impact of RAS mutations in MLL-rearranged acute lymphoblastic leukemia in infants. Haematologica June 2013;98(6):937–44. PMID:23403319. Pubmed Central PMCID: 3669451.

[46] Nakanishi H, Nakamura T, Canaani E, Croce CM. ALL1 fusion proteins induce deregulation of EphA7 and ERK phosphorylation in human acute leukemias. Proc Natl Acad Sci USA September 4, 2007;104(36):14442–7. PMID:17726105. Pubmed Central PMCID: PMC1964835. Epub 2007/08/30. eng.

[47] Mandanas RA, Leibowitz DS, Gharehbaghi K, Tauchi T, Burgess GS, Miyazawa K, et al. Role of p21 RAS in p210 bcr-abl transformation of murine myeloid cells. Blood September 15, 1993;82(6):1838–47. PMID:7691239.

[48] Pendergast AM, Quilliam LA, Cripe LD, Bassing CH, Dai Z, Li N, et al. BCR-ABL-induced oncogenesis is mediated by direct interaction with the SH2 domain of the GRB-2 adaptor protein. Cell October 8, 1993;75(1):175–85. PMID:8402896. Epub 1993/10/08. eng.

[49] Irving J, Matheson E, Minto L, Blair H, Case M, Halsey C, et al. Ras pathway mutations are prevalent in relapsed childhood acute lymphoblastic leukemia and confer sensitivity to MEK inhibition. Blood November 27, 2014;124(23):3420–30. PMID:25253770. Pubmed Central PMCID: 4246039.

[50] Yan J, Jiang N, Huang G, Tay JL, Lin B, Bi C, et al. Deregulated MIR335 that targets MAPK1 is implicated in poor outcome of paediatric acute lymphoblastic leukaemia. Br J Haematol October 2013;163(1):93–103. PMID:23888996.

[51] Schubbert S, Shannon K, Bollag G. Hyperactive Ras in developmental disorders and cancer. Nat Rev Cancer April 2007;7(4):295–308. PMID:17384584. Epub 2007/03/27. eng.

[52] Sabnis AJ, Cheung LS, Dail M, Kang HC, Santaguida M, Hermiston ML, et al. Oncogenic Kras initiates leukemia in hematopoietic stem cells. PLoS Biol March 17, 2009;7(3):e59. PMID:19296721. Pubmed Central PMCID: PMC2656550. Epub 2009/03/20. eng.

[53] Wang J, Liu Y, Tan LX, Lo JC, Du J, Ryu MJ, et al. Distinct requirements of hematopoietic stem cell activity and Nras G12D signaling in different cell types during leukemogenesis. Cell Cycle September 1, 2011;10(17):2836–9. PMID:21857161. Pubmed Central PMCID: PMC3218596. Epub 2011/08/23. eng.

[54] Kindler T, Cornejo MG, Scholl C, Liu J, Leeman DS, Haydu JE, et al. K-RasG12D-induced T-cell lymphoblastic lymphoma/leukemias harbor Notch1 mutations and are sensitive to gamma-secretase inhibitors. Blood October 15, 2008;112(8):3373–82. PMID:18663146. Pubmed Central PMCID: 2569181.

[55] Zhang J, Wang J, Liu Y, Sidik H, Young KH, Lodish HF, et al. Oncogenic Kras-induced leukemogeneis: hematopoietic stem cells as the initial target and lineage-specific progenitors as the potential targets for final leukemic transformation. Blood February 5, 2009;113(6):1304–14. PMID:19066392. Pubmed Central PMCID: PMC2637193. Epub 2008/12/11. eng.

[56] Wang J, Liu Y, Li Z, Wang Z, Tan LX, Ryu MJ, et al. Endogenous oncogenic Nras mutation initiates hematopoietic malignancies in a dose- and cell type-dependent manner. Blood July 14, 2011;118(2):368–79. PMID:21586752. Pubmed Central PMCID: PMC3138689. Epub 2011/05/19. eng.

[57] Lauchle JO, Kim D, Le DT, Akagi K, Crone M, Krisman K, et al. Response and resistance to MEK inhibition in leukaemias initiated by hyperactive Ras. Nature September 17, 2009;461(7262):411–4. PMID:19727076. Epub 2009/09/04. eng.

[58] Xu D, Liu X, Yu WM, Meyerson HJ, Guo C, Gerson SL, et al. Non-lineage/stage-restricted effects of a gain-of-function mutation in tyrosine phosphatase Ptpn11 (Shp2) on malignant transformation of hematopoietic cells. J Exp Med September 26, 2011;208(10):1977–88. PMID:21930766. Pubmed Central PMCID: 3182060.

[59] Kong G, Du J, Liu Y, Meline B, Chang YI, Ranheim EA, et al. Notch1 gene mutations target KRAS G12D-expressing CD8+ cells and contribute to their leukemogenic transformation. J Biol Chem June 21, 2013;288(25):18219–27. PMID:23673656. Pubmed Central PMCID: 3689964.

[60] Dail M, Li Q, McDaniel A, Wong J, Akagi K, Huang B, et al. Mutant Ikzf1, KrasG12D, and Notch1 cooperate in T lineage leukemogenesis and modulate responses to targeted agents. Proc Natl Acad Sci USA March 16, 2010;107(11):5106–11. PMID:20194733. Pubmed Central PMCID: 2841878.

[61] Van Meter ME, Diaz-Flores E, Archard JA, Passegue E, Irish JM, Kotecha N, et al. K-RasG12D expression induces hyperproliferation and aberrant signaling in primary hematopoietic stem/progenitor cells. Blood May 1, 2007;109(9):3945–52. PMID:17192389. Pubmed Central PMCID: PMC1874575. Epub 2006/12/29. eng.

[62] Li Q, Bohin N, Wen T, Ng V, Magee J, Chen SC, et al. Oncogenic Nras has bimodal effects on stem cells that sustainably increase competitiveness. Nature December 5, 2013;504(7478):143–7. PMID:24284627.

[63] Mullighan CG, Zhang J, Kasper LH, Lerach S, Payne-Turner D, Phillips LA, et al. CREBBP mutations in relapsed acute lymphoblastic leukaemia. Nature March 10, 2011;471(7337):235–9. PMID:21390130. Pubmed Central PMCID: 3076610.

[64] Tamai H, Miyake K, Takatori M, Miyake N, Yamaguchi H, Dan K, et al. Activated K-Ras protein accelerates human MLL/AF4-induced leukemo-lymphomogenicity in a transgenic mouse model. Leukemia 2011;25(5):888–91.

[65] Chen W, Li Q, Hudson WA, Kumar A, Kirchhof N, Kersey JH. A murine Mll-AF4 knock-in model results in lymphoid and myeloid deregulation and hematologic malignancy. Blood July 15, 2006;108(2):669–77. PMID:16551973. Pubmed Central PMCID: PMC1895483. Epub 2006/03/23. eng.

[66] Davidsson J, Paulsson K, Lindgren D, Lilljebjorn H, Chaplin T, Forestier E, et al. Relapsed childhood high hyperdiploid acute lymphoblastic leukemia: presence of preleukemic ancestral clones and the secondary nature of microdeletions and RTK-RAS mutations. Leukemia May 2010;24(5):924–31. PMID:20237506.

[67] Garza AS, Miller AL, Johnson BH, Thompson EB. Converting cell lines representing hematological malignancies from glucocorticoid-resistant to glucocorticoid-sensitive: signaling pathway interactions. Leuk Res May 2009;33(5):717–27. PMID:19012965. Epub 2008/11/18. eng.

[68] McCubrey JA, Steelman LS, Chappell WH, Abrams SL, Wong EW, Chang F, et al. Roles of the Raf/MEK/ERK pathway in cell growth, malignant transformation and drug resistance. Biochimica Biophysica Acta August 2007;1773(8):1263–84. PMID:17126425. Pubmed Central PMCID: 2696318. Epub 2006/11/28. eng.

[69] Yokota S, Nakao M, Horiike S, Seriu T, Iwai T, Kaneko H, et al. Mutational analysis of the N-ras gene in acute lymphoblastic leukemia: a study of 125 Japanese pediatric cases. Int J Hematol June 1998;67(4):379–87. PMID:9695411.

[70] Moorman AV, Richards SM, Martineau M, Cheung KL, Robinson HM, Jalali GR, et al. Outcome heterogeneity in childhood high-hyperdiploid acute lymphoblastic leukemia. Blood October 15, 2003;102(8):2756–62. PMID:12829593.

[71] Liang DC, Shih LY, Fu JF, Li HY, Wang HI, Hung IJ, et al. K-Ras mutations and N-Ras mutations in childhood acute leukemias with or without mixed-lineage leukemia gene rearrangements. Cancer February 15, 2006;106(4):950–6. PMID:16404744. Epub 2006/01/13. eng.

[72] Moorman AV, Irving J, Enshaei A, Parker CA, Sutton R, Kuiper R, et al., editors. Composite index for risk prediction in relapsed childhood acute lymphoblastic leukaemia. Vienna: European Haematology Association; 2015.

[73] Goemans BF, Zwaan CM, Harlow A, Loonen AH, Gibson BE, Hahlen K, et al. In vitro profiling of the sensitivity of pediatric leukemia cells to tipifarnib: identification of T-cell ALL and FAB M5 AML as the most sensitive subsets. Blood November 15, 2005;106(10):3532–7. PMID:16051737. Epub 2005/07/30. eng.

[74] Karp JE, Lancet JE, Kaufmann SH, End DW, Wright JJ, Bol K, et al. Clinical and biologic activity of the farnesyltransferase inhibitor R115777 in adults with refractory and relapsed acute leukemias: a phase 1 clinical-laboratory correlative trial. Blood June 1, 2001;97(11):3361–9. PMID:11369625.

[75] Widemann BC, Arceci RJ, Jayaprakash N, Fox E, Zannikos P, Goodspeed W, et al. Phase 1 trial and pharmacokinetic study of the farnesyl transferase inhibitor tipifarnib in children and adolescents with refractory leukemias: a report from the Children's Oncology Group. Pediatr Blood Cancer February 2011;56(2):226–33. PMID:20860038. Pubmed Central PMCID: 3271115.

[76] Downward J. Targeting RAS signalling pathways in cancer therapy. Nat Rev Cancer January 2003;3(1):11–22. PMID:12509763. Epub 2003/01/02. eng.

[77] Gilmartin AG, Bleam MR, Groy A, Moss KG, Minthorn EA, Kulkarni SG, et al. GSK1120212 (JTP-74057) is an inhibitor of MEK activity and activation with favorable pharmacokinetic properties for sustained in vivo pathway inhibition. Clin Cancer Res March 1, 2011;17(5): 989–1000. PMID:21245089. Epub 2011/01/20. eng.

[78] Kim K, Kong SY, Fulciniti M, Li X, Song W, Nahar S, et al. Blockade of the MEK/ERK signalling cascade by AS703026, a novel selective MEK1/2 inhibitor, induces pleiotropic anti-myeloma activity in vitro and in vivo. Br J Haematol May 2010;149(4):537–49. PMID:20331454. Epub 2010/03/25. eng.

[79] Bennouna J, Lang I, Valladares-Ayerbes M, Boer K, Adenis A, Escudero P, et al. A Phase II, open-label, randomised study to assess the efficacy and safety of the MEK1/2 inhibitor AZD6244 (ARRY-142886) versus capecitabine monotherapy in patients with colorectal cancer who have failed one or two prior chemotherapeutic regimens. Invest New Drugs 2011;29. PMID:20127139. Epub 2010/02/04. eng.

[80] Davies BR, Logie A, McKay JS, Martin P, Steele S, Jenkins R, et al. AZD6244 (ARRY-142886), a potent inhibitor of mitogen-activated protein kinase/extracellular signal-regulated kinase kinase 1/2 kinases: mechanism of action in vivo, pharmacokinetic/pharmacodynamic relationship, and potential for combination in preclinical models. Mol Cancer Ther August 2007;6(8):2209–19. PMID:17699718. Epub 2007/08/19. eng.

[81] Barrett SD, Bridges AJ, Dudley DT, Saltiel AR, Fergus JH, Flamme CM, et al. The discovery of the benzhydroxamate MEK inhibitors CI-1040 and PD 0325901. Bioorg Med Chem Lett December 15, 2008;18(24):6501–4. PMID:18952427.

[82] Yeh TC, Marsh V, Bernat BA, Ballard J, Colwell H, Evans RJ, et al. Biological characterization of ARRY-142886 (AZD6244), a potent, highly selective mitogen-activated protein kinase kinase 1/2 inhibitor. Clin Cancer Res March 1, 2007;13(5):1576–83. PMID:17332304. Epub 2007/03/03. eng.

[83] Meng XW, Chandra J, Loegering D, Van Becelaere K, Kottke TJ, Gore SD, et al. Central role of Fas-associated death domain protein in apoptosis induction by the mitogen-activated protein kinase kinase inhibitor CI-1040 (PD184352) in acute lymphocytic leukemia cells in vitro. J Biol Chem November 21, 2003;278(47):47326–39. PMID:12963734. Epub 2003/09/10. eng.

[84] Lock RB, Liem N, Farnsworth ML, Milross CG, Xue C, Tajbakhsh M, et al. The nonobese diabetic/severe combined immunodeficient (NOD/SCID) mouse model of childhood acute lymphoblastic leukemia reveals intrinsic differences in biologic characteristics at diagnosis and relapse. Blood June 1, 2002;99(11):4100–8. PMID:12010813.

[85] Janne PA, Shaw AT, Pereira JR, Jeannin G, Vansteenkiste J, Barrios C, et al. Selumetinib plus docetaxel for KRAS-mutant advanced non-small-cell lung cancer: a randomised, multicentre, placebo-controlled, phase 2 study. Lancet Oncol January 2013;14(1):38–47. PMID:23200175.

[86] Anuradha Banerjee RJ, Onar-Thomas A, Shengjie W, Nicolaides T, Turner D, Richardson S, et al. J Clin Oncol 2014;32. 5s abstr 10065.

[87] Shieh A, Ward AF, Donlan KL, Harding-Theobald ER, Xu J, Mullighan CG, et al. Defective K-Ras oncoproteins overcome impaired effector activation to initiate leukemia in vivo. Blood June 13, 2013;121(24):4884–93. PMID:23637129. Pubmed Central PMCID: PMC3682340. Epub 2013/05/03. eng.

[88] Engelman JA, Chen L, Tan X, Crosby K, Guimaraes AR, Upadhyay R, et al. Effective use of PI3K and MEK inhibitors to treat mutant Kras G12D and PIK3CA H1047R murine lung cancers. Nat Med December 2008;14(12):1351–6. PMID:19029981. Pubmed Central PMCID: 2683415.

[89] Haagensen EJ, Kyle S, Beale GS, Maxwell RJ, Newell DR. The synergistic interaction of MEK and PI3K inhibitors is modulated by mTOR inhibition. Br J Cancer March 13, 2012;106. PMID:22415236. Epub 2012/03/15. eng.

[90] Corcoran RB, Cheng KA, Hata AN, Faber AC, Ebi H, Coffee EM, et al. Synthetic lethal interaction of combined BCL-XL and MEK inhibition promotes tumor regressions in KRAS mutant cancer models. Cancer Cell January 14, 2013;23(1):121–8. PMID:23245996. Pubmed Central PMCID: 3667614.

[91] Rambal AA, Panaguiton ZL, Kramer L, Grant S, Harada H. MEK inhibitors potentiate dexamethasone lethality in acute lymphoblastic leukemia cells through the pro-apoptotic molecule BIM. Leukemia October 2009;23(10):1744–54. PMID:19404317. Pubmed Central PMCID: PMC2761998. Epub 2009/05/01. eng.

[92] Zhang W, Konopleva M, Burks JK, Dywer KC, Schober WD, Yang JY, et al. Blockade of mitogen-activated protein kinase/extracellular signal-regulated kinase kinase and murine double minute synergistically induces Apoptosis in acute myeloid leukemia via BH3-only proteins Puma and Bim. Cancer Res March 15, 2010;70(6):2424–34. PMID:20215498. Pubmed Central PMCID: PMC2840060. Epub 2010/03/11. eng.

[93] Brown P, Levis M, Shurtleff S, Campana D, Downing J, Small D. FLT3 inhibition selectively kills childhood acute lymphoblastic leukemia cells with high levels of FLT3 expression. Blood January 15, 2005;105(2):812–20. PMID:15374878.

[94] Liu W, Yu B, Xu G, Xu WR, Loh ML, Tang LD, et al. Identification of cryptotanshinone as an inhibitor of oncogenic protein tyrosine phosphatase SHP2 (PTPN11). J Med Chem September 26, 2013;56(18):7212–21. PMID:23957426.

[95] Ostrem JM, Peters U, Sos ML, Wells JA, Shokat KM. K-Ras(G12C) inhibitors allosterically control GTP affinity and effector interactions. Nature November 28, 2013;503(7477):548–51. PMID:24256730. Epub 2013/11/22. eng.

[96] Neubauer A, Maharry K, Mrozek K, Thiede C, Marcucci G, Paschka P, et al. Patients with acute myeloid leukemia and RAS mutations benefit most from postremission high-dose cytarabine: a Cancer and Leukemia Group B study. J Clin Oncol October 1, 2008;26(28):4603–9. PMID:18559876. Pubmed Central PMCID: 2653132.

[97] Meyer M, Rubsamen D, Slany R, Illmer T, Stabla K, Roth P, et al. Oncogenic RAS enables DNA damage- and p53-dependent differentiation of acute myeloid leukemia cells in response to chemotherapy. PLoS One 2009;4(11):e7768. PMID:19890398. Pubmed Central PMCID: 2767509. Epub 2009/11/06. eng.

CHAPTER 5

Oncogenic KRAS and the Inflammatory Micro-Environment in Pancreatic Cancer

H.-H. Chang[1], A. Schmidt[1,2], G. Eibl[1]

[1]*David Geffen School of Medicine at UCLA, Los Angeles, CA, United States;*
[2]*Universitätsklinikum Freiburg, Freiburg, Germany*

CONTENTS

ONCOGENIC KRAS DRIVES PANCREATIC CANCER DEVELOPMENT

Pancreatic cancer is one of the most deadly human diseases with overall 5-year survival rate of only about 5% and a median survival period of 4–6 months. It is now the fourth leading cause of cancer mortality in both men and women [1]. Due to the continuing lack of early diagnostic tools and absence of efficient therapeutic strategies, pancreatic cancer deaths are projected to become the second leading cause of cancer-related deaths by 2030 [3]. Among various malignancies in the pancreas, infiltrating pancreatic ductal adenocarcinoma (PDAC) is the predominant histo-pathological form, which accounts for up to 95% of all pancreatic tumors. Similar to many other epithelial cancers, PDAC is generally believed to originate from precursor lesions (pancreatic intra-epithelial neoplasias, PanINs) through a stepwise process with increasing degrees of morphologic atypia and accumulating genetic alterations [2]. Remarkably, activating mutations in *KRAS* are detected in >90% of human pancreatic tumors [3], and in ~40% of earliest-stage, lowest-grade PanIN-1 lesions [4], suggesting a key role of mutant KRAS in early pancreatic tumorigenesis. Among a variety of oncogenic *KRAS* mutations in human PDAC, point mutations in codon 12 occur most frequently [5]. Similarly in carcinogen-induced hamster PDAC, a well-known animal model sharing many characteristics with the human disease [6], *Kras* codon 12 mutation is a particularly frequent and early event.

Over the past 10–15 years, our understanding of oncogenic Kras in pancreatic tumorigenesis has been expanded because of several advances, including the development of conditional mutant Kras mouse models, which faithfully recapitulate many key aspects of human PDAC. These models generally involve pancreas-restricted expression of the constitutively active $Kras^{G12D}$ (or $Kras^{G12V}$) allele at its endogenous locus to maintain physiological expression levels. In the most basic form, the targeted expression of oncogenic Kras is achieved by

Conquering RAS. http://dx.doi.org/10.1016/B978-0-12-803505-4.00005-9

crossing the mouse strain harboring a latent *LSL-Kras*^{G12D} allele, with transgenic Cre mice that express Cre recombinase driven by pancreas-specific promoters [eg, Pdx1, Ptf-1a-p48 (p48), and elastase promoters]. The resulting mice, which are usually referred to as KC mice for "Kras/Cre," reproduce the full spectrum of progressive development of precursor lesions (PanIN 1–3) as seen in humans [7]. When additional genetic alterations are introduced, such as defects in tumor suppressor genes frequently mutated in human PDAC (eg, *Tp53, Ink4a/Arf*, and *Smad4*), the onset and incidence of invasive PDACs in this model are markedly promoted [8–10]. Importantly, Kras-driven murine PDACs also display an inflammatory desmoplastic stroma, a hallmark feature observed clinically [11]. Thus, the conditional Kras^{G12D} mouse model, with or without additional modifications, has allowed for detailed studies of the development of autochthonous pancreatic cancer with relevant micro-environmental alterations.

As revealed by the KC models in which Kras^{G12D} expression is targeted to mouse pancreatic progenitor cells (*Pdx1-Cre*– or *p48-Cre*–mediated), oncogenic Kras certainly plays a critical role in PanIN formation, an initial event in PDAC pathogenesis [7,9]. Besides PanIN initiation in early tumorigenesis, oncogenic Kras is also important in the maintenance of established PanINs, as well as in the development and sustenance of invasive PDACs. This notion is demonstrated by utilizing mouse models with inducible and reversible expression of Kras^{G12D} in the pancreas, with or without heterozygous knockout of the *Tp53* tumor suppressor gene [12,13]. In another study using a similar inducible model [14], oncogenic Kras^{G12D} was shown to indispensably support the continuous growth of pancreatic tumors by enhancing glycolic flux and diverting glucose metabolism into biosynthetic pathways, such as hexosamine biosynthesis pathway and the non-oxidative arm of the pentose phosphate pathway. The Kras^{G12D}-mediated metabolic reprogramming is achieved by transcriptional regulation of the key metabolic genes, which is mediated through mitogen-activated protein kinases (MAPK) and Myc. Notably, these studies also provided evidence that oncogenic Kras in PDAC is required for supporting the desmoplastic stroma [12,14], which is thought to play an important role in tumor progression and chemo-resistance [15]. Overall, oncogenic *Kras* mutation, found in early PanIN lesions and essentially all invasive pancreatic tumors, is believed to drive tumor initiation as well as to sustain tumor progression cell-autonomously or cell-non-autonomously.

THE IMPORTANCE OF INFLAMMATION IN PDAC DEVELOPMENT

Another notable feature of PDAC is the inflammatory micro-environment, which in this case is relevant both as a risk factor and as a key player in cancer progression. Epidemiologically, the association between inflammation and pancreatic cancer has been well established. For instance, patients with chronic pancreatitis (CP), a progressively destructive, inflammatory, and fibrotic condition, are at higher risk of developing pancreatic cancer [16–19]. In particular, among individuals with hereditary pancreatitis (a rare cause of CP), the risk for pancreatic cancer is tremendously increased, with a lifetime risk of 40–55% [19,20]. Interestingly, the correlation between pancreatitis and pancreatic cancer appears even stronger when the temporal history of previous pancreatitis is shorter [18,21], suggesting that in some cases pancreatitis could be an early manifestation in the process of pancreatic tumorigenesis and might be antecedently misdiagnosed. Indeed, an important hallmark of PDAC is the inflammatory desmoplastic stroma characterized by an extensive fibrosis and inflammatory response, resembling that of CP [15,22]. Infiltrating immune cells and stromal elements as components of a robust fibro-inflammatory reaction, along with initiated neoplastic cells, are thought to orchestrate the tumor micro-environment to foster proliferation, survival, metastasis, and immunosuppression, through the production of mediators such as cytokines, chemokines, and prostaglandins [23,24]. More detailed molecular mechanisms linking inflammation, oncogenic Kras, and PDAC progression will be discussed in the later sections.

There is a considerable amount of evidence from animal studies supporting the importance of inflammation in PDAC development and progression. In the KC mice mentioned earlier (*LSL-KRras*$^{G12D/+}$; *Pdx-1-Cre* or *p48-Cre*), with mutant Kras expressed throughout the pancreatic epithelium starting at embryogenic stages, pancreatic inflammation is one of the earliest sign as manifested by leukocyte infiltration and up-regulation of pro-inflammatory genes such as cyclooxygenase-2 (COX-2) in the PanIN lesions [7,25,26]. Of note, without the additional genetic alterations mentioned previously, the development of invasive PDAC in this model usually occurs very late (>12 months) and at low frequency (5–10%) [7], implicating the requirement of an extra insult. Consistent with this view, studies have shown that *Kras* oncogene-driven PanIN or PDAC progression in mice is accelerated under conditions of chronic or acute pancreatitis induced experimentally by cerulein [27–30], a cholecystokinin analogue that induces acinar cell death. Strikingly, CP is shown as a permissive factor for PanIN formation in certain mouse models normally refractory to oncogenic Kras activation alone, such as inducible models with mutant Kras expressed in differentiated adult pancreatic cells [27,28].

Interestingly, cerulein-induced pancreatitis in KC mice not only accelerates PanIN formation but also promotes epithelial–mesenchymal transition and leads to a marked increase of circulating pancreatic cells (CPCs) [31], which are the key parameters in the metastatic dissemination of transformed cells. Importantly, administration of dexamethasone, an anti-inflammatory steroid, almost completely eliminates Kras^{G12D}-induced PanIN lesions in the pancreas and significantly reduced CPCs in the blood [31]. Similar PanIN regression is also observed in the conditional Kras mice treated with non-steroidal anti-inflammatory drugs (NSAIDs) [32,33]. Overall, these studies highlight the prominent roles of inflammation in promoting neoplastic transformation in situ, as well as facilitating invasion and dissemination, which are key components of cancer metastasis. Moreover, inflammation may also induce the process of pancreatic acinar-to-ductal metaplasia (ADM) [27,34,35], a reprogramming event recognized as the mechanism preceding PanIN formation in PDAC initiation [30], supposing the presence of additional oncogenic stimulation. Taken together, the aforementioned animal studies underscore the pro-tumorigenic effect of injury and inflammation in essentially all stages of PDAC progression and implicates a cooperative interaction between oncogenic Kras and inflammatory cues.

INTER-RELATION BETWEEN ONCOGENIC KRAS AND INFLAMMATION IN PANCREATIC CANCER

Oncogenic Kras Induces Inflammatory Pathways in Pancreatic Cancer

The molecular connections between activated Kras and inflammation have mostly been elucidated in Kras mouse models. For example, using a transgenic model expressing higher than endogenous levels of $Kras^{G12V}$ in adult acinar cells, Logsdon et al. [36] demonstrated that hyperactivity of Kras led to a rapid development of progressive pancreatitis characterized by leukocyte infiltration, collagen deposition replacing the stroma, continued loss of acinar cells, and ADM. All of these findings closely resemble the clinical course of CP. Besides inducing pancreatic inflammation, enhanced Kras activity in acinar cells eventually resulted in spontaneous development of PDAC, which was not observed in mice expressing endogenous levels of mutant Kras unless accompanied by p53 ablation. Further, when examining the local inter-cellular mediators in the Kras-induced inflammation, transcriptional up-regulation of interleukins (IL) 1β and 6, Gro-α (CXCL1), and granulocyte-macrophage colony-stimulating factor (GM-CSF) was detected [36]. These inflammatory mediators are known to be associated with tumorigenesis, angiogenesis, and immune modulation through the activation of signaling pathways such as nuclear factor-kappa B (NF-κB) and Janus kinase/signal transducers and activators of transcription (JAK/STAT) cascades, which can further promote cytokine and chemokine production [23,37–41]. Interestingly, the pancreatic pathology caused by $Kras^{G12V}$ transgene over-expression was associated with cellular senescence [36], assessed by the up-regulation of the well-known senescence markers β-galactosidase and DEC1 (deleted in esophageal cancer 1), with increased expression of tumor suppressors p53, p15, and p16, known to be involved in oncogene-induced senescence. This phenomenon was initially shown in vitro that oncogenic

Kras induces cellular senescence associated with secretory properties, which is referred to as senescence-associated secretory phenotype (SASP) and is characterized by the secretion of a set of inflammatory cytokines including IL-6, Gro-α, and GM-CSF [42]. Oncogene-induced senescence has been validated in vivo in early stages of tumorigenesis in pancreatic and other cancers [43], acting as a barrier to prohibit growth and full transformation, while creating an inflammatory micro-environment. However, in high-grade PanIN lesions (PanIN-3) and advanced pancreatic cancer, evasion of senescence occurs with permissive mechanisms such as loss of tumor suppressors and pancreatitis-induced inflammation [33,36,44]. Collectively, oncogenic Kras-induced senescence, which eventually gets overcome during cancer progression, is an important component in establishing a tumor-promoting inflammatory environment by its secretory functions.

Further elaborating the pathways involved in oncogenic Kras-driven, cancer-related inflammatory responses, Ling et al. [45] demonstrated that the NF-κB pathway is a critical module in pancreatic cancer as also seen in other malignancies [46]. Pancreas-specific blockade of the NF-κB pathway by deletion of the inhibitor of nuclear factor kappa-B kinase subunit beta (IKKβ) led to inhibition of the key pro-inflammatory cytokines in a murine Kras model, along with improved histological outcomes in terms of inflammation and PanIN/PDAC progression. Further, $Kras^{G12D}$-induced over-expression of IL-1α, an NF-κB target gene known to exert strong pro-inflammatory activities, was hereby found to be a mediator between Kras and constitutive NF-κB activation in a forward feedback loop. Up-regulated by Kras-induced activator protein-1, IL-1α activates NF-κB, which in turn targets the genes of IL-1α and p62, both reciprocally sustaining NF-κB activation via a feedback loop [45].

In addition to triggering an NF-κB-activating feedback loop, IL-1α can also mediate Kras-induced inflammation through regulating SASP [47], which contributes essentially to inflammation and progression of metaplasia, as mentioned earlier. Using a murine model of pancreatic cancer (harboring a *Kras*^{G12D} mutation), it was shown that Kras-induced senescence depends on the histone deacetylase-associated protein SIN3B, whose levels are strongly correlated with Kras-induced IL-1α, implicating SIN3B-mediated regulation of IL-1α expression [44]. Deletion of *SIN3B* led to significantly decreased IL-1α, impaired senescence, a mitigated inflammatory response, and delays in the progression of $Kras^{G12D}$-driven pancreatic lesions. Importantly, similar correlation among SIN3B, IL-1α, and senescence-associated inflammatory response was also found in human tissues [44].

Activated by cytokines, STAT3 is another well-known critical mediator linking inflammation and neoplasia [48]. Studies using a $Kras^{G12D}$ mouse model have shown that it contributes to pancreatic cancer initiation and metaplasia-associated inflammation through matrix metalloproteinase (MMP)-7 [49]. Again, the inflammatory response was attenuated when STAT3 was knocked out in pancreatic epithelial cells. However, it is unclear if there is direct signaling linking Kras activity to STAT3 activation. In general, an inflammatory environment caused by oncogenic Kras attracts myeloid cells, which release IL-6. Secreted IL-6 in turn triggers STAT3 activation in pancreatic cells, enhancing the inflammatory response. Consequently, this amplification loop accelerates PanIN progression and pancreatic cancer development [50].

In addition, there are effects of high Kras activity on the cellular immune response. Oncogenic Kras attracts macrophages through up-regulation of a chemo-attractant, inter-cellular adhesion molecule-1 (ICAM-1) in acinar cells. Those infiltrating macrophages induce pancreatic inflammation and extracellular matrix remodeling by secreting tumor necrosis factor α (TNF-α) and MMP-9, which contribute to oncogenic Kras-driven ADM and progression to PanINs [26]. In addition, the IL-17 signaling axis, shown to mediate the cross talk between host immunity and pancreatic neoplastic cells, is another recently recognized response to oncogenic Kras. Using a conditional mouse model with tamoxifen-mediated $Kras^{G12D}$ activation, it was demonstrated that oncogenic Kras stimulates expression of IL-17 receptors on PanIN epithelial cells, as well as enhances recruitment of IL-17-producing T cells to the pancreatic micro-environment [51]. Importantly, genetic or pharmacological blockade of IL-17 signaling markedly down-regulates the downstream IL-6/Stat3 pathway, delays PanIN progression, and mitigates fibro-inflammatory responses in the pancreas. Based on this study, the "hematopoietic-to-epithelial" IL-17 signaling axis represents a prominent mediator of Kras-driven inflammation contributing to pancreatic neoplasia [51].

Inflammation Reciprocally Promotes Kras Over-activation

Contrary to the notion that mutant Kras is constitutively activated and spontaneously over-activates downstream pathways [eg, MAPK and phosphoinositide 3-kinase (PI3K)], *Kras* mutations are found in healthy, non-cancer-bearing individuals as well, where it does not exert a pro-tumorigenic property [52]. In fact, it was reported that certain stimuli may functionally augment the activity of mutant Kras beyond a critical threshold, and therefore promote Kras-driven neoplasia [53]. In pancreatic cancer, inflammation has been implicated as an important trigger to enhance Ras activity [36]. In line with this view, adult mice bearing a *Kras* mutation do not necessarily develop PanIN lesions or PDAC, unless CP co-exists [27]. Studies have shown that mutant Kras is able to activate an NF-κB-dependent positive feedback loop involving COX-2, which in turn amplifies Kras activity and therefore sustains itself [54]. Together, mutant Kras is able to create a pro-inflammatory environment via NF-κB, which not only helps the progression of pancreatic lesions but also supports a persistent strong Kras activity.

The inflammatory environment may also lead to oncogenic mutations of *KRAS*. Typically, activated inflammatory cells can foster an environment rich in reactive oxygen species (ROS) and growth factors, which may favor accumulation of DNA damage in rapidly proliferating parenchymal/neoplastic cells [23,55]. As in many other inflammatory disorders, ROS over-production and oxidative DNA adducts have been observed and pathologically linked to acute and chronic pancreatitis, as well as pancreatic cancer [56,57]. *KRAS* mutations were detected in ductal lesions in 27% of patients with CP without evidence of neoplasia and mutated p53 protein [58]. In addition, the presence and duration of CP is positively associated with the occurrence of *KRAS* mutations, implicating a mutagenic potential of inflammation [4,59]. Thus it is conceivable that an inflammatory environment in the pancreas can facilitate oncogenic mutagenesis of *KRAS*, as well as the accumulation of other spontaneous genetic aberrations (eg, in *CDKN2A*, *TP53*, and *SMAD4*).

Taken together, the interplay between oncogenic Kras and inflammation is likely multi-faceted. As a critical driver of pancreatic tumorigenesis, oncogenic Kras induces a variety of inflammatory responses, including secretion of cytokines, activation of key transcriptional factors, and modulation of immune cells. The inflammatory mediators also lead to functional and mutational stimulation of Kras, creating an amplification loop.

OBESITY-RELATED INFLAMMATION PROMOTES PANCREATIC CANCER DEVELOPMENT

One of the physiological conditions associated with low-grade, systemic chronic inflammation is obesity [60], which is a major modifiable risk factor [61–66] and unfavorable prognostic factor [67] for pancreatic cancer. Considering the rapid increase in the prevalence of obesity generally attributable to Western-style diets [68,69], understanding the complex relationship between obesity and pancreatic cancer will help identify targets for preventive or therapeutic strategies for this deadly disease. Although the biological mechanisms driving this association are still not clearly defined, accumulating evidence has pointed to the pro-inflammatory state associated with excess adiposity [70]. In obese individuals, there is an increase of systemic or local adipose tissue inflammation, characterized by immune cell infiltration and elevated levels of inflammatory mediators such as cytokines TNF-α and IL-6, and COX-2-derived prostaglandin E2 (PGE_2) [60,71–76].

The pro-tumorigenic effects of diet-driven obesity and associated inflammation have been more deeply studied using well-established animal models of pancreatic cancer development. In conditional KC mice high-fat diets (HFDs) are shown to accelerate PanIN development with enhanced pancreatic inflammation [77–79]. The promotional effect of the HFD may be associated with the pro-inflammatory TNF-α signaling, as PanIN development was abrogated on a TNF-α receptor-defective background (*TNFR1*$^{-/-}$) [77]. In a faithful model of obesity-associated Kras^{G12D}-driven pancreatic neoplasia, where a high-fat, high-calorie diet (HFCD) was given to conditional KC mice, we observed a higher percentage of advanced PanIN lesions [78], as well as increased PDAC incidence. Importantly, the development of PanINs and PDACs in the HFCD-fed KC mice was accompanied by robust signs of pancreatic inflammation characterized by acinar cell loss, inflammatory cell infiltrates, and stromal fibrosis [78]. This is consistent with other studies demonstrating marked infiltration of leukocytes with immuno-suppressive properties (eg, myeloid-derived suppressor cells, tumor-associated macrophages, and regulatory T cells) during early development of pancreatic cancer [25,80]. Although HFCD alone did not induce PanIN formation in wild-type mice, it led to a higher CP index reflecting minor loss of acinar cells, a moderated infiltration of inflammatory cells, and weak stromal fibrosis in wild-type mice. In addition, in mice fed the HFCD we observed substantial changes in cytokine/chemokine levels in the pancreas, including TNF-α and IL-6 [78]. Besides inflammatory signs in the pancreas, there was also an increased number of inflammatory foci in the visceral adipose tissues of HFCD-fed mice, especially in the peri-pancreatic region, which seems to be distinct from other visceral fat depots in terms of histology and cytokine profile [80a]. It is plausible that, in response to HFCD, inflammation is initiated in the expanded peri-pancreatic (or intra-pancreatic [81]) fat, which may in turn facilitate inflammation and neoplastic progression in the adjacent pancreas. Such effects may be mediated by secreted factors from adipocytes or adipose-infiltrating immune cells although detailed mechanistic studies are needed.

Philip et al. confirmed the results of HFD-promoted adiposity, PanIN lesions, pancreatic inflammation, and fibrosis in mice with either developmental or adulthood activation of Kras^{G12D} in the pancreas. Interestingly, the authors described a model, where mutant Kras can be further activated by an environmental stimulus such as COX-2-mediated inflammation induced by an HFD, thereby promoting PanIN progression and PDAC development [79]. These data strongly suggest a cross talk between oncogenic Kras and obesity-induced inflammation. Collectively, these studies demonstrate the significance of obesity-associated chronic inflammation in pancreatic tumorigenesis and highlight the links between obesity, inflammation, and oncogenic Kras signaling, involving diverse inflammatory mediators and a range of cell types in the tumor micro-environment.

POTENTIAL TARGETS FOR INTERVENTIONS

Despite its pivotal role in pancreatic cancer development, oncogenic KRAS is (or at least has been) widely considered an un-druggable target. Thus, targeting its effectors or regulators has been considered a clinically more effective therapeutic approach. For example, inhibition of the downstream mitogen-activated protein kinase kinase (MEK) has emerged as a promising therapeutic strategy, as demonstrated pre-clinically in combination with gemcitabine, an established chemotherapeutic agent for pancreatic cancer [82]. The PI3K/PDK1 signaling, another key effector of oncogenic Kras, has also been validated as a potential target in mouse models of Kras-driven PDAC [83]. However, clinical data for MEK inhibitors in combination with gemcitabine have been disappointing so far, at least in patients with advanced PDAC [84]. Targeting PI3K or downstream AKT/mammalian target of rapamycin (mTOR) signaling is also challenging because of the development of drug resistance as a common cause of treatment failure. It is therefore important to consider multi-target approaches when blocking signaling molecules to avoid compensatory over-activation of pro-tumorigenic pathways [85]. Based on the previous discussion, the inflammatory environment, which is associated with oncogenic Kras, may represent an intriguing target for therapy and prevention. To mitigate Kras-driven inflammatory responses, neutralizing antibodies can potentially be used against secreted or membrane-bound inflammatory mediators (eg, IL-6 [86], IL-1α, TNF-α, chemokine receptor CXCR2 [87], IL-17 or its receptor [88], and ICAM-1 [26]). Other approaches include blocking NF-κB activation and its downstream pathways [89] by IKK inhibitors, or other compounds with a broader range of activities such as anti-inflammatory drugs, phyto-chemicals (eg, curcumin [90]), proteasome inhibitors (eg, bortezomib [91,92]), and histone deacetylase inhibitors [93,94]. Similar to NF-κB, the transcription factor STAT3 is a convergence point for various pro-inflammatory and oncogenic pathways in pancreatic cancer, and therefore another attractive target [95]. The JAK/STAT pathway is an important signaling pathway, through which cytokines, hormones, and other soluble factors exert their functions, especially during inflammation [96]. A study carried out by Hurwitz et al. showed that patients with PDAC and high-grade systemic inflammation had an improvement in survival when receiving the JAK1/2 inhibitor ruxolitinib in combination with standard capecitabine chemotherapy. It is noteworthy that a significance between the treatment groups was only observed when a clear sign of systemic inflammation (defined as C-reactive protein >13 mg/L) was present. Patients with pancreatic tumors but no signs of inflammation did not benefit from this treatment. There are a few ongoing trials involving JAK1 or JAK2 inhibitors [84]. In addition, targeting obesity-related inflammation, especially in the visceral adipose tissue, is a promising route for prevention. Based on the studies mentioned previously, COX-2 is a key component of a positive feedback loop sustaining strong Kras activity [54], which can be attenuated by COX-2 inhibitors or

NSAIDs to prevent HFD-promoted pancreatic inflammation and tumorigenesis [79]. Thus, in addition to dietary control, approaches lowering COX-2 derived PGE_2 (a critical pro-inflammatory lipid mediator) may be pursued to prevent pancreatic cancer in obese individuals who are at high risk. Of note, scientific attention has been focused on metformin, which is the most widely prescribed drug for type-2 diabetes, an obesity-related metabolic disorder also linked to pancreatic cancer. Although the anti-tumor activities of metformin can largely be attributed to AMPK (AMP-activated protein kinase)-dependent inhibition of mTOR pathway and signaling cross talks [97,98], this reagent has also been shown to suppress NF-κB and STAT3 inflammatory pathways in mouse models of PDAC and other cancers [99,100]. The immune system can also be a target for prevention and therapy of pancreatic cancer. There have been attempts to activate elements of the adaptive immune system to eliminate tumor cells expressing cancer-specific antigens, including oncogenic Kras. A vaccine against mutant Kras has been developed and showed promising results with minimal side effects [101–103]. However, a strong immuno-suppressive micro-environment in pancreatic cancer may exclude and negate the effects of cytotoxic T cells. Therefore, strategies to overcome tumor immune evasion, which are now being actively pursued, should also be considered.

CONCLUSION

In summary, our understanding of oncogenic Kras-mediated effects in PDAC development, including its close linkage with inflammation, has been considerably improved owing to the valuable genetically engineered Kras animal models. The knowledge should be exploited for the development of novel efficacious intervention strategies disrupting the vicious connection between oncogenic Kras and inflammatory pathways. The potential mechanistic targets summarized earlier, and repurposing traditional drugs (eg, metformin and NSAIDs) as anti-cancer agents, await further clinical validation. Moreover, the complexity and regulation of immune responses in the pancreatic tumor micro-environment, and the detailed mechanisms of obesity-driven PDAC via inflammatory pathways, need to be further explored.

List of Acronyms and Abbreviations

ADM Acinar-to-ductal metaplasia
AMPK AMP-activated protein kinase
AP-1 Activator protein-1
COX-2 Cyclooxygenase-2
CP Chronic pancreatitis
CPC Circulating pancreatic cells
CXCL Chemokine (C-X-C motif) ligand

DEC1 Deleted in esophageal cancer 1
EMT Epithelial-mesenchymal transition
GM-CSF Granulocyte-macrophage colony-stimulating factor
HFCD High-fat, high-calorie diet
HFD High-fat diet
ICAM-1 Inter-cellular adhesion molecule-1
IKKβ Inhibitor of nuclear factor kappa-B kinase subunit beta
IL Interleukin
JAK Janus kinase
Kras/Cre KC
MAPK Mitogen-activated protein kinase
MEK Mitogen-activated protein kinase kinase
MMP Matrix metalloproteinase
MTOR Mammalian target of rapamycin
NF-κB Nuclear factor-kappa B
NSAID Non-steroidal anti-inflammatory drug
PanIN Pancreatic intra-epithelial neoplasia
PDAC Pancreatic ductal adenocarcinoma
PDK 3-Phosphoinositide-dependent protein kinase-1
Pdx1 Pancreatic and duodenal homeobox 1
PGE_2 Prostaglandin E2
PI3K Phosphoinositide 3-kinase
Ptf-1a Pancreas-specific transcription factor 1a
ROS Reactive oxygen species
SASP Senescence-associated secretory phenotype
STAT Signal transducers and activators of transcription
TNF-α Tumor necrosis factor α

References

[1] Siegel RL, Miller KD, Jemal A. Cancer statistics, 2015. CA Cancer J Clin 2015;65(1):5–29.

[2] Hezel AF, Kimm AC, elman, Stanger BZ, Bardeesy N, Depinho RA. Genetics and biology of pancreatic ductal adenocarcinoma. Genes Dev 2006;20(10):1218–49.

[3] Almoguera C, Shibata D, Forrester K, Martin J, Arnheim N, Perucho M. Most human carcinomas of the exocrine pancreas contain mutant c-K-ras genes. Cell 1988;53(4):549–54.

[4] Lohr M, Kloppel G, Maisonneuve P, Lowenfels AB, Luttges J. Frequency of K-ras mutations in pancreatic intraductal neoplasias associated with pancreatic ductal adenocarcinoma and chronic pancreatitis: a meta-analysis. Neoplasia 2005;7(1):17–23.

[5] Smit VT, Boot AJ, Smits AM, Fleuren GJ, Cornelisse CJ, Bos JL. KRAS codon 12 mutations occur very frequently in pancreatic adenocarcinomas. Nucleic Acids Res 1988;16(16):7773–82.

[6] Cerny WL, Mangold KA, Scarpelli DG. K-ras mutation is an early event in pancreatic duct carcinogenesis in the Syrian golden hamster. Cancer Res 1992;52(16):4507–13.

[7] Hingorani SR, Petricoin EF, Maitra A, Rajapakse V, King C, Jacobetz MA, et al. Preinvasive and invasive ductal pancreatic cancer and its early detection in the mouse. Cancer Cell 2003;4(6):437–50.

[8] Hingorani SR, Wang L, Multani AS, Combs C, Deramaudt TB, Hruban RH, et al. Trp53R172H and KrasG12D cooperate to promote chromosomal instability and widely metastatic pancreatic ductal adenocarcinoma in mice. Cancer Cell 2005;7(5):469–83.

[9] Aguirre AJ, Bardeesy N, Sinha M, Lopez L, Tuveson DA, Horner J, et al. Activated Kras and Ink4a/Arf deficiency cooperate to produce metastatic pancreatic ductal adenocarcinoma. Genes Dev 2003;17(24):3112–26.

[10] Bardeesy N, Cheng KH, Berger JH, Chu GC, Pahler J, Olson P, et al. Smad4 is dispensable for normal pancreas development yet critical in progression and tumor biology of pancreas cancer. Genes Dev 2006;20(22):3130–46.

[11] Guerra C, Barbacid M. Genetically engineered mouse models of pancreatic adenocarcinoma. Mol Oncol 2013;7(2):232–47.

[12] Collins MA, Bednar F, Zhang Y, Brisset JC, Galban S, Galban CJ, et al. Oncogenic Kras is required for both the initiation and maintenance of pancreatic cancer in mice. J Clin Invest 2012;122(2):639–53.

[13] Collins MA, Brisset JC, Zhang Y, Bednar F, Pierre J, Heist KA, et al. Metastatic pancreatic cancer is dependent on oncogenic Kras in mice. PLoS One 2012;7(12):e49707.

[14] Ying H, Kimmelman AC, Lyssiotis CA, Hua S, Chu GC, Fletcher-Sananikone E, et al. Oncogenic Kras maintains pancreatic tumors through regulation of anabolic glucose metabolism. Cell 2012;149(3):656–70.

[15] Neesse A, Michl P, Frese KK, Feig C, Cook N, Jacobetz MA, et al. Stromal biology and therapy in pancreatic cancer. Gut 2011;60(6):861–8.

[16] Lowenfels AB, Maisonneuve P, Cavallini G, Ammann RW, Lankisch PG, Andersen JR, et al. Pancreatitis and the risk of pancreatic cancer. International Pancreatitis Study Group. N Engl J Med 1993;328(20):1433–7.

[17] Malka D, Hammel P, Maire F, Rufat P, Madeira I, Pessione F, et al. Risk of pancreatic adenocarcinoma in chronic pancreatitis. Gut 2002;51(6):849–52.

[18] Duell EJ, Lucenteforte E, Olson SH, Bracci PM, Li D, Risch HA, et al. Pancreatitis and pancreatic cancer risk: a pooled analysis in the International Pancreatic Cancer Case-Control Consortium (PanC4). Ann Oncol 2012;23(11):2964–70.

[19] Yadav D, Lowenfels AB. The epidemiology of pancreatitis and pancreatic cancer. Gastroenterology 2013;144(6):1252–61.

[20] Weiss FU. Pancreatic cancer risk in hereditary pancreatitis. Front Physiol 2014;5:70.

[21] Bracci PM, Wang F, Hassan MM, Gupta S, Li D, Holly EA. Pancreatitis and pancreatic cancer in two large pooled case-control studies. Cancer Causes Control 2009;20(9):1723–31.

[22] Logsdon CD, Ji B. Ras activity in acinar cells links chronic pancreatitis and pancreatic cancer. Clin Gastroenterol Hepatol 2009;7(11 Suppl.):S40–3.

[23] Grivennikov SI, Greten FR, Karin M. Immunity, inflammation, and cancer. Cell 2010;140(6):883–99.

[24] Waghray M, Yalamanchili M, di Magliano MP, Simeone DM. Deciphering the role of stroma in pancreatic cancer. Curr Opin Gastroenterol 2013;29(5):537–43.

[25] Clark CE, Hingorani SR, Mick R, Combs C, Tuveson DA, Vonderheide RH. Dynamics of the immune reaction to pancreatic cancer from inception to invasion. Cancer Res 2007;67(19):9518–27.

[26] Liou GY, Doppler H, Necela B, Edenfield B, Zhang L, Dawson DW, et al. Mutant KRAS-induced expression of ICAM-1 in pancreatic acinar cells causes attraction of macrophages to expedite the formation of precancerous lesions. Cancer Discov 2015;5(1):52–63.

[27] Guerra C, Schuhmacher AJ, Canamero M, Grippo PJ, Verdaguer L, Perez-Gallego L, et al. Chronic pancreatitis is essential for induction of pancreatic ductal adenocarcinoma by K-Ras oncogenes in adult mice. Cancer Cell 2007;11(3):291–302.

[28] Gidekel Friedlander SY, Chu GC, Snyder EL, Girnius N, Dibelius G, Crowley D, et al. Context-dependent transformation of adult pancreatic cells by oncogenic K-Ras. Cancer Cell 2009;16(5):379–89.

[29] Carriere C, Young AL, Gunn JR, Longnecker DS, Korc M. Acute pancreatitis markedly accelerates pancreatic cancer progression in mice expressing oncogenic Kras. Biochem Biophys Res Commun 2009;382(3):561–5.

[30] Kopp JL, von Figura G, Mayes E, Liu FF, Dubois CL, Morris 4th JP, et al. Identification of Sox9-dependent acinar-to-ductal reprogramming as the principal mechanism for initiation of pancreatic ductal adenocarcinoma. Cancer Cell 2012;22(6):737–50.

[31] Rhim AD, Mirek ET, Aiello NM, Maitra A, Bailey JM, McAllister F, et al. EMT and dissemination precede pancreatic tumor formation. Cell 2012;148(1–2):349–61.

[32] Funahashi H, Satake M, Dawson D, Huynh NA, Reber HA, Hines OJ, et al. Delayed progression of pancreatic intraepithelial neoplasia in a conditional Kras(G12D) mouse model by a selective cyclooxygenase-2 inhibitor. Cancer Res 2007;67(15):7068–71.

[33] Guerra C, Collado M, Navas C, Schuhmacher AJ, Hernandez-Porras I, Canamero M, et al. Pancreatitis-induced inflammation contributes to pancreatic cancer by inhibiting oncogene-induced senescence. Cancer Cell 2011;19(6):728–39.

[34] Morris 4th JP, Cano DA, Sekine S, Wang SC, Hebrok M. Beta-catenin blocks Kras-dependent reprogramming of acini into pancreatic cancer precursor lesions in mice. J Clin Invest 2010;120(2):508–20.

[35] Liou GY, Doppler H, Necela B, Krishna M, Crawford HC, Raimondo M, et al. Macrophage-secreted cytokines drive pancreatic acinar-to-ductal metaplasia through NF-kappaB and MMPs. J Cell Biol 2013;202(3):563–77.

[36] Ji B, Tsou L, Wang H, Gaiser S, Chang DZ, Daniluk J, et al. Ras activity levels control the development of pancreatic diseases. Gastroenterology 2009;137(3):1072–82. 1082 e1-6.

[37] Roshani R, McCarthy F, Hagemann T. Inflammatory cytokines in human pancreatic cancer. Cancer Lett 2014;345(2):157–63.

[38] Vonderheide RH, Bayne LJ. Inflammatory networks and immune surveillance of pancreatic carcinoma. Curr Opin Immunol 2013;25(2):200–5.

[39] Pylayeva-Gupta Y, Lee KE, Hajdu CH, Miller G, Bar-Sagi D. Oncogenic Kras-induced GM-CSF production promotes the development of pancreatic neoplasia. Cancer Cell 2012;21(6):836–47.

[40] Bayne LJ, Beatty GL, Jhala N, Clark CE, Rhim AD, Stanger BZ, et al. Tumor-derived granulocyte-macrophage colony-stimulating factor regulates myeloid inflammation and T cell immunity in pancreatic cancer. Cancer Cell 2012;21(6):822–35.

[41] Baumgart S, Ellenrieder V, Fernandez-Zapico ME. Oncogenic transcription factors: cornerstones of inflammation-linked pancreatic carcinogenesis. Gut 2013;62(2):310–6.

[42] Coppe JP, Patil CK, Rodier F, Sun Y, Munoz DP, Goldstein J, et al. Senescence-associated secretory phenotypes reveal cell-nonautonomous functions of oncogenic RAS and the p53 tumor suppressor. PLoS Biol 2008;6(12):2853–68.

[43] Collado M, Serrano M. Senescence in tumours: evidence from mice and humans. Nat Rev Cancer 2010;10(1):51–7.

[44] Rielland M, Cantor DJ, Graveline R, Hajdu C, Mara L, Diaz Bde D, et al. Senescence-associated SIN3B promotes inflammation and pancreatic cancer progression. J Clin Invest 2014;124(5):2125–35.

[45] Ling J, Kang Y, Zhao R, Xia Q, Lee DF, Chang Z, et al. KrasG12D-induced IKK2/beta/NF-kappaB activation by IL-1alpha and p62 feedforward loops is required for development of pancreatic ductal adenocarcinoma. Cancer Cell 2012;21(1):105–20.

[46] Staudt LM. Oncogenic activation of NF-kappaB. Cold Spring Harb Perspect Biol 2010;2(6):a000109.

[47] Orjalo AV, Bhaumik D, Gengler BK, Scott GK, Campisi J. Cell surface-bound IL-1alpha is an upstream regulator of the senescence-associated IL-6/IL-8 cytokine network. Proc Natl Acad Sci USA 2009;106(40):17031–6.

[48] Yu H, Pardoll D, Jove R. STATs in cancer inflammation and immunity: a leading role for STAT3. Nat Rev Cancer 2009;9(11):798–809.

[49] Fukuda A, Wang SC, Morris 4th JP, Folias AE, Liou A, Kim GE, et al. Stat3 and MMP7 contribute to pancreatic ductal adenocarcinoma initiation and progression. Cancer Cell 2011;19(4):441–55.

[50] Lesina M, Kurkowski MU, Ludes K, Rose-John S, Treiber M, Kloppel G, et al. Stat3/Socs3 activation by IL-6 transsignaling promotes progression of pancreatic intraepithelial neoplasia and development of pancreatic cancer. Cancer Cell 2011;19(4):456–69.

[51] McAllister F, Bailey JM, Alsina J, Nirschl CJ, Sharma R, Fan H, et al. Oncogenic Kras activates a hematopoietic-to-epithelial IL-17 signaling axis in preinvasive pancreatic neoplasia. Cancer Cell 2014;25(5):621–37.

[52] Parsons BL, Meng F. K-RAS mutation in the screening, prognosis and treatment of cancer. Biomark Med 2009;3(6):757–69.

[53] Huang H, Daniluk J, Liu Y, Chu J, Li Z, Ji B, et al. Oncogenic K-Ras requires activation for enhanced activity. Oncogene 2014;33(4):532–5.

[54] Daniluk J, Liu Y, Deng D, Chu J, Huang H, Gaiser S, et al. An NF-kappaB pathway-mediated positive feedback loop amplifies Ras activity to pathological levels in mice. J Clin Invest 2012;122(4):1519–28.

[55] Hussain SP, Hofseth LJ, Harris CC. Radical causes of cancer. Nat Rev Cancer 2003; 3(4):276–85.

[56] Kodydkova J, Vavrova L, Stankova B, Macasek J, Krechler T, Zak A. Antioxidant status and oxidative stress markers in pancreatic cancer and chronic pancreatitis. Pancreas 2013;42(4):614–21.

[57] Wang M, Abbruzzese JL, Friess H, Hittelman WN, Evans DB, Abbruzzese MC, et al. DNA adducts in human pancreatic tissues and their potential role in carcinogenesis. Cancer Res 1998;58(1):38–41.

[58] Luttges J, Diederichs A, Menke MA, Vogel I, Kremer B, Kloppel G. Ductal lesions in patients with chronic pancreatitis show K-ras mutations in a frequency similar to that in the normal pancreas and lack nuclear immunoreactivity for p53. Cancer 2000;88(11):2495–504.

[59] Lohr M, Maisonneuve P, Lowenfels AB. K-Ras mutations and benign pancreatic disease. Int J Pancreatol 2000;27(2):93–103.

[60] Gregor MF, Hotamisligil GS. Inflammatory mechanisms in obesity. Annu Rev Immunol 2011;29:415–45.

[61] Bracci PM. Obesity and pancreatic cancer: overview of epidemiologic evidence and biologic mechanisms. Mol Carcinog 2012;51(1):53–63.

[62] Berrington de Gonzalez A, Sweetland S, Spencer E. A meta-analysis of obesity and the risk of pancreatic cancer. Br J Cancer 2003;89(3):519–23.

[63] Larsson SC, Orsini N, Wolk A. Body mass index and pancreatic cancer risk: a meta-analysis of prospective studies. Int J Cancer 2007;120(9):1993–8.

[64] Arslan AA, Helzlsouer KJ, Kooperberg C, Shu XO, Steplowski E, Bueno-de-Mesquita HB, et al. Anthropometric measures, body mass index, and pancreatic cancer: a pooled analysis from the Pancreatic Cancer Cohort Consortium (PanScan). Arch Intern Med 2010;170(9):791–802.

[65] Genkinger JM, Spiegelman D, Anderson KE, Bernstein L, van den Brandt PA, Calle EE, et al. A pooled analysis of 14 cohort studies of anthropometric factors and pancreatic cancer risk. Int J Cancer 2011;129(7):1708–17.

[66] Aune D, Greenwood DC, Chan DS, Vieira R, Vieira AR, Navarro Rosenblatt DA, et al. Body mass index, abdominal fatness and pancreatic cancer risk: a systematic review and non-linear dose-response meta-analysis of prospective studies. Ann Oncol 2012;23(4):843–52.

[67] Yuan C, Bao Y, Wu C, Kraft P, Ogino S, Ng K, et al. Prediagnostic body mass index and pancreatic cancer survival. J Clin Oncol 2013;31(33):4229–34.

[68] An R. Prevalence and trends of adult obesity in the US, 1999–2012. ISRN Obes 2014;2014:185132.

[69] Ng M, Fleming T, Robinson M, Thomson B, Graetz N, Margono C, et al. Global, regional, and national prevalence of overweight and obesity in children and adults during 1980–2013: a systematic analysis for the Global Burden of Disease Study 2013. Lancet 2014;384(9945):766–81.

[70] Khandekar MJ, Cohen P, Spiegelman BM. Molecular mechanisms of cancer development in obesity. Nat Rev Cancer 2011;11(12):886–95.

[71] Weisberg SP, McCann D, Desai M, Rosenbaum M, Leibel RL, Ferrante Jr AW. Obesity is associated with macrophage accumulation in adipose tissue. J Clin Invest 2003;112(12):1796–808.

[72] Kern PA, Ranganathan S, Li C, Wood L, Ranganathan G. Adipose tissue tumor necrosis factor and interleukin-6 expression in human obesity and insulin resistance. Am J Physiol Endocrinol Metab 2001;280(5):E745–51.

[73] Bahceci M, Gokalp D, Bahceci S, Tuzcu A, Atmaca S, Arikan S. The correlation between adiposity and adiponectin, tumor necrosis factor alpha, interleukin-6 and high sensitivity C-reactive protein levels. Is adipocyte size associated with inflammation in adults? J Endocrinol Invest 2007;30(3):210–4.

[74] Subbaramaiah K, Morris PG, Zhou XK, Morrow M, Du B, Giri D, et al. Increased levels of COX-2 and prostaglandin E2 contribute to elevated aromatase expression in inflamed breast tissue of obese women. Cancer Discov 2012;2(4):356–65.

[75] Morris PG, Zhou XK, Milne GL, Goldstein D, Hawks LC, Dang CT, et al. Increased levels of urinary PGE-M, a biomarker of inflammation, occur in association with obesity, aging, and lung metastases in patients with breast cancer. Cancer Prev Res (Phila) 2013;6(5):428–36.

[76] Chang HH, Young SH, Sinnett-Smith J, Chou CE, Moro A, Hertzer KM, et al. Prostaglandin E2 activates the mTORC1 pathway through an EP_4/cAMP/PKA and EP_1/Ca^{2+}-mediated mechanism in the human pancreatic carcinoma cell line PANC-1. Am J Physiol Cell Physiol 2015;309(10):C639–49.

[77] Khasawneh J, Schulz MD, Walch A, Rozman J, Hrabe de Angelis M, Klingenspor M, et al. Inflammation and mitochondrial fatty acid beta-oxidation link obesity to early tumor promotion. Proc Natl Acad Sci USA 2009;106(9):3354–9.

[78] Dawson DW, Hertzer K, Moro A, Donald G, Chang HH, Go VL, et al. High-fat, high-calorie diet promotes early pancreatic neoplasia in the conditional KrasG12D mouse model. Cancer Prev Res (Phila) 2013;6(10):1064–73.

[79] Philip B, Roland CL, Daniluk J, Liu Y, Chatterjee D, Gomez SB, et al. A high-fat diet activates oncogenic Kras and COX2 to induce development of pancreatic ductal adenocarcinoma in mice. Gastroenterology 2013;145(6):1449–58.

[80] Clark CE, Beatty GL, Vonderheide RH. Immunosurveillance of pancreatic adenocarcinoma: insights from genetically engineered mouse models of cancer. Cancer Lett 2009; 279(1):1–7.

[80a] Hertzer KM, Xu M, Moro A, Dawson DW, Du L, Li G, et al. Robust early inflammation of the peri-pancreatic visceral adipose tissue during diet-induced obesity in the KrasG12D model of pancreatic cancer. Pancreas 2016;45(3):458–65.

[81] Rebours V, Gaujoux S, d'Assignies G, Sauvanet A, Ruszniewski P, Levy P, et al. Obesity and fatty pancreatic infiltration are risk factors for pancreatic precancerous lesions (PanIN). Clin Cancer Res 2015;21(15):3522–8.

[82] Vena F, Li Causi E, Rodriguez-Justo M, Goodstal S, Hagemann T, Hartley JA, et al. The Mek1/2 inhibitor pimasertib enhances gemcitabine efficacy in pancreatic cancer models by altering protein levels of ribonucleotide reductase subunit-1 (Rrm1). Clin Cancer Res 2015;21.

[83] Eser S, Reiff N, Messer M, Seidler B, Gottschalk K, Dobler M, et al. Selective requirement of PI3K/PDK1 signaling for Kras oncogene-driven pancreatic cell plasticity and cancer. Cancer Cell 2013;23(3):406–20.

[84] Garrido-Laguna I, Hidalgo M. Pancreatic cancer: from state-of-the-art treatments to promising novel therapies. Nat Rev Clin Oncol 2015;12(6):319–34.

[85] Rozengurt E, Soares HP, Sinnet-Smith J. Suppression of feedback loops mediated by PI3K/mTOR induces multiple overactivation of compensatory pathways: an unintended consequence leading to drug resistance. Mol Cancer Ther 2014;13(11):2477–88.

[86] Ancrile B, Lim KH, Counter CM. Oncogenic Ras-induced secretion of IL6 is required for tumorigenesis. Genes Dev 2007;21(14):1714–9.

[87] Hertzer KM, Donald GW, Hines OJ. CXCR2: a target for pancreatic cancer treatment? Expert Opin Ther Targets 2013;17(6):667–80.

[88] Wu HH, Hwang-Verslues WW, Lee WH, Huang CK, Wei PC, Chen CL, et al. Targeting IL-17B-IL-17RB signaling with an anti-IL-17RB antibody blocks pancreatic cancer metastasis by silencing multiple chemokines. J Exp Med 2015;212(3):333–49.

[89] Carbone C, Melisi D. NF-kappaB as a target for pancreatic cancer therapy. Expert Opin Ther Targets 2012;16(Suppl. 2):S1–10.

[90] Kanai M. Therapeutic applications of curcumin for patients with pancreatic cancer. World J Gastroenterol 2014;20(28):9384–91.

[91] Nawrocki ST, Sweeney-Gotsch B, Takamori R, McConkey DJ. The proteasome inhibitor bortezomib enhances the activity of docetaxel in orthotopic human pancreatic tumor xenografts. Mol Cancer Ther 2004;3(1):59–70.

[92] Lee JK, Ryu JK, Yang KY, Woo SM, Park JK, Yoon WJ, et al. Effects and mechanisms of the combination of suberoylanilide hydroxamic acid and bortezomib on the anticancer property of gemcitabine in pancreatic cancer. Pancreas 2011;40(6):966–73.

[93] Chien W, Lee DH, Zheng Y, Wuensche P, Alvarez R, Wen DL, et al. Growth inhibition of pancreatic cancer cells by histone deacetylase inhibitor belinostat through suppression of multiple pathways including HIF, NFκB, and mTOR signaling in vitro and in vivo. Mol Carcinog 2014;53(9):722–35.

[94] Bai J, Demirjian A, Sui J, Marasco W, Callery MP. Histone deacetylase inhibitor trichostatin A and proteasome inhibitor PS-341 synergistically induce apoptosis in pancreatic cancer cells. Biochem Biophys Res Commun 2006;348(4):1245–53.

[95] Corcoran RB, Contino G, Deshpande V, Tzatsos A, Conrad C, Benes CH, et al. STAT3 plays a critical role in KRAS-induced pancreatic tumorigenesis. Cancer Res 2011;71(14):5020–9.

[96] Villarino AV, Kanno Y, Ferdinand JR, O'Shea JJ. Mechanisms of Jak/STAT signaling in immunity and disease. J Immunol 2015;194(1):21–7.

[97] Kisfalvi K, Eibl G, Sinnett-Smith J, Rozengurt E. Metformin disrupts crosstalk between G protein-coupled receptor and insulin receptor signaling systems and inhibits pancreatic cancer growth. Cancer Res 2009;69(16):6539–45.

[98] Kisfalvi K, Moro A, Sinnett-Smith J, Eibl G, Rozengurt E. Metformin inhibits the growth of human pancreatic cancer xenografts. Pancreas 2013;42(5):781–5.

[99] Tan XL, Bhattacharyya KK, Dutta SK, Bamlet WR, Rabe KG, Wang E, et al. Metformin suppresses pancreatic tumor growth with inhibition of NFkappaB/STAT3 inflammatory signaling. Pancreas 2015;44(4):636–47.

[100] Hirsch HA, Iliopoulos D, Struhl K. Metformin inhibits the inflammatory response associated with cellular transformation and cancer stem cell growth. Proc Natl Acad Sci USA 2013;110(3):972–7.

[101] Gjertsen MK, Bakka A, Breivik J, Saeterdal I, Gedde-Dahl 3rd T, Stokke KT, et al. Ex vivo ras peptide vaccination in patients with advanced pancreatic cancer: results of a phase I/II study. Int J Cancer 1996;65(4):450–3.

[102] Carbone DP, Ciernik IF, Kelley MJ, Smith MC, Nadaf S, Kavanaugh D, et al. Immunization with mutant p53- and K-ras-derived peptides in cancer patients: immune response and clinical outcome. J Clin Oncol 2005;23(22):5099–107.

[103] Gjertsen MK, Bakka A, Breivik J, Saeterdal I, Solheim BG, Soreide O, et al. Vaccination with mutant ras peptides and induction of T-cell responsiveness in pancreatic carcinoma patients carrying the corresponding RAS mutation. Lancet 1995;346(8987):1399–400.

CHAPTER 6

Activation of Ras by Post-Translational Modifications

S. Xiang[1], W. Bai[1], G. Bepler[2], X. Zhang[2]

[1]University of South Florida, Tampa, FL, United States; [2]Karmanos Cancer Institute, Detroit, MI, United States

CONTENTS

INTRODUCTION

The three human Ras genes encode four proteins: H-Ras, N-Ras, and two K-Ras splicing variants: K-Ras4A and K-Ras4B [1–4]. Each Ras protein is a 21-KDa guanine nucleotide–binding protein with an intrinsic GTPase activity. Ras is considered as a molecular switch for a wide variety of signaling pathways, including cell proliferation and apoptosis. In the GTP-bound state, Ras is active and can interact with effectors to trigger signaling pathways in the cell. In the GDP-bound state, Ras becomes inactive and losses its ability to bind to its effectors. Because both the intrinsic GTPase activity and GTP-binding ability of Ras proteins are relatively low, these properties of Ras are greatly facilitated by accessory proteins. In general, Ras proteins are activated by guanine nucleotide exchange factors (GEFs), which release GDP from the guanine nucleotide–binding site of Ras and promote GTP-binding to Ras leading to Ras interacting with various downstream effector proteins, such as rapidly accelerated fibrosarcoma (Raf-1), phosphoinositide 3-kinase (PI3K), and Ras-like guanine nucleotide dissociation stimulator (RalGDS). Conversely, Ras is inactivated by GTPase-activating proteins (GAPs), which enhance the intrinsic GTPase activity of Ras and increase GTP hydrolysis to GDP. Oncogenic Ras mutations were mostly found at amino acids G12, G13, and Q61. These Ras mutants have impaired intrinsic GTPase activities leading to the accumulation of sustainable activated Ras proteins [5]. In addition to Ras mutations, loss of GAP function leads to Ras activation and tumorigenesis in von Recklinghausen neurofibrosis [6].

Ras mutations are found in >30% of human cancers, and Ras isoforms are considered as drivers for oncogenesis [3,7,8]. Of these, K-Ras accounts for most of the Ras mutations in human cancers. Intense efforts have been focused on developing new drugs to efficiently inhibit activities of Ras by prevention of Ras-GTP formation, Ras-effector interactions, and Ras membrane localization.

Conquering RAS. http://dx.doi.org/10.1016/B978-0-12-803505-4.00006-0

[9]. However, the mutant Ras proteins are very difficult to target by anti-cancer drugs [10]. Therefore, understanding of the regulation of Ras activity is very important for future drug design.

Post-translational modification (PTM) of proteins is highly involved in cell signaling pathways. With these delicate modifications, cells could react to signals from the outside environment. For example, protein phosphorylation can act as an on and off switch for the propagation of intra-cellular signals [11]. All four Ras proteins consist of a highly homologous G-domain (1–168 or 1–169 aa) and a C-terminal 20 amino acids termed the hyper-variable region (HVR) [4,12]. In addition to regulation of Ras activity by the direct GTP-binding, Ras possesses multiple PTMs, including farnesylation, palmitoylation, geranylgeranylation, ubiquitination, phosphorylation, nitrosylation, and acetylation. These modifications can affect Ras trafficking, localization, and activation [13,14]. In particular, Ras proteins are required to be farnesylated and palmitoylated at its CAAX terminus to create a hydrophobic lipid domain for the inner plasma membrane attachment. These two lipidations are essential for the correct function of Ras proteins. In addition, several other PTMs including lysine acetylation and ubiquitination have been reported to influence the GTP exchange rate, which in turn affects the function of Ras. Here, we summarize most of the PTMs of Ras proteins and the enzymes that are responsible for these PTMs.

RAS EFFECTORS

Ras regulates many cellular functions, including gene expression, actin-cytoskeletal integrity, cell proliferation, differentiation, cell adhesion and migration, and apoptosis, through interaction with a myriad of downstream effectors. These effectors, including Rafs, PI3Ks, and RalGDS, all contain a conserved Ras-binding domain (RBD), which binds to an N-terminal effector loop on Ras. Raf-1 was identified as the first known Ras effector through genetic studies in *Drosophila melanogaster* and *Caenorhabditis elegans* [15,16], and it is the first Ser/Thr kinase in the Raf-MEK (MAPK/ERK kinase)-ERK (extracellular signal-regulated kinase) MAPK (mitogen-activated protein kinase) signaling cascade. MAPKs are highly conserved and are key players for the mitogen-stimulated cell proliferation. Many proteins in this pathway like epidermal growth factor receptor, Ras, and Raf are mutated in human cancers [17].

PI3Ks, which are lipid kinases for the synthesis of PI-3, 4, 5-triphosphate using PI-4, 5 phosphates as substrates, are also Ras effectors [18,19]. The Src homology-2 (SH2) domain of PI3Ks binds to the phospho-Tyr presented on activated Tyr kinases. Furthermore, the RBD on PI3Ks binds to Ras, which resides on the inner plasma membrane. Both bindings bring PI3Ks to the membrane facilitating the interaction of PI3Ks with substrate lipids. In the case of oncogenic Ras, this recruitment occurs in the absence of upstream stimuli. PI3K activity is known for the activation of Akt family Ser/Thr kinases, which inhibit cell apoptosis and promote cell survival. One of Ral GEFs, Ral-GDS, is also a direct effector of Ras–GTP [20,21]. The carboxyl-terminal region in Ral-GDS contains an RBD. Binding of Ras releases Ral-GDS from its inhibitory status. Ral-GDS is a Ral-specific guanine nucleotide GEF, stimulating GDP release and GTP loading and activation of Ral GTPase. Ral proteins, including RalA and RalB, also function as molecular switches to activate a number of cellular processes, including anchorage-independent cell growth, vesicle trafficking, and cytoskeletal organization [22,23]. Another important Ras effector, T lymphoma invasion and metastasis-inducing 1 (TIAM1), a Rac-specific GEF that stimulates GDP–GTP exchange of Rac, activates Rac, a Rho family GTPase, and connects extracellular signals to cytoskeletal activities [24]. In summary, Ras is a prolific signaling molecule that is involved both in normal cellular homeostasis and in pathologic conditions.

POST-TRANSLATIONAL MODIFICATIONS OF RAS

Ras is transiently expressed on the endoplasmic reticulum (ER) and stably present on the Golgi apparatus. Ras is also reported to reside on endosomes [25]. Matsuda et al. reported that Ras is only activated at the plasma membrane in response to epidermal growth factor (EGF) treatment [26]. By contrast, another small GTPase, Ras-related protein 1 (Rap1), is active in the center of the cell on EGF stimulation. It makes sense that RasGAP activity is very high near the ER and the Golgi apparatus, but is low under the plasma membrane.

Can Ras also be activated on non-plasma membranes? If these organelle membranes can serve as important sites for interaction between Ras and its effectors, these membranes could act as an environment for activating Ras for certain functions, for example, gene transcription, cell proliferation, and cell apoptosis [27–29]. Chiu et al. [27] noticed that oncogenic H-Ras and N-Ras interact with Raf-1 on the Golgi and that endogenous Ras and un-palmitoylated H-Ras are activated in response to mitogens on the Golgi and ER. Interestingly, Ras activation is restricted to the Golgi and is undetectable at the plasma membrane after T-cell receptor stimulation [30], mainly because Ras guanyl releasing protein 1 (RasGRP1) is a key exchange factor at Golgi for Ras activation [31,32]. Grp-null mice have marked defects in thymocyte differentiation, proliferation, and diacylglycerol-dependent Ras signaling [33]. Overall, Ras signaling from Golgi seems particularly prominent in lymphocytes.

We now understand that post-translational lipid modifications including farnesylation and palmitoylation direct Ras proteins to multiple membranes

[34,35]. Nascent Ras is a globular, hydrophilic protein, and its association with membranes is regulated by a variety of PTMs. The dynamic changes of these modifications could affect Ras trafficking and functions. Membrane association posits Ras in two dimensions and facilitates its interaction with GEFs, GAPs, and its downstream effectors. In addition to prenylation, other conditional modifications have been reported, including phosphorylation, ubiquitination, and acetylation. These conditional PTMs add another layer for regulating the activity of Ras.

Prenylation

Prenylation of CAAX proteins involves covalent addition of either farnesyl (15-carbon) or geranylgeranyl (20-carbon) isoprenoids to the conserved cysteine residues at or near the C-terminus of proteins. The enzymes that catalyze farnesylation and geranylgeranylation were identified in the 1990s and termed farnesyltransferase (FTase) and geranylgeranyltransferase type I (GGTase-I), respectively. The mating factor in fungi was first found to be farnesylated at a Cys residue in the late 1970s. Later, nuclear envelope protein, Lamin B, was identified as the first farnesylated mammalian protein, followed by Ras, which controls signaling transduction for cell growth and differentiation [36–38]. In addition to farnesylation, the 20-carbon geranylgeranyl isoprenoid can be attached to proteins [39]. In fact, geranylgeranyl is the predominant isoprenoid found in cellular proteins. Comparison of the amino acid sequence in the C-terminus of these proteins and factors have showed that all these proteins contain a Cys residue at the fourth position from the end, to form the so-called CAAX motif. A myriad of proteins containing the CAAX motif have been identified through searching the protein database. Many of those belong to the Ras superfamily of GTP-binding proteins. In particular, Ras prenylation is required for its oncogenic transformation in the fibroblast cells [40].

The CAAX Processing

Newly synthesized Ras is a globular and hydrophilic protein. Its association with the cell membrane requires a series of PTMs at its C-terminal HVR. Ras is a member of CAAX proteins that terminate with a CAAX sequence, in which C is a Cys, A is an aliphatic amino acid, and X is any amino acid. The correct processing and modification of the C-terminal CAAX motif is essential for the transport of Ras from Golgi to the plasma membrane. Ras first undergoes farnesylation at its C-terminal HVR region. Three enzymes including FTase, Ras converting enzyme 1 (RCE1), and isoprenylcysteine carboxyl methyltransferase (ICMT) work sequentially at the CAAX sequence for farnesylation, proteolysis, and methylation (Fig. 6.1) [35,41]. The FTase irreversibly adds a 15-carbon farnesyl lipid at the Cys residue in the cytosol, leading to RAS binding to the membrane of ER. Then RCE1 on ER cleaves the AAX amino acids to make the farnesylcysteine as a new C-terminus [42]. Ras farnesylation is a

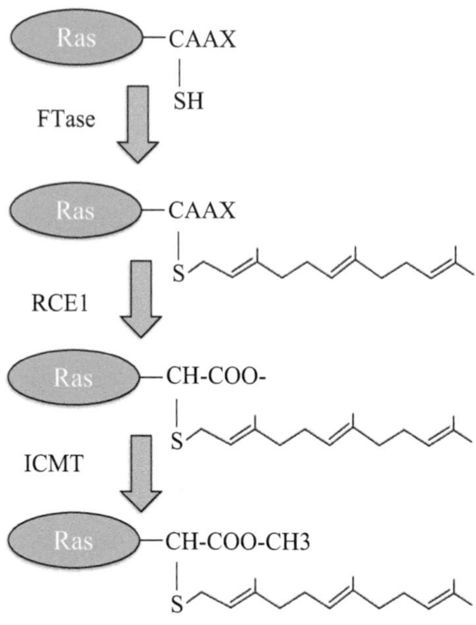

FIGURE 6.1 Farnesylation of Ras.

Farnesylation is the processes by which a cysteine residue in the C-terminal hyper-variable region of Ras proteins is first post-translationally modified with an isoprenoid lipid and then RCE1, a CAAX endoprotease, removes AAX triple amino acid residues. At last, the exposed carboxyl group is methylated by ICMT.

pre-requisite for the RCE1-mediated cleavage. Following the cleavage, another enzyme ICMT on ER catalyzes the methylesterfication of the α-carboxyl group of the farnesylcysteine [43]. The carboxyl methylation reaction is reversible, representing another possibility of regulating Ras localization and activity. However, a Ras-specific methylesterase has not yet been identified. These three modifications transform the C-terminal of Ras to a hydrophobic region. The newly farnesylated C-terminus inserts into the membrane with a modest affinity and is required for the biological activity of Ras proteins [44].

Not all CAAX proteins undergo farnesylation. If the amino acid in the X position is a Leu, for example, most of the Rho family GTPases and γ-subunits of heterotrimeric G proteins have a Leu at their C-terminus, the Cys residue will be added to a 20-carbon polyisoprene lipid by geranylgeranyltransferase type I (GGTase-I) via a stable thioester bond. As in Ras farnesylation, geranylgeranylation helps Rho proteins associate with the membrane. On the other hand, the geranylgeranyl lipid of Rho proteins can insert into the hydrophobic pocket of a cytosolic chaperone, Rho-specific GDP dissociation inhibitor (RhoGDI), which can extract Rho proteins from membranes and solubilize them in the cytosol [45]. Therefore, the geranylgeranyl-binding proteins like RhoGDI can regulate the protein trafficking.

The Enzymes for Prenylation

Farnesyltransferase

FTase is a heterodimer containing an α and a β subunit (48 and 46 KDa, respectively) in mammalian cells. The activity of FTase requires both Zn^{2+} and Mg^{2+} [46]. FTase can bind to farnesyl pyrophosphate (FPP) and peptide substrate independently. The β subunit of FTase binds to FPP with a very high affinity without covalent bond formation. It has been shown that both FPP and protein substrate bind to the β subunit of FTase. However, FTase was predicted to first bind to FPP, then rapidly reacts with the peptide substrate to form product, followed by releasing the farnesylated products [47].

One important feature of FTase is that it can recognize short peptides containing CAAX motifs and farnesylate those peptides [48]. However, substitution of an amino acid at the second A position by an aromatic residue Phe (F) makes the peptide CVFM become a competitive inhibitor, which is the basis for designing the peptidomimetic inhibitors of FTase [49].

Geranylgeranyltransferase Type I

GGTase-I catalyzes the addition of a geranylgeranyl group from geranylgeranylpyrophosphate to proteins ending with CAAL (the C-terminal amino acid usually is leucine or phenylalanine) [50]. The structure of GGTase-I has been resolved, and it is similar to that of FTase. It also contains alpha and beta two subunits, which share the same alpha subunit with FTase [51]. Some CAAX proteins, such as RhoA, cell division cycle 42 (CDC42), and RAP1, are geranylgeranylated by the protein GGTase-I. However, mammalian K-Ras and N-Ras, which have a carboxyl terminal Met, can be farnesylated by FTase in normal conditions and geranylgeranylated under conditions when FTase is inhibited in vivo [52–54].

Ras Converting Enzyme 1

RCE1 mediates proteolytic trimming of the C-terminal AAX tripeptide from CAAX proteins that have undergone isoprenylation. Medically relevant targets of RCE1 include Ras and Ras-related GTPases (eg, N-Ras, H-Ras, K-Ras, Rho) which are often mutated in cancer. RCE1 orthologs are present in all branches of life, but only eukaryotic orthologs are known to interact with isoprenylated substrates. A common feature of RCE1 proteins is that they are integral membrane proteins having multiple membrane spans. Eukaryotic RCE1 integrates into the ER, whereas prokaryotic orthologs are presumably located on the plasma membrane. The crystal structure of a *Methanococcus maripaludis* homolog of RCE1, whose endopeptidase specificity for farnesylated peptides mimics that of eukaryotic Rce1, has been resolved [42]. RCE1 is a founding member of a novel family of Glu-dependent intra-membrane proteases. Inactivation of Rce1 results in disassociation of Ras proteins from the plasma membrane leading to inhibition of Ras-induced transformation of fibroblasts, but progression of K-Ras-induced myeloproliferative disease [55–57]. Rce1-deficient mice develop lethal dilated cardiomyopathy [58], and RCE1 is also essential for the survival of photoreceptor cells [59].

Isoprenylcysteine Carboxyl Methyltransferase

ICMT catalyzes the final step of the CAAT processing. It was first identified in yeast as STE14P, which is responsible for methylation of the alpha-factor-mating pheromone at its prenylated C-terminus [60,61]. Later, ICMT was cloned from mammalian cells, exhibiting a similar characteristic as its yeast homologue [62]. The enzyme only recognizes substrates with prenylation, including both farnesyl and geranylgeranyl moieties, and the enzymatic activity localizes solely in the membrane fractions [63]. The crystal structure analysis of ICMT has showed that it is a unique methyltransferase that utilizes S-adenosyl-L-methionine (SAM) as a co-factor [43]. Briefly, ICMT comprises a core of five trans-membrane α helices and a co-factor-binding pocket enclosed within a highly conserved C-terminal catalytic subdomain. A tunnel linking the reactive methyl group of SAM to the inner membrane provides access for the prenyl lipid substrate.

Targeting Ras Prenylation

Inhibition of the farnesyl transferase by farnesyl transferase inhibitors (FTIs) has been examined to block Ras signaling in human cancers [64–67]. FTase has a binding site for recognizing the CAAX box of Ras proteins and another one for binding to FPP. Based on the crystal structure of FTase, the inhibitors were designed to successfully suppress the FTase activity [68,69]. However, an alternative prenylation called geranyl-geranylation was found in N-Ras and K-Ras, but not in H-Ras, when farnesylation is blocked by FTIs [53,70]. This modification is a 20-carbon isoprenoid group transferred by GGTase-I. Geranyl-geranylated Ras is associated with the plasma membrane and active, rendering cancer cells driven by N-Ras and K-Ras become tolerated with FTIs [17,52–54,71]. Because geranyl-geranylated proteins are much more abundant than farnesylated proteins, targeting GGTases by GGTase-Is may result in toxicity. It has been shown that GGTase-I deficiency reduces tumor formation and improves survival in mice with K-Ras-induced lung cancer [72]. Animal studies showed that targeting both FTase and GGTase-I in mice effectively reduces K-Ras-induced lung carcinogenesis, and extends the lifespan of these mice considerably more than either FTase or GGTase-I deficiency alone, suggesting that dual inhibition of FTase and GGTase-I may be therapeutically beneficial for cancer patients [73]. Therefore, designing an inhibitor targeting both enzymes could be a better strategy to block the K-Ras signaling pathway [74].

RCE1 has been proved to be important for Ras translocation [56]. Depletion of RCE1 in fibroblasts inhibits cell growth and reduces Ras-induced transformation [57]. But only modest inhibitory effect was shown when inactivating RCE1 in cancer cells. Substrate analogs like modified CAAX peptides have been developed to inhibit RCE1 activity [75]. However, the mice study showed that depletion of RCE1 exacerbates the K-Ras4B-driven myeloproliferative disease, indicating that this enzyme may not be a good anti-cancer drug target. It also suggests that Ras is one of many substrates of this enzyme.

The effect of inhibition of the third enzyme ICMT has been reported to dramatically reduce the cell growth rate and inhibit K-Ras-induced oncogenic transformation [76,77]. The anti-cancer drug methotrexate induces higher levels of homocysteine that causes hypomethylation in cells and reduces Ras methylation by almost 90%, leading to Ras mis-localization and a decreased activity of ERK1/2 and Akt [78]. New small molecule inhibitors for ICMT were synthesized to induce apoptosis and reduce tumor growth in a variety of model systems [79].

Palmitoylation and De-palmitoylation Cycle

Palmitoylation of Ras proteins is served as a secondary signal for plasma membrane targeting. Ras can be palmitoylated with the 16-carbon palmitoyl chain at cysteine residues just upstream the CAAX sequence on the Golgi apparatus. H-Ras (C181 and C184), N-Ras (C181), and K-Ras4A (C180) can all be palmitoylated by palmitoyl acyltransferases (PATs) [80,81]. The PAT responsible for Ras palmitoylation was identified as the DHHC9-GPC16 (DHHC domain containing 9-Golgi complex associated protein of 16 KDa) complex. DHHC9 is a trans-membrane protein that resides in the Golgi apparatus where palmitoylation of Ras occurs. In addition to DHH9-GPC16, other PATs may also involve in Ras palmitoylation. The palmitoylation of Ras proteins is required for the transport of Ras from Golgi to plasma membrane. The farnesylated Ras has only a modest affinity for membrane [44], but palmitoylation of Ras further increases the affinity more than 100-fold [82,83]. Altogether, both farnesylation and palmitoylation target Ras proteins to the cell membrane for activation.

It was found that H-Ras and N-Ras can be de-palmitoylated by acyl-protein thioesterase (APT1), which reduces the affinity of Ras to the plasma membrane, and release Ras proteins from the plasma membrane, which then are diffused back to the Golgi apparatus [84,85]. Thus, the palmitoylation–de-palmitoylation cycle of N-Ras and H-Ras regulates the Ras trafficking between the Golgi and plasma membrane. Inhibition of APT enzymes has been exploited to disrupt Ras trafficking and attenuates oncogenic growth signaling [86,87].

In addition, FKBP12, a *cis-trans* prolyl isomerase, regulates the palmitoylation/de-palmitoylation cycle of H-Ras by binding to H-Ras in a palmitoylation-dependent fashion and promoting de-palmitoylation through a mechanism that requires the prolyl–isomerase activity of FKBP12 on P179 of H-Ras [88].

Phosphorylation

Unlike other Ras isoforms, K-Ras4B cannot be palmitoylated because of the lack of cysteine residues upstream of the CAAX sequence. Instead, through a unique poly-lysine sequence that is positively charged in the HVR, K-Ras4B forms an electrostatic interaction with the negative head groups of the inner leaflets of the plasma membrane. However, the activity of K-Ras4B proteins can be attenuated by phosphorylation of S181 by proteins kinase C (PKC) [89]. Phosphorylation of S181 in this region by PKC neutralizes the positive charge and promotes the dissociation of K-Ras4B from the plasma membrane to ER.

In 1987, phosphorylation of K-Ras4B was first discovered after simulation of PKC with phorbol-12-myristate-13-acetate (PMA) by Ballester and colleagues [90]. They found that K-Ras4B is a direct substrate of PKC and predicted that the possible phosphorylation site was S181, which is located just upstream of the CAAX motif. At that time, Ras had not been characterized as a CAAX protein, which can be farnesylated at the Cys residue. Therefore, the significance of S181 phosphorylation was not recognized till 2006 when Philips and colleagues confirmed the existence of S181 phosphorylation in K-Ras4B by activation of PKC with bryostatins and PMA [89]. Interestingly, their results showed that after transient stimulation of PKC-α, a subset of K-Ras4B rapidly became phosphorylated at S181, which induced a fast translocation of the phosphorylated K-Ras4B from plasma membrane to internal membranes including ER, Golgi, and mitochondria. In addition, translocation of the phosphorylated K-Ras4B to the outer membrane of mitochondria also promoted cell apoptosis via a pathway that requires the interaction between K-Ras and Bcl-XL. Since phosphorylation of K-Ras4B at S181 induced cell death, bryostatin-1, a PKC agonist, indeed inhibited K-Ras-driven tumor growth in an S181-dependent manner. The cells with phosphorylation mimetic mutant K-RasG12V/S181E were sensitive to bryostatin treatment, but the cells with K-RasG12V/S181A, a non-phosphorylation mimic, were resistant to bryostatin treatment. Thus, the agents that promote S181 phosphorylation could be used for K-Ras mutant patients.

In contrast, depletion of the PKC activity by chronic PMA treatment in mouse fibroblasts has been previously reported to induce K-Ras-dependent cell apoptosis [91]. Stimulation of PKC in COS (CV-1 in origin, transformed by SV40) cells led to activation of Ras and formation of Ras-Raf-1 complexes containing active Raf-1 [92]. Expression of activated v-H-Ras in Jurkat cells, or expression of v-K-Ras in murine fibroblasts, induces apoptosis while suppression of the PKC activity [93]. Therefore, both stimulation and suppression of PKC activity may lead to K-Ras-dependent apoptosis, demonstrating that Ras controls apoptosis in a context-dependent manner.

Moreover, calmodulin (CaM) was found to prevent Ras activation by PKC in 3T3 fibroblasts, thus down-regulating the Ras-Raf-MEK-ERK pathway. In 3T3 fibroblast cells, activation of PKC by 12-O-tetradecanoylphorbol-13-acetate (TPA) does not induce Ras activation; however, inhibition of CaM together with TPA treatment induces Ras activation by PKC. It was shown that CaM inhibits Ras activation through direct binding of GTP-bound K-Ras4B when S181 is not phosphorylated [94]. The phosphomimetic mutant S181D cannot bind to CaM, indicating the two events, CaM binding and S181 phosphorylation, are mutually exclusive. CaM does not interact with H-Ras and N-Ras, which have no such phosphorylation site. For K-Ras, the C-terminal HVR region including polybasic residues, S181 residue, and farnesyl group are all required for CaM binding. In vitro phosphorylation experiments showed that the phosphorylation of K-Ras by PKC was inhibited by CaM [95]. Overall, the results indicate that PKC phosphorylation of K-Ras somehow activates Ras activity and this effect could be inhibited by CaM binding to K-Ras.

In addition, S181 phosphorylation has also been found to inhibit the organization of K-Ras proteins into plasma membrane nanoclusters [96,97]. Using high-resolution immunoelectron microscopy, Sarah et al. found that Raf-1 was preferentially recruited to and retained in K-Ras-GTP nanoclusters. Thus, the formation of K-Ras nanoclusters is crucial for mitogen-activated protein kinase activation. Inhibition of formation of K-Ras nanoclusters could prevent Ras signaling transduction. The formation of K-Ras-GTP nanoclusters is disrupted by S181 phosphorylation, which is catalyzed by PKC. Cellular fractionation results showed that the S181 phosphorylation mimetic GFP-K-RasG12V-S181D is more cytosolic compared with GFP-K-RasG12V, indicating that S181 phosphorylation neutralizes the positive charge in the polylysine stretch and promotes the dissociation of K-Ras from the plasma membrane [98]. Interestingly, Galectin-3 (Gal3), a predominantly cytosolic protein, which contains a hydrophobic binding pocket for binding of the prenylation moieties and helping K-Ras nanocluster formation, was not regulated by S181 phosphorylation, suggesting they are two independent contributors for K-Ras nanocluster formation.

Ubiquitination

The F-box protein β-transducin repeats-containing protein (β-TrCP) promotes poly-ubiquitination and proteasome-mediated degradation of all Ras isoforms and inhibits the Ras transformation ability in cells [99]. In addition, Wnt signaling decreases β-TrCP–induced polyubiquitination of Ras, thereby enhancing Ras activities [100]. Smad ubiquitination regulatory factor 2 (SMURF2) and ubiquitin-conjugating enzyme E2D 1 (UBCH5) were identified as a critical

E3:E2 complex for β-TrCP protein degradation. Loss of SMURF2 increased the β-TrCP level and accelerated Ras degradation [101].

Ubiquitination of Ras was first discovered to restrict Ras activities. H-Ras and N-Ras are found di-ubiquitinated to a single lysine residue, but K-Ras is refractory to this modification [102]. H-RasC186S mutant in HVR region abrogates the ubiquitination, indicating this ubiquitination is HVR dependent. The endosomal E3 ligase, Rabex-5/RabGEF1, the mammalian ortholog of yeast Vps9p, and a GEF for Rab5, is responsible for mono- and di-ubiquitination of H-Ras and N-Ras [103]. Ubiquitination of Ras restricts endosomal Ras anchoring and suppresses downstream ERK activation [103]. It was also shown that Ras ubiquitination by Rabex-5 restricts Ras signaling in *Drosophila* in vivo to establish proper organ size, wing vein pattern, and eye versus antennal fate [104]. It is uncertain which specific lysine site is involved although the authors generated an H-Ras8KR mutant in which all eight lysine sites were mutated to arginine and found this mutant was enriched in the Golgi compartment.

Interestingly, Ras can also be activated by ubiquitination. Sasaki et al. reported that K-Ras4B is mono-ubiquitinated on K104 and K147 [105]. The mono-ubiquitination of K-Ras at K147 promotes its activity through increased affinity for Raf and PI3K, enhancing the oncogenic functions of K-Ras. Mono-ubiquitinated K-Ras predominantly in the GTP-loaded state and unconjugated K-Ras predominantly in the GDP-loaded state and, therefore, ubiquitination may enhance GTP-loading of K-Ras. Mutation of K147 to arginine (R) in the K-Ras-G12V mutant reduced its binding to Raf and PI3K and decreased tumor size and weight in a 3T3 nude mice xenograft model. The structural analysis showed that K147 mono-ubiquitination in Ras severely inhibits the GAP-mediated hydrolysis, leading to Ras activation. Similarly, the ubiquitination sites at K117, K147, and K170 in H-Ras were also detected [88]. However, H-Ras is mainly activated by ubiquitination at K117, which accelerates an intrinsic nucleotide exchange, thereby promoting GTP loading. Interestingly, even a small percentage of K147 mono-ubiquitination of K-Ras could lead to meaningful biological consequences. These findings demonstrated that different Ras isoforms are mono-ubiquitinated at different sites, with distinct mechanisms for their activation. It is unknown which E3 ligases and de-ubiquitinating enzymes (DUBs) are responsible for K-Ras K147 and H-Ras K117 ubiquitination and de-ubiquitination, respectively. Overall, the two different outcomes after Ras ubiquitination indicate that various mechanisms are implicated in Ras regulation in different cell types and tissues for different Ras isoforms.

Acetylation

Besides ubiquitination, it was shown that both wild-type and mutant K-Ras are acetylated at K104 [106]. Acetylation of K104 attenuates the K-Ras transforming activity by interfering with GEF-induced nucleotide exchange. Both HDAC6 and SIRT2 de-acetylate K-Ras in cancer cells, and knockdown of HDAC6 or SirT2 reduces the viability in NIH3T3 cells expressing K-RasG12V, but not in cells expressing K-RasG12V/K104A [107]. These results suggest that therapeutic targeting of HDAC6 and/or SIRT2 may benefit the patients with cancers expressing mutant forms of K-Ras. Whether other Ras family proteins can be regulated by acetylation and de-acetylation remains to be determined.

S-Nitrosylation

Nitric oxide (NO) regulates cell division, tumorigenesis, cardiovascular signaling, DNA repair, apoptosis, and neurodegenerative diseases [108–111] and serves as a positive regulator of carcinogenesis and cancer development [112,113]. S-nitrosylation is a redox-mediated PTM through covalently attaching NO to protein cysteine residues. NO can be synthesized in vivo by various nitric oxide synthase (NOS) isoforms.

Ras proteins can be nitrosylated at C118 when exposed to NO [114,115]. This site is located in the nucleotide-binding region, which is highly conserved among all Ras isoforms. C118 can directly interact with nitrogen dioxide radical or with glutathionyl radical through its thio group. This modification increases guanine nucleotide exchange and promotes more efficient Ras activation, activating downstream MAPK pathways [116,117]. S-nitrosylation of wild-type Ras has been shown to promote pancreatic tumor growth without Ras oncogenic mutation. Activation of endothelial NOS by AKT through S1177 increases NO production, which in turn maintains pancreatic tumor growth by activating wild-type Ras through S-nitrosylation of C118 [118]. In estrogen receptor negative breast cancer, NO has been reported to activate the Ras/MEK/ERK signaling pathway through Ras S-nitrosylation, which phosphorylates and activates Ets-1 transcriptional activity [119]. The Ets-1 transcription factor further regulates the expression of genes involved in tumor progression and metastasis.

In lung cancer, wild-type Ras was also found to be S-nitrosylated and activated by nitrosative stress. It was also shown that Ras nitrosylation was de-nitrosylated by S-nitrosoglutathione (GSNO) reductase. De-nitrosylation of Ras decreases its activity. Interestingly, the activity and expression level of GSNO reductase are decreased in human lung cancer. In addition, GSNO reductase is abnormally distributed in the perinuclear region in cancer cell lines. Therefore, the decreased activity of GSNO reductase can promote the development of lung cancer, especially for smokers [120].

Bacterial Toxins and Exoenzymes

It has been reported that modifications by bacterial toxins inhibit Ras signaling to MAPKs. The Ras superfamily proteins seem to be a favorite target of bacterial virulence factors, especially for the Rho family proteins. However, there are also bacterial enzymes that are active against Ras. Bacterial toxin Exoenzymes S (ExoS) from *Pseudomonas aeruginosa*, an ADP-ribosyl transferase, can add ADP-ribose to R41 and R128 of Ras in vitro and in vivo [121]. The mutant Ras proteins including H-Ras, N-Ras, and K-Ras are modified in a manner qualitatively similar to their wild-type counterparts. ExoS disrupts the interaction between Ras and its effector Raf-1, thus inhibiting the Ras signaling. Another type of toxin is from *Clostridium sordelli*, which induces antibiotic-associated diarrhea. These toxins are monoglucosyltransferases that modify Ras as well as Rho GTPases by transferring the glucose moiety from UDP-glucose to T35, which is located in the effector region, transmitting signals to downstream effector molecules. Thus this modification stabilizes the effector region in its inactive GDP-bound state and impairs Ras signaling to the Raf/MEK/ERK, RalGEF/Ral, and PI3K-Akt signaling pathways, resulting in apoptotic cell death [122].

CONCLUSIONS

Ras proteins (H-Ras, N-Ras, and K-Ras) are small GTPases that mediate the transduction of signals from diverse extracellular stimuli to intra-cellular signaling pathways. It is now well established that Ras proteins dynamically localize to different cellular membranes via their PTMs. Although all Ras isoforms are found on the plasma membrane, H-Ras and N-Ras are also present at the Golgi and endosomes, and K-Ras can be found in the ER and the outer mitochondrial membrane [14,123]. Specifically, Ras proteins in the Golgi pool activate the MAPK pathway that is essential in the development of T-cell population [124]. It seems that activation of Ras in the plasma membrane is fast and transient, whereas activation of Ras in the Golgi is delayed and sustained [27]. Finally, the mitochondrial K-Ras is phosphorylated and induces apoptosis through its association with Bcl-XL [89]. These functions of Ras are all closely related to the PTM of Ras proteins. Ras has been shown to form stable homo-dimers or hetero-dimers [125–129] and activates the MAPK pathway, which contrasts the earlier belief that Ras signals as a monomeric GTPase or Ras forms a cluster with five to eight monomers. Consistent with the previous data, the intact HVR region is required for Ras dimerization and activation, indicating prenylation is important for Ras dimerization. It is unknown whether other PTMs are involved in dimer formation.

Direct pharmacologic inhibition of Ras has been challenging because of a lack of suitable binding pockets on the surface of Ras proteins. An alternative approach is to target the PTM of Ras, which prevents the attachment of Ras from the cell membrane. FTIs and GGTIs are designed to inhibit FTase and GGTase-I, which catalyze the farnesylation and geranylgeranylation on Ras, respectively. It was shown that in the presence of FTIs, the mutant K-Ras and N-Ras are still activated because of the activation of Ras by new geranylgeranylation. Thus, the development of dual inhibitors targeting both FTase and GGTase-I is highly demanded for inactivating Ras. In addition, correct localization and signaling by farnesylated K-Ras is regulated by the prenyl-binding protein photoreceptor cGMP phosphodiesterase δ (PDEδ), which augments K-Ras and H-Ras signaling by enriching Ras in the plasma membrane [130,131]. Therefore, targeting the interaction between Ras and prenyl-binding protein PDEδ has been proposed as an alternative strategy to inactivate Ras [132].

The post-translational modifications discussed in this chapter have been summarized in Table 6.1. Through mass spectrometry analysis, many new modified sites of Ras have been identified (see www.phosphosite.com). These new sites await further investigation for their effects on Ras activation. Overall, a deeper

Table 6.1 The Post-Translational Modifications of Ras Proteins

Modification	Site	Enzyme	Function
Farnesylation	H-Ras,C186, CVLS	FTase	The first signal targeting Ras to plasma membrane
	N-Ras, C186, CVVM	FTase	
	K-Ras4A, Cys186, CIIM	FTase	
	K-Ras4B, Cys185, CVIM	FTase	
Geranylgeranylation	N-Ras, C186, CVVM,	GGTase-I	The first signal targeting Ras to plasma membrane
	K-Ras4A, C186, CIIM,	GGTase-I	
	K-Ras4B, C185, CVIM,	GGTase-I	
Palmitoylation	H-Ras, C181, C184	DHHC9-GPC16	The second signal targeting Ras to plasma membrane; transporting Ras from Golgi to plasma membrane
	N-Ras, C181		
	K-Ras4A, C180		
Phosphorylation	KRas4B, S181	PKCα	Ras trafficking and cell death
Ubiquitination	H/N/K-Ras (unknown sites)	β-TrCP	Ras protein degradation
	H/N-Ras (unknown sites)	Rabex-5/RabGEF1	Inhibiting Ras
	K-Ras4B, K104, K147	Unknown E3 ligase	Activating Ras
	H-Ras, K117, K147, K170		
Acetylation	K-Ras4B, K104	HDAC6, SirT2	Attenuating K-Ras transforming activity
S-nitrosylation	All Ras isoforms, C118	NOS	Activating Ras
ADP-ribosylation	All Ras isoforms, R41, R128	ExoS	Inhibiting Ras signaling
Glucosylation	All Ras isoforms, T35	Monoglucosyl-transferases	Impairing Ras signaling

DHHC9-GPC16, *DHHC domain containing 9-Golgi complex associated protein of 16KDa;* ExoS, *exoenzymes S;* FTase, *farnesyltransferase;* GGTase-I, *geranylgeranyltransferase type I;* NOS, *nitric oxide synthase;* PKCα, *proteins kinase Cα;* β-TrCP, *β-transducin repeat-containing protein.*

understanding of Ras modifications will definitely help design efficient inhibitors for blocking Ras signaling.

List of Acronyms and Abbreviations

APT Acyl-protein thioesterase
CaM Calmodulin
CDC42 Cell division cycle 42
DAG Diacylglycerol
DHHC9-GPC16 DHHC domain containing 9-Golgi complex associated protein of 16 KDa
DUB De-ubiquitinating enzyme
ER Endoplasmic reticulum
FTase Farnesyltransferase
FTI Farnesyltransferase inhibitor
GAP GTPase-activating protein
GEF Guanine nucleotide exchange factor
GGTase Geranylgeranyltransferase
GRP1 Guanyl releasing protein 1
GSNO S-Nitrosoglutathione
HVR Hyper-variable region
ICMT Isoprenylcyteine carboxylmethyltransferase
IMP Intramembrane protease
NF Neurofibrosis
NO Nitric oxide
PAT Palmitoyl acyltransferase
PDEδ Photoreceptor cGMP phosphodiesterase δ
PI3K Phosphoinositide 3-kinase
PKC Proteins kinase C
PMA Phorbol-12-myristate-13-acetate
PTM Post-translational modification
Ral-GDS Ral guanine nucleotide dissociation stimulator
RBD Ras-binding domain
RCE1 Ras converting enzyme 1
RhoGDI Rho-specific GDP dissociation inhibitor
SAM S-adenosyl-L-methionine
SH2 Src homology-2
SMURF2 Smad ubiquitination regulatory factor 2
SOS Son of sevenless
TCR T-cell receptor
TIAM1 T lymphoma invasion and metastasis-inducing 1
TPA 12-O-tetradecanoylphorbol-13-acetate
β-TrCP β-transducin repeat-containing protein

Acknowledgment

This work was funded by an NIH grant R01CA164147 to X.Z.

References

[1] Fernandez-Medarde A, Santos E. Ras in cancer and developmental diseases. Genes Cancer 2011;2(3):344–58.

[2] Der CJ, Krontiris TG, Cooper GM. Transforming genes of human bladder and lung carcinoma cell lines are homologous to the Ras genes of Harvey and Kirsten sarcoma viruses. Proc Natl Acad Sci USA 1982;79(11):3637–40.

[3] Parada LF, Tabin CJ, Shin C, Weinberg RA. Human EJ bladder carcinoma oncogene is homologue of Harvey sarcoma virus Ras gene. Nature 1982;297(5866):474–8.

[4] Ehrhardt A, Ehrhardt GR, Guy X, Schrader JW. Ras and relatives–job sharing and networking keep an old family together. Exp Hematol 2002;30(10):1089–106.

[5] Gremer L, Merbitz-Zahradnik T, Dvorsky R, Cirstea IC, Kratz CP, Zenker M, et al. Germline KRAS mutations cause aberrant biochemical and physical properties leading to developmental disorders. Hum Mutat 2011;32(1):33–43.

[6] Le LQ, Parada LF. Tumor microenvironment and neurofibromatosis type I: connecting the GAPs. Oncogene 2007;26(32):4609–16.

[7] Prior IA, Lewis PD, Mattos C. A comprehensive survey of Ras mutations in cancer. Cancer Res 2012;72(10):2457–67.

[8] Pylayeva-Gupta Y, Grabocka E, Bar-Sagi D. RAS oncogenes: weaving a tumorigenic web. Nat Rev Cancer 2011;11(11):761–74.

[9] Gysin S,Z, Salt M, Young A, McCormick F. Therapeutic strategies for targeting Ras proteins. Genes Cancer 2011;2(3):359–72.

[10] Spiegel J, Cromm PM, Zimmermann G, Grossmann TN, Waldmann H. Small-molecule modulation of Ras signaling. Nat Chem Biol 2014;10(8):613–22.

[11] Karin M, Hunter T. Transcriptional control by protein phosphorylation: signal transmission from the cell surface to the nucleus. Curr Biol 1995;5(7):747–57.

[12] Willumsen BM, Christensen A, Hubbert NL, Papageorge AG, Lowy DR. The p21 Ras C-terminus is required for transformation and membrane association. Nature 1984;310(5978):583–6.

[13] Ahearn IM, Haigis K, Bar-Sagi D, Philips MR. Regulating the regulator: post-translational modification of RAS. Nat Rev Mol Cell Biol 2012;13(1):39–51.

[14] Hancock JF. Ras proteins: different signals from different locations. Nat Rev Mol Cell Biol 2003;4(5):373–84.

[15] Nishida Y, Hata M, Ayaki T, Ryo H, Yamagata M, Shimizu K, et al. Proliferation of both somatic and germ cells is affected in the *Drosophila* mutants of raf proto-oncogene. EMBO J 1988;7(3):775–81.

[16] Watari Y, Kariya K, Shibatohge M, Liao Y, Hu CD, Goshima M, et al. Identification of Ce-AF-6, a novel *Caenorhabditis elegans* protein, as a putative Ras effector. Gene 1998;224(1–2):53–8.

[17] Roberts PJ, Der CJ. Targeting the Raf-MEK-ERK mitogen-activated protein kinase cascade for the treatment of cancer. Oncogene 2007;26(22):3291–310.

[18] Castellano E, Downward J. RAS interaction with PI3K: more than just another effector pathway. Genes Cancer 2011;2(3):261–74.

[19] Mendoza MC, Er EE, Blenis J. The Ras-ERK and PI3K-mTOR pathways: cross-talk and compensation. Trends Biochem Sci 2011;36(6):320–8.

[20] Matsubara K, Kishida S, Matsuura Y, Kitayama H, Noda M, Kikuchi A. Plasma membrane recruitment of RalGDS is critical for Ras-dependent Ral activation. Oncogene 1999;18(6):1303–12.

[21] Neel NF, Martin TD, Stratford JK, Zand TP, Reiner DJ, Der CJ. The RalGEF-Ral effector signaling network: the road less traveled for anti-Ras drug discovery. Genes Cancer 2011;2(3):275–87.

[22] Hamad NM, Elconin JH, Karnoub AE, Bai W, Rich JN, Abraham RT, et al. Distinct requirements for Ras oncogenesis in human versus mouse cells. Genes Dev 2002;16(16):2045–57.

[23] Lim KH, Baines AT, Fiordalisi JJ, Shipitsin M, Feig LA, Cox AD, et al. Activation of RalA is critical for Ras-induced tumorigenesis of human cells. Cancer Cell 2005;7(6):533–45.

[24] Lambert JM, Lambert QT, Reuther GW, Malliri A, Siderovsky DP, Sondek J, et al. Tiam1 mediates Ras activation of Rac by a PI(3)K-independent mechanism. Nat Cell Biol 2002;4(8):621–5.

[25] Lu A, Tebar F, Alvarez-Moya B, Lopez-Alcala C, Calvo M, Enrich C, et al. A clathrin-dependent pathway leads to KRas signaling on late endosomes en route to lysosomes. J Cell Biol 2009;184(6):863–79.

[26] Ohba Y, Kurokawa K, Matsuda M. Mechanism of the spatio-temporal regulation of Ras and Rap1. EMBO J 2003;22(4):859–69.

[27] Chiu VK, Bivona T, Hach A, Sajous JB, Silletti J, Wiener H, et al. Ras signalling on the endoplasmic reticulum and the Golgi. Nat Cell Biol 2002;4(5):343–50.

[28] Bivona TG, Philips MR. Ras pathway signaling on endomembranes. Curr Opin Cell Biol 2003;15(2):136–42.

[29] Rebollo A, Perez-Sala D, Martinez AC. Bcl-2 differentially targets K-, N-, and H-Ras to mitochondria in IL-2 supplemented or deprived cells: implications in prevention of apoptosis. Oncogene 1999;18(35):4930–9.

[30] Bivona TG, Perez De Castro I, Aheran IM, Grana TM, Chiu VK, Lockyer PJ, et al. Phospholipase Cgamma activates Ras on the Golgi apparatus by means of RasGRP1. Nature 2003;424(6949):694–8.

[31] Ebinu JO, Stang SL, Teixeira C, Bottorff DA, Hooton J, Blumberg PM, et al. RasGRP links T-cell receptor signaling to Ras. Blood 2000;95(10):3199–203.

[32] Caloca MJ, Zugaza JL, Bustelo XR. Exchange factors of the RasGRP family mediate Ras activation in the Golgi. J Biol Chem 2003;278(35):33465–73.

[33] Dower NA, Stang SL, Bottorff DA, Ebinu JO, Dickie P, Ostergaard HL, et al. RasGRP is essential for mouse thymocyte differentiation and TCR signaling. Nat Immunol 2000;1(4):317–21.

[34] Apolloni A, Prior IA, Lindsay M, Parton RG, Hancock JF. H-Ras but not K-Ras traffics to the plasma membrane through the exocytic pathway. Mol Cell Biol 2000;20(7):2475–87.

[35] Choy E, Chiu VK, Silletti J, Feoktistov M, Morimoto T, Michaelson D, et al. Endomembrane trafficking of Ras: the CAAX motif targets proteins to the ER and Golgi. Cell 1999;98(1):69–80.

[36] Wolda SL, Glomset JA. Evidence for modification of lamin B by a product of mevalonic acid. J Biol Chem 1988;263(13):5997–6000.

[37] Farnsworth CC, Wolda SL, Gelb MH, Glomset JA. Human lamin B contains a farnesylated cysteine residue. J Biol Chem 1989;264(34):20422–9.

[38] Casey PJ, Solski PA, Der CJ, Buss JE. p21ras is modified by a farnesyl isoprenoid. Proc Natl Acad Sci USA 1989;86(21):8323–7.

[39] Farnsworth CC, Gelb MH, Glomset JA. Identification of geranylgeranyl-modified proteins in HeLa cells. Science 1990;247(4940):320–2.

[40] Kohl NE, Mosser SD, deSolms SJ, Giuliani EA, Pompliano DL, Graham SL, et al. Selective inhibition of Ras-dependent transformation by a farnesyltransferase inhibitor. Science 1993;260(5116):1934–7.

[41] Wright LP, Philips MR. Thematic review series: lipid posttranslational modifications. CAAX modification and membrane targeting of Ras. J Lipid Res 2006;47(5):883–91.

[42] Manolaridis I, Kulkarni K, Dodd RB, Ogasawara S, Zhang Z, Bineva G, et al. Mechanism of farnesylated CAAX protein processing by the intramembrane protease Rce1. Nature 2013;504(7479):301–5.

[43] Yang J, Kulkarni K, Manolaridis I, Zhang Z, Dodd RB, Mas-Droux C, et al. Mechanism of isoprenylcysteine carboxyl methylation from the crystal structure of the integral membrane methyltransferase ICMT. Mol Cell 2011;44(6):997–1004.

[44] Silvius JR, l'Heureux F. Fluorimetric evaluation of the affinities of isoprenylated peptides for lipid bilayers. Biochemistry 1994;33(10):3014–22.

[45] Dovas A, Couchman JR. RhoGDI: multiple functions in the regulation of Rho family GTPase activities. Biochem J 2005;390(Pt 1):1–9.

[46] Huang CC, Casey PJ, Fierke CA. Evidence for a catalytic role of zinc in protein farnesyltransferase. Spectroscopy of Co2+-farnesyltransferase indicates metal coordination of the substrate thiolate. J Biol Chem 1997;272(1):20–3.

[47] Park HW, Boduluri SR, Moomaw JF, Casey PJ, Beese LS. Crystal structure of protein farnesyltransferase at 2.25 angstrom resolution. Science 1997;275(5307):1800–4.

[48] Boutin JA, Marande W, Goussard M, Loynel A, Canet E, Fauchere JL. Chromatographic assay and peptide substrate characterization of partially purified farnesyl- and geranylgeranyltransferases from rat brain cytosol. Arch Biochem Biophys 1998;354(1):83–94.

[49] Goldstein JL, Brown MS, Stradley SJ, Reiss Y, Gierasch LM. Nonfarnesylated tetrapeptide inhibitors of protein farnesyltransferase. J Biol Chem 1991;266(24):15575–8.

[50] Zhang FL, Diehl RE, Kohl NE, Gibbs JB, Giros B, Casey PJ, et al. cDNA cloning and expression of rat and human protein geranylgeranyltransferase type-I. J Biol Chem 1994;269(5):3175–80.

[51] Seabra MC, Reiss Y, Casey PJ, Brown MS, Goldstein JL. Protein farnesyltransferase and geranylgeranyltransferase share a common alpha subunit. Cell 1991;65(3):429–34.

[52] Rowell CA, Kowalczyk JJ, Lewis MD, Garcia AM. Direct demonstration of geranylgeranylation and farnesylation of Ki-Ras in vivo. J Biol Chem 1997;272(22):14093–7.

[53] Whyte DB, Kirschmeier P, Hockenberry TN, Nunez-Oliva I, James L, Catino JJ, et al. K- and N-Ras are geranylgeranylated in cells treated with farnesyl protein transferase inhibitors. J Biol Chem 1997;272(22):14459–64.

[54] Lerner EC, Zhang TT, Knowles DB, Qian Y, Hamilton AD, Sebti SM. Inhibition of the prenylation of K-Ras, but not H- or N-Ras, is highly resistant to CAAX peptidomimetics and requires both a farnesyltransferase and a geranylgeranyltransferase I inhibitor in human tumor cell lines. Oncogene 1997;15(11):1283–8.

[55] Wahlstrom AM, Cutts BA, Karlsson C, Andersson KM, Liu M, Sjogren AK, et al. Rce1 deficiency accelerates the development of K-RAS-induced myeloproliferative disease. Blood 2007;109(2):763–8.

[56] Kim E, Ambroziak P, Otto JC, Taylor B, Ashby M, Shannon K, et al. Disruption of the mouse Rce1 gene results in defective Ras processing and mislocalization of Ras within cells. J Biol Chem 1999;274(13):8383–90.

[57] Bergo MO, Ambroziak P, Gregory C, George A, Otto JC, Kim E, et al. Absence of the CAAX endoprotease Rce1: effects on cell growth and transformation. Mol Cell Biol 2002;22(1):171–81.

[58] Bergo MO, Lieu HD, Gavino BJ, Ambroziak P, Otto JC, Casey PJ, et al. On the physiological importance of endoproteolysis of CAAX proteins: heart-specific RCE1 knockout mice develop a lethal cardiomyopathy. J Biol Chem 2004;279(6):4729–36.

[59] Christiansen JR, Kolandaivelu S, Bergo MO, Ramamurthy V. RAS-converting enzyme 1-mediated endoproteolysis is required for trafficking of rod phosphodiesterase 6 to photoreceptor outer segments. Proc Natl Acad Sci USA 2011;108(21):8862–6.

[60] Hrycyna CA, Clarke S. Farnesyl cysteine C-terminal methyltransferase activity is dependent upon the STE14 gene product in *Saccharomyces cerevisiae*. Mol Cell Biol 1990;10(10): 5071–6.

[61] Hrycyna CA, Sapperstein SK, Clarke S, Michaelis S. The *Saccharomyces cerevisiae* STE14 gene encodes a methyltransferase that mediates C-terminal methylation of a-factor and RAS proteins. EMBO J 1991;10(7):1699–709.

[62] Stephenson RC, Clarke S. Identification of a C-terminal protein carboxyl methyltransferase in rat liver membranes utilizing a synthetic farnesyl cysteine-containing peptide substrate. J Biol Chem 1990;265(27):16248–54.

[63] Romano JD, Schmidt WK, Michaelis S. The *Saccharomyces cerevisiae* prenylcysteine carboxyl methyltransferase Ste14p is in the endoplasmic reticulum membrane. Mol Biol Cell 1998;9(8):2231–47.

[64] Siegel-Lakhai WS, Crul M, Zhang S, Sparidans RW, Pluim D, Howes A, et al. Phase I and pharmacological study of the farnesyltransferase inhibitor tipifarnib (Zarnestra, R115777) in combination with gemcitabine and cisplatin in patients with advanced solid tumours. Br J Cancer 2005;93(11):1222–9.

[65] Sparano JA, Moulder S, Kazi A, Vahdat L, Li T, Pellegrino C, et al. Targeted inhibition of farnesyltransferase in locally advanced breast cancer: a phase I and II trial of tipifarnib plus dose-dense doxorubicin and cyclophosphamide. J Clin Oncol 2006;24(19):3013–8.

[66] Sparano JA, Moulder S, Kazi A, Coppola D, Negassa A, Vahdat L, et al. Phase II trial of tipifarnib plus neoadjuvant doxorubicin-cyclophosphamide in patients with clinical stage IIB-IIIC breast cancer. Clin Cancer Res 2009;15(8):2942–8.

[67] Omer CA, Chen Z, Diehl RE, Conner MW, Chen HY, Trumbauer ME, et al. Mouse mammary tumor virus-Ki-rasB transgenic mice develop mammary carcinomas that can be growth-inhibited by a farnesyl:protein transferase inhibitor. Cancer Res 2000;60(10): 2680–8.

[68] Lerner EC, Qian Y, Blaskovich MA, Fossum RD, Vogt A, Sun J, et al. Ras CAAX peptidomimetic FTI-277 selectively blocks oncogenic Ras signaling by inducing cytoplasmic accumulation of inactive Ras-Raf complexes. J Biol Chem 1995;270(45):26802–6.

[69] Sun J, Qian Y, Hamilton AD, Sebti SM. Ras CAAX peptidomimetic FTI 276 selectively blocks tumor growth in nude mice of a human lung carcinoma with K-Ras mutation and p53 deletion. Cancer Res 1995;55(19):4243–7.

[70] Crespo NC, Ohkanda J, Yen TJ, Hamilton AD, Sebti SM. The farnesyltransferase inhibitor, FTI-2153, blocks bipolar spindle formation and chromosome alignment and causes prometaphase accumulation during mitosis of human lung cancer cells. J Biol Chem 2001;276(19):16161–7.

[71] Sun J, Qian Y, Hamilton AD, Sebti SM. Both farnesyltransferase and geranylgeranyltransferase I inhibitors are required for inhibition of oncogenic K-Ras prenylation but each alone is sufficient to suppress human tumor growth in nude mouse xenografts. Oncogene 1998;16(11):1467–73.

[72] Sjogren AK, Andersson KM, Liu M, Cutts BA, Karlsson C, Wahlstrom AM, et al. GGTase-I deficiency reduces tumor formation and improves survival in mice with K-RAS-induced lung cancer. J Clin Invest 2007;117(5):1294–304.

[73] Liu M, Sjogren AK, Karlsson C, Ibrahim MX, Andersson KM, Olofsson FJ, et al. Targeting the protein prenyltransferases efficiently reduces tumor development in mice with K-RAS-induced lung cancer. Proc Natl Acad Sci USA 2010;107(14):6471–6.

[74] Lobell RB, Liu D, Buser CA, Davide JP, De Puy E, Hamilton K, et al. Preclinical and clinical pharmacodynamic assessment of L-778,123, a dual inhibitor of farnesyl:protein transferase and geranylgeranyl:protein transferase type-I. Mol Cancer Ther 2002;1(9):747–58.

[75] Hollander IJ, Frommer E, Aulabaugh A, Mallon R. Human Ras converting enzyme endoproteolytic specificity at the P2′ and P3′ positions of K-Ras-derived peptides. Biochim Biophys Acta 2003;1649(1):24–9.

[76] Wahlstrom AM, Cutts BA, Liu M, Lindskog A, Karlsson C, Sjogren AK, et al. Inactivating ICMT ameliorates K-RAS-induced myeloproliferative disease. Blood 2008;112(4):1357–65.

[77] Bergo MO, Gavino BJ, Hong C, Beigneux AP, McMahon M, Casey PJ, et al. Inactivation of ICMT inhibits transformation by oncogenic K-Ras and B-Raf. J Clin Invest 2004;113(4):539–50.

[78] Winter-Vann AM, Kamen BA, Bergo MO, Young SG, Melnyk S, James SJ, et al. Targeting Ras signaling through inhibition of carboxyl methylation: an unexpected property of methotrexate. Proc Natl Acad Sci USA 2003;100(11):6529–34.

[79] Wang M, Hossain MS, Tan W, Coolman B, Zhou J, Liu S, et al. Inhibition of isoprenylcysteine carboxylmethyltransferase induces autophagic-dependent apoptosis and impairs tumor growth. Oncogene 2010;29(35):4959–70.

[80] Rocks O, Gerauer M, Vartak N, Koch S, Huang ZP, Pechlivanis M, et al. The palmitoylation machinery is a spatially organizing system for peripheral membrane proteins. Cell 2010;141(3):458–71.

[81] Swarthout JT, Lobo S, Farh L, Croke MR, Greentree WK, Deschenes RJ, et al. DHHC9 and GCP16 constitute a human protein fatty acyltransferase with specificity for H- and N-Ras. J Biol Chem 2005;280(35):31141–8.

[82] Shahinian S, Silvius JR. Doubly-lipid-modified protein sequence motifs exhibit long-lived anchorage to lipid bilayer membranes. Biochemistry 1995;34(11):3813–22.

[83] Schroeder H, Leventis R, Rex S, Schelhaas M, Nagele E, Waldmann H, et al. S-Acylation and plasma membrane targeting of the farnesylated carboxyl-terminal peptide of N-Ras in mammalian fibroblasts. Biochemistry 1997;36(42):13102–9.

[84] Rocks O, Peyker A, Kahms M, Verveer PJ, Koerner C, Lumbierres M, et al. An acylation cycle regulates localization and activity of palmitoylated Ras isoforms. Science 2005;307(5716):1746–52.

[85] Goodwin JS, Drake KR, Rogers C, Wright L, Lippincott-Schwartz J, Philips MR, et al. Depalmitoylated Ras traffics to and from the Golgi complex via a nonvesicular pathway. J Cell Biol 2005;170(2):261–72.

[86] Dekker FJ, Rocks O, Vartak N, Menninger S, Hedberg C, Balamurugan R, et al. Small-molecule inhibition of APT1 affects Ras localization and signaling. Nat Chem Biol 2010;6(6):449–56.

[87] Davda D, Martin BR. Acyl protein thioesterase inhibitors as probes of dynamic S-palmitoylation. Medchemcomm 2014;5(3):268–76.

[88] Baker R, Wilkerson EM, Sumita K, Isom DG, Sasaki AT, Dohlman HG, et al. Differences in the regulation of K-Ras and H-Ras isoforms by monoubiquitination. J Biol Chem 2013;288(52):36856–62.

[89] Bivona TG, Quatela SE, Bodemann BO, Ahearn IM, Soskis MJ, Mor A, et al. PKC regulates a farnesyl-electrostatic switch on K-Ras that promotes its association with Bcl-XL on mitochondria and induces apoptosis. Mol Cell 2006;21(4):481–93.

[90] Ballester R, Furth ME, Rosen OM. Phorbol ester- and protein kinase C-mediated phosphorylation of the cellular Kirsten Ras gene product. J Biol Chem 1987;262(6):2688–95.

[91] Xia S, Forman LW, Faller DV. Protein kinase C delta is required for survival of cells expressing activated p21RAS. J Biol Chem 2007;282(18):13199–210.

[92] Marais R, Light Y, Mason C, Paterson H, Olson MF, Marshall CJ. Requirement of Ras-GTP-Raf complexes for activation of Raf-1 by protein kinase C. Science 1998;280(5360):109–12.

[93] Chen CY, Faller DV. Direction of p21ras-generated signals towards cell growth or apoptosis is determined by protein kinase C and Bcl-2. Oncogene 1995;11(8):1487–98.

[94] Alvarez-Moya B, Lopez-Alcala C, Drosten M, Bachs O, Agell N. K-Ras4B phosphorylation at Ser181 is inhibited by calmodulin and modulates K-Ras activity and function. Oncogene 2010;29(44):5911–22.

[95] Villalonga P, Lopez-Alcala C, Chiloeches A, Gil J, Marais R, Bachs O, et al. Calmodulin prevents activation of Ras by PKC in 3T3 fibroblasts. J Biol Chem 2002;277(40):37929–35.

[96] Barcelo C, Paco N, Beckett AJ, Alvarez-Moya B, Garrido E, Gelabert M, et al. Oncogenic K-Ras segregates at spatially distinct plasma membrane signaling platforms according to its phosphorylation status. J Cell Sci 2013;126(Pt 20):4553–9.

[97] Plowman SJ, Ariotti N, Goodall A, Parton RG, Hancock JF. Electrostatic interactions positively regulate K-Ras nanocluster formation and function. Mol Cell Biol 2008;28(13):4377–85.

[98] Sidhu RS, Clough RR, Bhullar RP. Ca2+/calmodulin binds and dissociates K-RasB from membrane. Biochem Biophys Res Commun 2003;304(4):655–60.

[99] Kim SE, Yoon JY, Jeong WJ, Jeon SH, Park Y, Yoon JB, et al. H-Ras is degraded by Wnt/beta-catenin signaling via beta-TrCP-mediated polyubiquitylation. J Cell Sci 2009;122(Pt 6):842–8.

[100] Jeong WJ, Yoon J, Park JC, Lee SH, Lee SH, Kaduwal S, et al. Ras stabilization through aberrant activation of Wnt/beta-catenin signaling promotes intestinal tumorigenesis. Sci Signal 2012;5(219):ra30.

[101] Shukla S, Allam US, Ahsan A, Chen G, Krishnamurthy PM, Marsh K, et al. KRAS protein stability is regulated through SMURF2: UBCH5 complex-mediated beta-TrCP1 degradation. Neoplasia 2014;16(2):115–28.

[102] Jura N, Scotto-Lavino E, Sobczyk A, Bar-Sagi D. Differential modification of Ras proteins by ubiquitination. Mol Cell 2006;21(5):679–87.

[103] Xu L, Lubkov V, Taylor LJ, Bari-Sagi D. Feedback regulation of Ras signaling by Rabex-5-mediated ubiquitination. Curr Biol 2010;20(15):1372–7.

[104] Yan H, Jahanshahi M, Hovarth EA, Liu HY, Pfleger CM. Rabex-5 ubiquitin ligase activity restricts Ras signaling to establish pathway homeostasis in *Drosophila*. Curr Biol 2010;20(15):1378–82.

[105] Sasaki AT, Carracedo A, Locasale JW, Anastasiou D, Takeuchi K, Kahoud ER, et al. Ubiquitination of K-Ras enhances activation and facilitates binding to select downstream effectors. Sci Signal 2011;4(163):ra13.

[106] Yang MH, Nickerson S, Kim ET, Liot C, Laurent G, Spang R, et al. Regulation of RAS oncogenicity by acetylation. Proc Natl Acad Sci USA 2012;109(27):10843–8.

[107] Yang MH, Laurent G, Bause AS, Spang R, German N, Haigis MC, et al. HDAC6 and SIRT2 regulate the acetylation state and oncogenic activity of mutant K-RAS. Mol Cancer Res 2013;11(9):1072–7.

[108] Halloran M, Parakh S, Atkin JD. The role of s-nitrosylation and s-glutathionylation of protein disulphide isomerase in protein misfolding and neurodegeneration. Int J Cell Biol 2013;2013:797914.

[109] Iyer AK, Azad N, Wang L, Rojanasakul Y. Role of S-nitrosylation in apoptosis resistance and carcinogenesis. Nitric Oxide 2008;19(2):146–51.

[110] Tang CH, Wei W, Liu L. Regulation of DNA repair by S-nitrosylation. Biochim Biophys Acta 2012;1820(6):730–5.

[111] Foster MW, Hess DT, Stamler JS. Protein S-nitrosylation in health and disease: a current perspective. Trends Mol Med 2009;15(9):391–404.

[112] Aranda E, Lopez-Pedreera C, De La Haba-Rodriguez JR, Rodriguez-Ariza A. Nitric oxide and cancer: the emerging role of S-nitrosylation. Curr Mol Med 2012;12(1):50–67.

[113] Monteiro HP, Costa PE, Reis AK, Stern A. Nitric oxide: protein tyrosine phosphorylation and protein S-nitrosylation in cancer. Biomed J 2015;38(5):380–8.

[114] Lander HM, Hajjar DP, Hempstead BL, Mirza UA, Chait BT, Campbell S, et al. A molecular redox switch on p21(Ras). Structural basis for the nitric oxide-p21(Ras) interaction. J Biol Chem 1997;272(7):4323–6.

[115] Williams JG, Pappu K, Campbell SL. Structural and biochemical studies of p21Ras S-nitrosylation and nitric oxide-mediated guanine nucleotide exchange. Proc Natl Acad Sci USA 2003;100(11):6376–81.

[116] Lander HM, Hirata H. Redox regulation of cell signalling. Nature 1996;381(6581):380–1.

[117] Heo J, Campbell SL. Mechanism of p21Ras S-nitrosylation and kinetics of nitric oxide-mediated guanine nucleotide exchange. Biochemistry 2004;43(8):2314–22.

[118] Lim KH, Ancrile BB, Kashatus DF, Counter CM. Tumour maintenance is mediated by eNOS. Nature 2008;452(7187):646–9.

[119] Switzer CH, Cheng RY, Ridnour LA, Glynn SA, Ambs S, Wink DA. Ets-1 is a transcriptional mediator of oncogenic nitric oxide signaling in estrogen receptor-negative breast cancer. Breast Cancer Res 2012;14(5):R125.

[120] Marozkina NV, Wei C, Yemen S, Wallrabe H, Nagji AS, Liu L, et al. S-nitrosoglutathione reductase in human lung cancer. Am J Respir Cell Mol Biol 2012;46(1):63–70.

[121] Ganesan AK, Vincent TS, Olson JC, Barbieri JT. *Pseudomonas aeruginosa* exoenzyme S disrupts Ras-mediated signal transduction by inhibiting guanine nucleotide exchange factor-catalyzed nucleotide exchange. J Biol Chem 1999;274(31):21823–9.

[122] Just I, Selzer J, Hofmann F, Green GA, Aktories K. Inactivation of Ras by *Clostridium sordellii* lethal toxin-catalyzed glucosylation. J Biol Chem 1996;271(17):10149–53.

[123] Mor A, Philips MR. Compartmentalized Ras/MAPK signaling. Annu Rev Immunol 2006;24:771–800.

[124] Daniels MA, Teixeiro E, Gill J, Hausmann B, Roubaty D, Holmberg K, et al. Thymic selection threshold defined by compartmentalization of Ras/MAPK signalling. Nature 2006;444(7120):724–9.

[125] Santos E. Dimerization opens new avenues into Ras signaling research. Sci Signal 2014; 7(324):pe12.

[126] Muratcioglu S, Chavan TS, Freed BC, Jang H, Khavrutskii L, Freed RN, et al. GTP-dependent K-Ras dimerization. Structure 2015;23(7):1325–35.

[127] Lin WC, Iversen L, Tu HL, Rhodes C, Christensen SM, Iwig JS, et al. H-Ras forms dimers on membrane surfaces via a protein-protein interface. Proc Natl Acad Sci USA 2014;111(8):2996–3001.

[128] Nan X, Tamguney TM, Collisson EA, Lin LJ, Pitt C, Galeas J, et al. Ras-GTP dimers activate the mitogen-activated protein kinase (MAPK) pathway. Proc Natl Acad Sci USA 2015;112(26):7996–8001.

[129] Inouye K, Mizutani S, Koide H, Kaziro Y. Formation of the Ras dimer is essential for Raf-1 activation. J Biol Chem 2000;275(6):3737–40.

[130] The KRAS-PDEdelta interaction is a therapeutic target. Cancer Discov 2013; 3(7):OF20.

[131] Chandra A, Grecco HE, Pisupati V, Perera D, Cassidy L, Skoulidis F, et al. The GDI-like solubilizing factor PDEdelta sustains the spatial organization and signalling of Ras family proteins. Nat Cell Biol 2012;14(2):148–58.

[132] Zimmermann G, Papke B, Ismail S, Vartak N, Chandra A, Hoffmann M, et al. Small molecule inhibition of the KRAS-PDEdelta interaction impairs oncogenic KRAS signalling. Nature 2013;497(7451):638–42.

CHAPTER 7

Cross Talk Between Snail and Mutant K-Ras Contributes to Pancreatic Cancer Progression

C.R. Chow[1], K. Ebine[1], H.Z. Hattaway[1], K. Kumar[1,2], H.G. Munshi[1,2]

[1]Northwestern University, Chicago, IL, United States; [2]Jesse Brown VA Medical Center, Chicago, IL, United States

CONTENTS

INTRODUCTION

The American Cancer Society's most recent estimates indicate that there will be over 53,000 new cases of pancreatic ductal adenocarcinoma (PDAC) and over 41,750 deaths from the disease in 2016 [1]. PDAC is currently the fourth leading cause of death from cancer in the United States, with a median survival of approximately 6 months and 1-year survival rate of approximately 20% [1]. Despite advances in surgery, radiation, and chemotherapy, the overall survival rate for the vast majority of patients with PDAC has not changed significantly in more than three decades [1–3]. The current overall 5-year survival rate among patients with pancreatic cancer is approximately 6% [1]. It is projected that by 2020 pancreatic cancer will likely become the second leading cause of death from cancer in the United States, second only to non–small cell lung cancer [1].

A number of different factors are thought to contribute to the aggressive nature of pancreatic cancer. Given the retroperitoneal location of the pancreas, patients are often asymptomatic during the initial stages of cancer development and progression [2–4]. Moreover, there are currently no effective screening tests available for pancreatic cancer [2–4]. Consequently, over 80% of the patients with PDAC present either with un-resectable locally advanced disease or with metastatic spread [2–4]. Human PDAC tumors also demonstrate evidence of epithelial–mesenchymal transition (EMT) [5,6], which has been identified as a key step in tumor invasion and metastasis, enabling cells to disrupt normal tissue architecture and invade surrounding structures [7–9]. EMT also contributes to the generation of fibro-inflammatory reaction [7–9], which is particularly pronounced in human PDAC tumors and can regulate cancer progression and mediate response to therapy [10,11]. Furthermore, mutant K-ras, which is present in over 90% of human PDAC tumors [12,13] and is required for

Conquering RAS. http://dx.doi.org/10.1016/B978-0-12-803505-4.00007-2

tumor development in mouse models of PDAC [14–16], has been challenging to target directly [17–19]. In addition, the current conventional therapies have not been able to successfully eradicate cancer stem cells [20,21], which can re-establish tumors following treatment.

K-RAS AND PANCREATIC CANCER PROGRESSION

Mutations in *KRAS* gene that result in a dominant active form of K-ras GTPase protein are seen in over 90% of human pancreatic tumors [12,13]. Although the majority of *KRAS* mutations occur in codon 12, rare mutations in codon 13 have also been identified in human pancreatic tumors [22]. Most mutations in pancreatic cancer change glycine at codon 12 to valine (G12V) or to aspartate (G12D) [22]. Although mutation to serine (G12S) is a common K-ras mutation in other tumor types, it is unusual in pancreatic cancer [22]. *KRAS* mutations have been identified in precursor pancreatic intra-epithelial neoplastic (PanIN) lesions [23], indicating that mutation in the *KRAS* gene is an early event in pancreatic cancer progression. Consistent with the findings in human tumors, genetically engineered mouse models have demonstrated that mutation in the *KRAS* gene is necessary for tumor development [14–16].

Human PDAC tumors are also associated with other signature mutations, including mutations in the *CDKN2A*, *TP53*, and *SMAD4* tumor suppressor genes [2–4]. Genetically engineered mouse models have clearly established a role for oncogenic K-ras in driving tumor initiation as well as enabling tumor progression in combination with these other signature mutations [14–16,24]. Mutant K-ras cooperates with loss of p53 to promote chromosomal instability to cause widely metastatic disease in mice [24]. Although Smad4 loss alone is not sufficient to cause pancreatic tumors [14,15], Smad4 loss in combination with mutant K-ras causes tumors that resemble human intra-ductal papillary mucinous neoplasms or mucinous cystic neoplasms [14,15]. Histologically, loss of Smad4 in these mouse models is associated with differentiated pancreatic tumors [14]. In contrast, pancreatic tumors with intact Smad4 frequently demonstrate evidence of EMT and higher propensity for metastasis [14]. These studies suggest that, although the mutational landscape of pancreatic cancer is complex, mutation of K-ras is a critical step in the pathway to pancreatic tumor development.

Although mutant K-ras has been challenging to target directly [17,18,22], down-regulating K-ras attenuates tumor progression. A short hairpin RNA–mediated down-regulation of K-rasG12D decreased migration and invasion of pancreatic cancer cells in vitro and reduced metastasis in the orthotopic mouse model of pancreatic cancer [25]. Knockdown of K-rasG12D resulted in increased E-cadherin expression that was associated with decreased levels of EMT-regulating transcription factors [25]. Inactivation of K-rasG12D in mice transgenic for inducible K-rasG12D led to regression of precursor lesions and also caused regression of established tumors [26]. Metastatic lesions in these mice were also shown to depend on K-ras for their ongoing tumor growth [27]. However, relapse following down-regulation of K-rasG12D can activate an alternative transcriptional program mediated by Yap1 that enables proliferation of K-rasG12D-independent tumor cells [28]. Interestingly, a similar transcriptional program has been shown to be present in the quasi-mesenchymal subset of human pancreatic tumors [28,29]. The quasi-mesenchymal tumors, which have increased expression of mesenchymal genes, are also less dependent on oncogenic K-ras compared with the classical subset of human pancreatic tumors [29].

EPITHELIAL-TO-MESENCHYMAL TRANSITION AND PANCREATIC CANCER PROGRESSION

EMT is a developmental process that imbues cells that are a part of a rigid architecture to remodel the extra-cellular matrix (ECM), become motile, and spread to distant sites [7–9]. As cells undergo EMT, they lose their epithelial features such as tight cell-to-cell contacts and apical-basal polarity and down-regulate E-cadherin [7–9]. They also develop a mesenchymal phenotype by taking on a spindle-like morphology, becoming motile, and expressing mesenchymal markers, eg, fibronectin and vimentin [7–9]. Although the features of EMT in cancer cells were initially characterized in vitro, mouse models have demonstrated the role of dynamic EMT in pancreatic cancer. Through lineage labeling of epithelial cells with yellow fluorescent protein (YFP), Rhim et al. convincingly demonstrated that cancer cells in the genetically engineered $Kras^{G12D}$/P53/PdxCre (KPC) mouse model undergo EMT during tumor progression [6]. Significantly, >40% of epithelial cells labeled with YFP demonstrated evidence of EMT through losing E-cadherin expression and/or up-regulating EMT transcription factors ZEB1 or Snail [6]. Interestingly, they demonstrated that EMT occurs in PanIN lesions, early in tumor progression before there is any histologic evidence of cancer [6] (Fig.7.1).

Moreover, studies from patients with a variety of cancers have also provided evidence for EMT in vivo. In human PDAC samples, fibronectin and vimentin are increased in high-grade tumors and within poorly differentiated areas of low-grade tumors [30]. This increase is associated with a corresponding decrease in E-cadherin expression. Significantly, patients whose tumors demonstrate EMT have poorer outcome [31]. In a study based on a rapid autopsy program for patients with pancreatic cancer, approximately 75% of the primary tumors with mesenchymal features developed metastasis to liver and lung [31].

Human and mouse PDAC tumors have also been shown to express EMT-regulating transcription factors. Tissue microarray analysis of PDAC tumors has shown an inverse relationship between ZEB1 and E-cadherin expression [32]. Down-regulating ZEB1 in PDAC cell lines increased E-cadherin expression and restored the epithelial phenotype [32]. ZEB1 was shown to be responsible for EMT and increased migration and invasion following activation of nuclear factor (NF)-κB signaling in PDAC cells [33]. ZEB1 was also shown to be up-regulated in cancer cells undergoing EMT in the KPC mouse model [6]. Importantly, ZEB1 expression in human PDAC specimens correlated with advanced tumor grade and worse outcomes [32–34].

The Snail family of transcription factors also functions to regulate EMT in pancreatic cancer cells. Snail expression is detected in cancer cells undergoing EMT in the KPC mouse model [6]. Snail is up-regulated in >75% of human PDAC tumors [35], whereas Slug (Snai2) is expressed in ~50% of human PDAC tumors [35]. Snail expression also inversely correlated with E-cadherin expression, with decreased E-cadherin expression associated with higher tumor grade. Although the role of Snail in pancreatic cancer is well established, the role of Slug in pancreatic cancer progression is less well understood. There is increasing evidence that Snail and Slug proteins can have both similar and differing roles in the development and progression of cancer [7]. Developmentally, Snail knockout mice die early in gestation, whereas Slug-deficient mice are viable with minimal abnormalities [36–38]. Structurally, the Slug protein is largely homologous to the Snail protein, but the Slug protein additionally has a unique SLUG domain [39]. The binding affinity for E-cadherin promoter is higher for Snail than for Slug [40], and Snail induces a more pronounced EMT in breast and skin cancer systems [41,42]. In breast cancer, Slug expression is associated with a semi-differentiated state with E-cadherin(+) carcinoma cells,

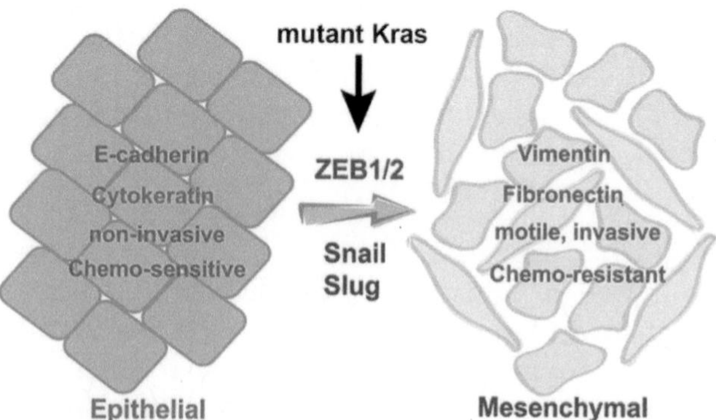

FIGURE 7.1 Schematic of epithelial-to-mesenchymal transition (EMT).
Cells that undergo EMT lose their epithelial markers, such as E-cadherin and cytokeratin, and sheet-like architecture and take on a mesenchymal phenotype with increased vimentin and fibronectin, along with single-cell, spindle-like morphology. These cells are invasive, have stem-cell-like properties, and demonstrate increased chemo-resistance.

whereas Snail expression is associated with vimentin(+)/E-cadherin(−) undifferentiated carcinoma [42,43]. It has been suggested that Slug and Snail may act sequentially to induce EMT, with Slug acting early to trigger EMT and Snail acting later to complete the EMT process [44].

INTERPLAY BETWEEN K-RAS AND SNAIL IN REGULATING PANCREATIC FIBROSIS

Pancreatic cancer is associated with a pronounced fibro-inflammatory reaction that can regulate cancer progression and mediate response to therapy. This fibrotic reaction consists of proliferating stromal cells together with collagen-rich ECM [10,12,45]. Significantly, the fibrotic reaction can account for >80% of the tumor mass [10,12]. Although collagen-rich ECM is well known to function as a barrier to invasion [10,46], analyses of human tumors have shown that increased collagen I expression can be associated with poor prognosis and increased metastases [47–49]. Importantly, mouse models have demonstrated that mutant K-ras contributes to pancreatic fibrosis. Expression of mutant K-ras in mouse models causes activation of stellate cells [13,50], which are the key mediators of the collagen-rich stromal reaction in vivo [51,52]. Interestingly, switching off mutant K-ras in the inducible K-ras mouse model causes regression of the stromal reaction [26], indicating that K-ras expression in the epithelial cells not only contributes to the stromal reaction but also helps to sustain the stromal reaction [26] (Fig. 7.2).

Snail can also promote fibrosis in vivo [53–55]. Transgenic over-expression of Snail in the kidney is sufficient to induce fibrosis in mice [53], whereas

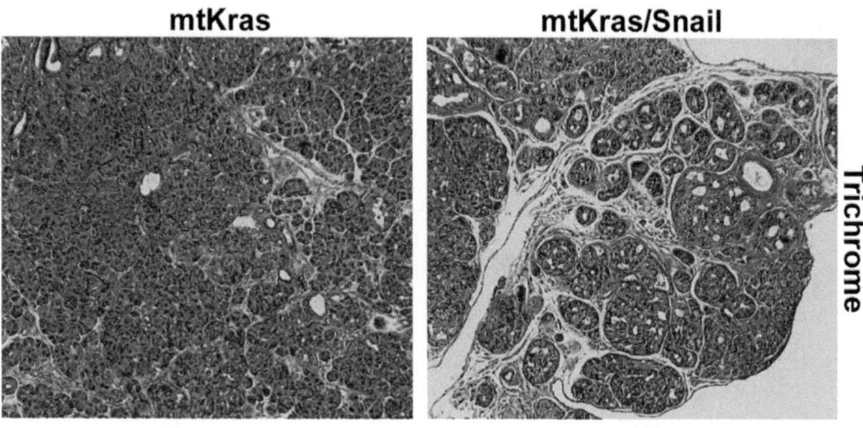

FIGURE 7.2 Snail co-expression with mutant K-ras mice promotes fibrosis.
Pancreatic sections from 3-month-old mice were analyzed for fibrosis using trichrome staining (blue, fibrosis) (see also Ref. [56]).

ablation of Snail in the liver attenuates chemical-induced fibrosis [54]. Snail also promotes the activation of liver stellate cells that are important for generation of the fibrotic response during liver injury [55]. In contrast to these reports, we have found that over-expression of Snail in the mouse pancreas did not result in changes in the mouse pancreas [56]. Snail expression also did not affect changes in the mouse pancreas following cerulein-induced pancreatitis [56]. However, Snail co-expression with mutant K-ras mice increased pancreatic fibrosis compared with mutant K-ras mice [56]. Snail expression in mouse pancreas also enhanced expression of α-SMA, indicating that Snail expression in epithelial cells caused activation of stellate cells in mutant K-ras mice [56]. We also showed that Snail expression in the epithelial cells in vitro and in vivo increased Smad2 phosphorylation in the stroma by inducing expression and secretion of transforming growth factor (TGF)-β2 by the epithelial cells [56].

Interestingly, not only do Ras and Snail cooperate to promote changes in the mouse pancreas but also Ras induce expression of Snail in cancer cells. Ras signaling plays an important role in regulating induction of Snail by TGF-β in cancer cells [57]. Knocking down mutant Ras in pancreatic cancer, Panc1 cells blocked TGF-β-induced Snail expression, whereas expression of constitutively active Ras enhanced TGF-β-mediated induction of Snail in HeLa cells. Together these results indicate that the cross talk between Snail and K-ras contributes to pancreatic fibrosis and tumor progression.

INTERPLAY BETWEEN K-RAS AND SNAIL IN REGULATING PANCREATIC INFLAMMATION

Inflammation also plays a significant role in PDAC development and progression [58,59]. Chronic pancreatitis, which is associated with ongoing inflammation and fibrosis [60], is a risk factor for pancreatic cancer in humans and contributes to PDAC progression in mouse models [59,61,62]. In addition, acute pancreatitis can accelerate the progression of precursor PanIN lesions to PDAC in mutant K-ras-driven mouse models of pancreatic cancer [63,64]. Importantly, although expression of embryonic mutant K-ras is sufficient for tumor initiation in various mouse models of pancreatic cancer [13,65], expression of mutant K-ras in adult mouse pancreas does not result in any obvious phenotypic changes [59]. However, induction of chronic pancreatitis promotes PDAC development in adult mice expressing mutant K-ras [59,66].

Significantly, expression of mutant K-ras in epithelial cells in mouse pancreas increases inflammatory cell infiltration (eg, myeloid cells) in the stroma [63,64,67]. The myeloid cells in the pancreas can in turn accelerate PanIN progression and PDAC development by releasing interleukin-6 (IL-6), which then activates Stat3 signaling in the K-ras-expressing epithelial cells [63]. The increase in Stat3 activation supports persistent K-ras-driven cell proliferation that is required for ADM and PanIN development [26], and also increases matrix metalloproteinase-7 (MMP-7) levels, which contributes to tumor growth and metastasis [26]. In addition, activation of K-ras in pancreatic ductal cells induces expression of granulocyte macrophage colony-stimulating factor, which recruits and expands the number of immunosuppressive myeloid cells in the stroma, thereby restraining the anti-tumor immune response [68]. Interestingly, inactivation of K-ras in the inducible K-ras mouse model decreases the levels of phospho-Stat3, IL-6, and MMP-7, indicating that K-ras is required for initiating and sustaining the inflammatory response in vivo [26].

Inflammatory signaling can also increase EMT. Snail activity is increased via stabilization at the protein level in response to tumor necrosis factor α–driven NF-κB signaling [69]. Conversely, Snail can also modulate inflammatory signaling in vivo through up-regulation of chemokines and cytokines [54,70,71]. Snail over-expression in keratinocytes increases production of cytokines IL-6 and IL-8 and the chemokine CXCL1 [70]. Snail over-expression in epidermal keratinocytes in a transgenic mouse model promotes cutaneous inflammation that is associated with increased IL-6 production [71]. Snail ablation in turn attenuates inflammatory response in a chemical-induced liver fibrosis model [54]. Significantly, Snail expression is increased in macrophages at sites of injury in vivo and mediates migration of macrophages in vitro [72] (Fig. 7.3).

We have found that Snail can cooperate with mutant K-ras to enhance inflammatory response in mouse pancreas [73]. Snail expression in mutant K-ras mice enhanced the number of macrophages, Gr-1(+) cells, and mast cells in the

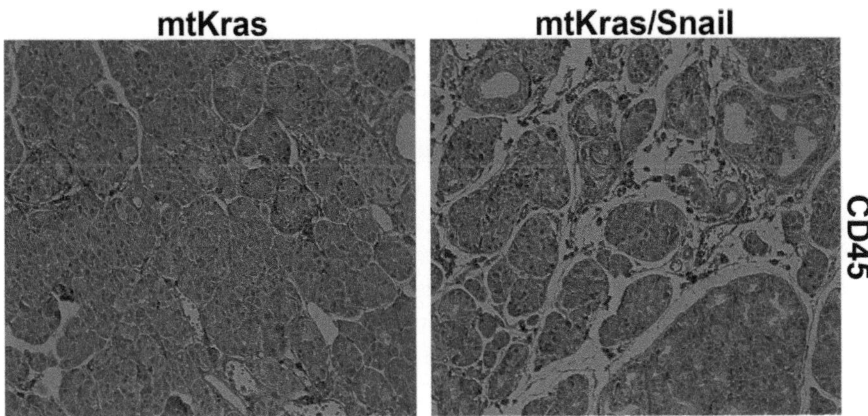

FIGURE 7.3 Snail co-expression with mutant K-ras mice promotes inflammation. Pancreatic tissue samples from 3-month-old mice were collected and stained for leukocytes using CD45 antibody (see also Ref. [73]).

pancreatic stroma [73]. Significantly, mast cells are increased in human PDAC tumors and contribute to PDAC progression. We have also found that Snail expression in human PDAC tumors correlated with expression of the mast cell activation marker tryptase and with the cytokine stem cell factor (SCF) [73]. Consistent with the human samples, we have found that Snail expression in the mutant K-ras mice increased expression of SCF. We also show that Snail expression in pancreatic cancer cells increased SCF production to increase mast cell migration [73]. Together these results indicate that the cross talk between Snail and mutant K-ras contributed not only to the pancreatic fibrosis but also to the pronounced inflammatory reaction that is present in human tumors.

ROLE OF K-RAS AND SNAIL IN REGULATING PANCREATIC CANCER STEM CELLS

There is increasing interest in a sub-population of cells within tumors that have stem cell-like properties. These cells, which usually number <1% of the total tumor cells, are frequently associated with metastatic foci and chemo-resistance [74–76]. These stem cell-like cells in pancreatic cancer were initially identified as being CD44+/CD24+/ESA+(epithelial specific antigen) cells and found to have the ability to differentiate into a heterogeneous tumor cell population and form tumors [77]. These triple-positive cells were shown to be over 100-fold more tumorigenic than unsorted cells. These stem cell-like cells also have increased aldehyde dehydrogenase (ALDH) activity, with the ALDH-high cells capable of producing tumors at very low numbers [31]. The ALDH-high cells are increased within pancreatic cancer metastases. Significantly, patients whose primary tumors exhibit high ALDH activity have increased propensity to develop metastases and have worse survival [31].

Mutant K-ras has also been shown to be important in regulating cancer stem-like cells. Down-regulation of mutant K-ras in pancreatic cancer cells decreased cancer stem cell markers and also decreased the sphere-forming ability of these cells [25]. K-ras knockdown also decreased the tumor-initiating capacity of the pancreatic cancer cells in xenograft mouse models [25]. In contrast, over-expression of mutant K-ras in colon cancer cells increased the sphere-forming ability of these cells [78]. Moreover, over-expression of K-rasG12D in the colonic epithelium facilitated the expansion of cancer stem cell-like cells in an APC model of colon cancer [78]. These results indicate that mutant K-ras in pancreatic tumors can regulate pancreatic cancer stem cells.

Expression of EMT transcription factors, including Snail, in human mammary epithelial cells induces stem cell-like cells with increased expression of stem cell markers and increased sphere-forming ability [76]. In contrast, down-regulating EMT-regulating transcription factor ZEB1 in pancreatic cancer cells decreased stem cell markers and also decreased sphere-forming ability [79]. Similarly, Snail down-regulation in pancreatic cancer cells decreased expression of ALDH and decreased sphere-forming ability [80]. Because down-regulation of Ras in pancreatic cancer cells decreases expression of EMT-activating transcription factors [25], it is possible that the effect of K-ras on cancer stem cells may be mediated in part through regulation of EMT.

CONCLUSION

There is significant interplay between mutant K-ras and Snail in pancreatic cancer progression. Cancer cells can up-regulate Snail and undergo EMT early in K-ras-driven tumorigenesis in the mouse models of pancreatic cancer. Snail can additionally cooperate with mutant K-ras to enhance fibrosis and inflammation in the mouse pancreas, and this cooperation may also contribute to the regulation of pancreatic cancer stem-like cells. In addition, other EMT-regulating transcription factors such as ZEB1 and Slug may also play a role in pancreatic cancer progression. However, more research needs to be performed to understand the full significance of expression of these proteins during pancreatic cancer progression.

Acknowledgments

This work was supported by grants R01CA126888 and R01CA186885 (to H.G. Munshi) from the NCI and a Merit award I01BX001363 (to H.G. Munshi) from the Department of Veterans Affairs. This research was also supported by the training grant T32CA070085 (to C.R. Chow) from the NCI.

References

[1] Siegel RL, Miller KD, Jemal A. Cancer statistics, 2015. CA Cancer J Clin 2016;65(1):7–30.

[2] Garrido-Laguna I, Hidalgo M. Pancreatic cancer: from state-of-the-art treatments to promising novel therapies. Nat Rev Clin Oncol 2015;12(6):319–34.

[3] Ryan DP, Hong TS, Bardeesy N. Pancreatic adenocarcinoma. N Engl J Med 2014;371(11):1039–49.

[4] Vincent A, Herman J, Schulick R, Hruban RH, Goggins M. Pancreatic cancer. Lancet 2011;378(9791):607–20.

[5] Krantz SB, Shields MA, Dangi-Garimella S, Munshi HG, Bentrem DJ. Contribution of epithelial-to-mesenchymal transition and cancer stem cells to pancreatic cancer progression. J Surg Res 2012;173(1):105–12.

[6] Rhim AD, Mirek ET, Aiello NM, Maitra A, Bailey JM, McAllister F, et al. EMT and dissemination precede pancreatic tumor formation. Cell 2012;148(1–2):349–61.

[7] Thiery JP, Acloque H, Huang RY, Nieto MA. Epithelial-mesenchymal transitions in development and disease. Cell 2009;139(5):871–90.

[8] Thiery JP. Epithelial-mesenchymal transitions in development and pathologies. Curr Opin Cell Biol 2003;15(6):740–6.

[9] Kalluri R, Weinberg RA. The basics of epithelial-mesenchymal transition. J Clin Invest 2009;119(6):1420–8.

[10] Shields MA, Dangi-Garimella S, Redig AJ, Munshi HG. Biochemical role of the collagen-rich tumor microenvironment in pancreatic cancer progression. Biochem J 2012;441(2):541–52.

[11] Chu GC, Kimmelman AC, Hezel AF, DePinho RA. Stromal biology of pancreatic cancer. J Cell Biochem 2007;101(4):887–907.

[12] Maitra A, Hruban RH. Pancreatic cancer. Annu Rev Pathol 2008;3:157–88.

[13] Hingorani SR, Petricoin EF, Maitra A, Rajapakse V, King C, Jacobetz MA, et al. Preinvasive and invasive ductal pancreatic cancer and its early detection in the mouse. Cancer Cell 2003;4(6):437–50.

[14] Bardeesy N, Cheng KH, Berger JH, Chu GC, Pahler J, Olson P, et al. Smad4 is dispensable for normal pancreas development yet critical in progression and tumor biology of pancreas cancer. Genes Dev 2006;20(22):3130–46.

[15] Izeradjene K, Combs C, Best M, Gopinathan A, Wagner A, Grady WM, et al. Kras(G12D) and Smad4/Dpc4 haploinsufficiency cooperate to induce mucinous cystic neoplasms and invasive adenocarcinoma of the pancreas. Cancer Cell 2007;11(3):229–43.

[16] Ijichi H, Chytil A, Gorska AE, Aakre ME, Fujitani Y, Fujitani S, et al. Aggressive pancreatic ductal adenocarcinoma in mice caused by pancreas-specific blockade of transforming growth factor-beta signaling in cooperation with active Kras expression. Genes Dev 2006;20(22): 3147–60.

[17] Gysin S, Salt M, Young A, McCormick F. Therapeutic strategies for targeting ras proteins. Genes Cancer 2011;2(3):359–72.

[18] Baines AT, Xu D, Der CJ. Inhibition of Ras for cancer treatment: the search continues. Future Med Chem 2011;3(14):1787–808.

[19] McCleary-Wheeler AL, McWilliams R, Fernandez-Zapico ME. Aberrant signaling pathways in pancreatic cancer: a two compartment view. Mol Carcinog 2012;51(1):25–39.

[20] Abel EV, Simeone DM. Biology and clinical applications of pancreatic cancer stem cells. Gastroenterology 2013;144(6):1241–8.

[21] Penchev VR, Rasheed ZA, Maitra A, Matsui W. Heterogeneity and targeting of pancreatic cancer stem cells. Clin Cancer Res 2012;18(16):4277–84.

[22] Stephen AG, Esposito D, Bagni RK, McCormick F. Dragging ras back in the ring. Cancer Cell 2014;25(3):272–81.

[23] Klimstra DS, Longnecker DS. K-ras mutations in pancreatic ductal proliferative lesions. Am J Pathol 1994;145(6):1547–50.

[24] Hingorani SR, Wang L, Multani AS, Combs C, Deramaudt TB, Hruban RH, et al. Trp53R172H and KrasG12D cooperate to promote chromosomal instability and widely metastatic pancreatic ductal adenocarcinoma in mice. Cancer Cell 2005;7(5):469–83.

[25] Rachagani S, Senapati S, Chakraborty S, Ponnusamy MP, Kumar S, Smith LM, et al. Activated KrasG(1)(2)D is associated with invasion and metastasis of pancreatic cancer cells through inhibition of E-cadherin. Br J Cancer 2011;104(6):1038–48.

[26] Collins MA, Bednar F, Zhang Y, Brisset JC, Galban S, Galban CJ, et al. Oncogenic Kras is required for both the initiation and maintenance of pancreatic cancer in mice. J Clin Invest 2012;122(2):639–53.

[27] Collins MA, Brisset JC, Zhang Y, Bednar F, Pierre J, Heist KA, et al. Metastatic pancreatic cancer is dependent on oncogenic Kras in mice. PLoS One 2012;7(12):e49707.

[28] Kapoor A, Yao W, Ying H, Hua S, Liewen A, Wang Q, et al. Yap1 activation enables bypass of oncogenic Kras addiction in pancreatic cancer. Cell 2014;158(1):185–97.

[29] Collisson EA, Sadanandam A, Olson P, Gibb WJ, Truitt M, Gu S, et al. Subtypes of pancreatic ductal adenocarcinoma and their differing responses to therapy. Nat Med 2011;17(4):500–3.

[30] Javle MM, Gibbs JF, Iwata KK, Pak Y, Rutledge P, Yu J, et al. Epithelial-mesenchymal transition (EMT) and activated extracellular signal-regulated kinase (p-Erk) in surgically resected pancreatic cancer. Ann Surg Oncol 2007;14(12):3527–33.

[31] Rasheed ZA, Yang J, Wang Q, Kowalski J, Freed I, Murter C, et al. Prognostic significance of tumorigenic cells with mesenchymal features in pancreatic adenocarcinoma. J Natl Cancer Inst 2010;102(5):340–51.

[32] Arumugam T, Ramachandran V, Fournier KF, Wang H, Marquis L, Abbruzzese JL, et al. Epithelial to mesenchymal transition contributes to drug resistance in pancreatic Cancer. Cancer Res 2009;69(14):5820–8.

[33] Maier HJ, Schmidt-Strassburger U, Huber MA, Wiedemann EM, Beug H, Wirth T. NF-kappaB promotes epithelial-mesenchymal transition, migration and invasion of pancreatic carcinoma cells. Cancer Lett 2010;295(2):214–28.

[34] Buck E, Eyzaguirre A, Barr S, Thompson S, Sennello R, Young D, et al. Loss of homotypic cell adhesion by epithelial-mesenchymal transition or mutation limits sensitivity to epidermal growth factor receptor inhibition. Mol Cancer Ther 2007;6(2):532–41.

[35] Hotz B, Arndt M, Dullat S, Bhargava S, Buhr HJ, Hotz HG. Epithelial to mesenchymal transition: expression of the regulators snail, slug, and twist in pancreatic cancer. Clin Cancer Res 2007;13(16):4769–76.

[36] Jiang R, Lan Y, Norton CR, Sundberg JP, Gridley T. The Slug gene is not essential for mesoderm or neural crest development in mice. Dev Biol 1998;198(2):277–85.

[37] Nieto MA, Sargent MG, Wilkinson DG, Cooke J. Control of cell behavior during vertebrate development by Slug, a zinc finger gene. Science 1994;264(5160):835–9.

[38] Carver EA, Jiang R, Lan Y, Oram KF, Gridley T. The mouse snail gene encodes a key regulator of the epithelial-mesenchymal transition. Mol Cell Biol 2001;21(23):8184–8.

[39] Nieto MA. The snail superfamily of zinc-finger transcription factors. Nat Rev Mol Cell Biol 2002;3(3):155–66.

[40] Bolos V, Peinado H, Perez-Moreno MA, Fraga MF, Esteller M, Cano A. The transcription factor Slug represses E-cadherin expression and induces epithelial to mesenchymal transitions: a comparison with Snail and E47 repressors. J Cell Sci 2003;116(Pt 3):499–511.

[41] Olmeda D, Montes A, Moreno-Bueno G, Flores JM, Portillo F, Cano A. Snai1 and Snai2 collaborate on tumor growth and metastasis properties of mouse skin carcinoma cell lines. Oncogene 2008;27(34):4690–701.

[42] Côme C, Magnino F, Bibeau F, De Santa Barbara P, Becker KF, Theillet C, et al. Snail and slug play distinct roles during breast carcinoma progression. Clin Cancer Res 2006;12(18):5395–402.

[43] Elloul S, Elstrand MB, Nesland JM, Trope CG, Kvalheim G, Goldberg I, et al. Snail, Slug, and Smad-interacting protein 1 as novel parameters of disease aggressiveness in metastatic ovarian and breast carcinoma. Cancer 2005;103(8):1631–43.

[44] Savagner P, Yamada KM, Thiery JP. The zinc-finger protein slug causes desmosome dissociation, an initial and necessary step for growth factor-induced epithelial-mesenchymal transition. J Cell Biol 1997;137(6):1403–19.

[45] Bardeesy N, DePinho RA. Pancreatic cancer biology and genetics. Nat Rev Cancer 2002;2(12):897–909.

[46] Kadler K. Extracellular matrix 1: fibril-forming collagens. Protein Profile 1995;2(5):491–619.

[47] Ramaswamy S, Ross KN, Lander ES, Golub TR. A molecular signature of metastasis in primary solid tumors. Nat Genet 2003;33(1):49–54.

[48] Tavazoie SF, Alarcon C, Oskarsson T, Padua D, Wang Q, Bos PD, et al. Endogenous human microRNAs that suppress breast cancer metastasis. Nature 2008;451(7175):147–52.

[49] Egeblad M, Rasch MG, Weaver VM. Dynamic interplay between the collagen scaffold and tumor evolution. Curr Opin Cell Biol 2010;22(5):697–706.

[50] Grippo PJ, Nowlin PS, Demeure MJ, Longnecker DS, Sandgren EP. Preinvasive pancreatic neoplasia of ductal phenotype induced by acinar cell targeting of mutant Kras in transgenic mice. Cancer Res 2003;63(9):2016–9.

[51] Apte MV, Haber PS, Applegate TL, Norton ID, McCaughan GW, Korsten MA, et al. Periacinar stellate shaped cells in rat pancreas: identification, isolation, and culture. Gut 1998;43(1):128–33.

[52] Bachem MG, Schneider E, Gross H, Weidenbach H, Schmid RM, Menke A, et al. Identification, culture, and characterization of pancreatic stellate cells in rats and humans. Gastroenterology 1998;115(2):421–32.

[53] Boutet A, De Frutos CA, Maxwell PH, Mayol MJ, Romero J, Nieto MA. Snail activation disrupts tissue homeostasis and induces fibrosis in the adult kidney. EMBO J 2006;25(23):5603–13.

[54] Rowe RG, Lin Y, Shimizu-Hirota R, Hanada S, Neilson EG, Greenson JK, et al. Hepatocyte-derived snail1 propagates liver fibrosis progression. Mol Cell Biol 2011;31(2):2392–403.

[55] Scarpa M, Grillo AR, Brun P, Macchi V, Stefani A, Signori S, et al. Snail1 transcription factor is a critical mediator of hepatic stellate cell activation following hepatic injury. Am J Physiol Gastrointest Liver Physiol 2011;300(2):G316–26.

[56] Shields MA, Ebine K, Sahai V, Kumar K, Siddiqui K, Hwang RF, et al. Snail cooperates with KrasG12D to promote pancreatic fibrosis. Mol Cancer Res 2013;11(9):1078–87.

[57] Horiguchi K, Shirakihara T, Nakano A, Imamura T, Miyazono K, Saitoh M. Role of Ras signaling in the induction of snail by transforming growth factor-beta. J Biol Chem 2009;284(1):245–53.

[58] Gidekel Friedlander SY, Chu GC, Snyder EL, Girnius N, Dibelius G, Crowley D, et al. Context-dependent transformation of adult pancreatic cells by oncogenic K-Ras. Cancer Cell 2009;16(5):379–89.

[59] Guerra C, Schuhmacher AJ, Cañamero M, Grippo PJ, Verdaguer L, Pérez-Gallego L, et al. Chronic pancreatitis is essential for induction of pancreatic ductal adenocarcinoma by K-Ras oncogenes in adult mice. Cancer Cell 2007;11(3):291–302.

[60] Braganza JM, Lee SH, McCloy RF, McMahon MJ. Chronic pancreatitis. Lancet 2011; 377(9772):1184–97.

[61] Lowenfels AB, Maisonneuve P, Cavallini G, Ammann RW, Lankisch PG, Andersen JR, et al. Pancreatitis and the risk of pancreatic cancer. International pancreatitis study group. N Engl J Med 1993;328(20):1433–7.

[62] Yadav D, Lowenfels AB. The epidemiology of pancreatitis and pancreatic cancer. Gastroenterology 2013;144(6):1252–61.

[63] Lesina M, Kurkowski MU, Ludes K, Rose-John S, Treiber M, Kloppel G, et al. Stat3/Socs3 activation by IL-6 transsignaling promotes progression of pancreatic intraepithelial neoplasia and development of pancreatic cancer. Cancer Cell 2011;19(4):456–69.

[64] Fukuda A, Wang SC, Morris 4th JP, Folias AE, Liou A, Kim GE, et al. Stat3 and MMP7 contribute to pancreatic ductal adenocarcinoma initiation and progression. Cancer Cell 2011;19(4):441–55.

[65] Aguirre AJ, Bardeesy N, Sinha M, Lopez L, Tuveson DA, Horner J, et al. Activated Kras and Ink4a/Arf deficiency cooperate to produce metastatic pancreatic ductal adenocarcinoma. Genes Dev 2003;17(24):3112–26.

[66] Guerra C, Collado M, Navas C, Schuhmacher AJ, Hernandez-Porras I, Canamero M, et al. Pancreatitis-induced inflammation contributes to pancreatic cancer by inhibiting oncogene-induced senescence. Cancer Cell 2011;19(6):728–39.

[67] Li N, Grivennikov SI, Karin M. The unholy trinity: inflammation, cytokines, and STAT3 shape the cancer microenvironment. Cancer Cell 2011;19(4):429–31.

[68] Pylayeva-Gupta Y, Lee KE, Hajdu CH, Miller G, Bar-Sagi D. Oncogenic Kras-induced GM-CSF production promotes the development of pancreatic neoplasia. Cancer Cell 2012;21(6):836–47.

[69] Wu Y, Deng J, Rychahou PG, Qiu S, Evers BM, Zhou BP. Stabilization of snail by NF-κB is required for inflammation-induced cell migration and invasion. Cancer Cell 2009;15(5):416–28.

[70] Lyons JG, Patel V, Roue NC, Fok SY, Soon LL, Halliday GM, et al. Snail up-regulates proinflammatory mediators and inhibits differentiation in oral keratinocytes. Cancer Res 2008;68(12):4525–30.

[71] Du F, Nakamura Y, Tan TL, Lee P, Lee R, Yu B, et al. Expression of snail in epidermal keratinocytes promotes cutaneous inflammation and hyperplasia conducive to tumor formation. Cancer Res 2010;70(24):10080–9.

[72] Hotz B, Visekruna A, Buhr HJ, Hotz HG. Beyond epithelial to mesenchymal transition: a novel role for the transcription factor Snail in inflammation and wound healing. J Gastrointest Surg 2010;14(2):388–97.

[73] Knab LM, Ebine K, Chow CR, Raza SS, Sahai V, Patel AP, et al. Snail cooperates with KrasG12D in vivo to increase stem cell factor and enhance mast cell infiltration. Mol Cancer Res 2014;12(10):1440–8.

[74] Santisteban M, Reiman JM, Asiedu MK, Behrens MD, Nassar A, Kalli KR, et al. Immune-induced epithelial to mesenchymal transition in vivo generates breast cancer stem cells. Cancer Res 2009;69(7):2887–95.

[75] Gupta PB, Chaffer CL, Weinberg RA. Cancer stem cells: mirage or reality? Nat Med 2009;15(9):1010–2.

[76] Mani SA, Guo W, Liao M-J, Eaton EN, Ayyanan A, Zhou AY, et al. The epithelial-mesenchymal transition generates cells with properties of stem cells. Cell 2008;133(4):704–15.

[77] Li C, Heidt DG, Dalerba P, Burant CF, Zhang L, Adsay V, et al. Identification of pancreatic cancer stem cells. Cancer Res 2007;67(3):1030–7.

[78] Moon BS, Jeong WJ, Park J, Kim TI, Min do S, Choi KY. Role of oncogenic K-Ras in cancer stem cell activation by aberrant Wnt/beta-catenin signaling. J Natl Cancer Inst 2014;106(2):djt373.

[79] Wellner U, Schubert J, Burk UC, Schmalhofer O, Zhu F, Sonntag A, et al. The EMT-activator ZEB1 promotes tumorigenicity by repressing stemness-inhibiting microRNAs. Nat Cell Biol 2009;11(12):1487–95.

[80] Zhou W, Lv R, Qi W, Wu D, Xu Y, Liu W, et al. Snail contributes to the maintenance of stem cell-like phenotype cells in human pancreatic cancer. PLoS One 2014;9(1):e87409.

2

SECTION

Novel Therapeutic Approaches Targeting RAS and Related Pathways

CHAPTER 8

Search for Inhibitors of Ras-Driven Cancers

A.B. Keeton[1,2], G.A. Piazza[1,2]
[1]*University of South Alabama Mitchell Cancer Institute, Mobile, AL, United States;*
[2]*ADT Pharmaceuticals, Inc., Orange Beach, AL, United States*

BACKGROUND

Cancer is one of the leading causes of death in the developed world, with over 1 million cases diagnosed each year and 500,000 deaths per year from cancer in the United States alone. Overall it is estimated that more than one in three people will develop some form of neoplastic disease during their lifetime. There are more than 200 different types of cancers, four of which, breast, lung, colorectal (CRC), and prostate, account for over half of all new cases. A significant fraction of these tumors arise from mutations in *ras* genes that activate Ras proteins, which drive critically important signaling pathways needed for cancer cell proliferation and/or survival.

Ras proteins are key regulators of several aspects of normal cell growth and malignant transformation, including cellular proliferation and survival, invasiveness, angiogenesis, and metastasis. Ras proteins are active in most human tumor cells because of activating mutations in *ras* genes or from alterations in up-stream or down-stream signaling components [1]. For example, numerous growth factor receptors that are over-expressed in tumors will signal mitogenesis through activation of Ras signaling. Certain molecular targeted therapies that inhibit RAS signaling pathways, therefore, would be expected to inhibit the growth, survival, and spread of tumor cells with activated Ras or mutant *ras*. This chapter will focus on the various experimental strategies that have been used to discover either direct or indirect inhibitors of Ras.

Human homologues of the transforming viral Rat sarcoma oncogene were identified over 30 years ago [2,3]. Such mutations result in the activation of several Ras isoforms, including H-Ras, N-Ras, or K-Ras, that drive complex signaling pathways leading to uncontrolled cell growth and tumor development. Mutations in *ras* occur de novo in approximately one-third of all human cancers and are especially prevalent in pancreatic, colorectal, and lung tumors [1]. Although

CONTENTS

Conquering RAS. http://dx.doi.org/10.1016/B978-0-12-803505-4.00008-4

Ras mutations are infrequent in other tumor types, for example, breast cancer, Ras can be pathologically activated by growth factor receptors that signal through Ras, connecting numerous effectors to the activation of downstream signaling pathways. Mutations in *ras* also develop in tumors that become resistant to chemotherapy and radiation, as well as targeted therapies, such as receptor tyrosine kinase inhibitors [4,5]. Despite the fact that *ras* mutations have been known for many years, there are no currently available cancer therapeutics that selectively suppress the growth of tumor cells with activated Ras. In fact, Ras has been described as "un-druggable" because of the lack of obvious binding pockets outside of the catalytic site, and the relative abundance of and high affinity for its substrate, GTP [5,6].

In the absence of *ras* mutations, Ras may be activated by growth factor receptors at the plasma membrane. Among the clinically important receptor tyrosine kinases (RTKs), the epidermal growth factor receptor is frequently hyper-activated in lung, colon, and other solid tumors [4,7]. Canonical RTK activators of Ras recruit a guanine nucleotide exchange factor (GEF), such as Sos, to its intra-cellular domain or to an adaptor protein, Grb2 [8], dissociating GDP from Ras at the plasma membrane, thus allowing binding of GTP to make the active, signaling competent form of Ras. In the case of wild-type Ras, this active state is transient, because GTPase activating proteins such as P120-Ras GTPase activating protein (RasGAP) or neurofibromin (NF1) bind and cooperate with residues within the switch regions and P-loop of Ras to form the catalytic site that hydrolyzes GTP to GDP [8]. Oncogenic mutations effect H-Ras, K-Ras, or N-Ras primarily in the amino acids at positions 12, 13, and 61 that interfere with GTPase activating protein (GAP)-mediated inactivation [1]. For activation of Ras to occur, it must be localized at the plasma membrane. This occurs primarily as a result of lipid modifications on the C-terminal region by the enzymatic addition of prenyl groups (farnesyl or geranylgeranyl moieties) within the CAAX motif, followed by proteolytic cleavage of N-terminal amino acids, methylation, addition of a palmitoyl chain, or through electrostatic interaction of a polybasic region with plasma membrane phospholipids, as in the case of K-Ras4B [5,9,10].

A large and growing list of downstream effectors and signaling pathways are engaged by activated Ras [11]. The canonical effectors of activated Ras are the C- and B-isoforms of Raf kinase, which directly bind via their Ras-binding domains (RBDs) [12]. Mutations in B-Raf, in particular, are a common driver in melanoma that can enable Ras-independent signaling. In either case, Raf phosphorylates its well-characterized substrates, mitogen/extracellular signal-regulated kinase (MEK)1/2 [13]. These highly selective, dual-specificity kinases are responsible for activation of extracellular signal-regulated kinase (ERK)1/2, which then translocates to the nucleus where its activity is associated with cell cycle progression, anabolic metabolism, reduction in cell death signaling, and increased genomic instability [13]. On the other hand, activated ERK is also associated with senescence, and thus potentially a problematic target [13].

Phosphoinositide 3-kinase (PI3K) can be activated by direct interaction with Ras, or indirectly through adaptor proteins recruited to activated RTKs [14]. In either case, the result is activation of survival pathways including downstream kinases phosphoinositide-dependent kinases, Akt, and mechanistic target of rapamycin (mTOR), all of which are important in the formation and maintenance of the tumor cell phenotype [14,15]. Another small GTPase, Ras like (Ral), lies downstream of activated Ras and is involved in a variety of membrane fusion, internalization, and trafficking processes [16]. This is driven by direct interaction between GTP-bound Ras and the Ral GEF (RalGDS). Exactly how important activation of Ral is to the oncogenic power of Ras has not been resolved, but clearly there is a correlation with anchorage-independent tumor cell growth, invasion, metastasis, and poor prognosis [16]. Numerous other effector pathways lie downstream of Ras, which may play greater or lesser roles in tumorigenesis or tumor maintenance, depending on the tumor type in question [11].

DIRECTLY TARGETING RAS

Ras Activation

The catalytic activity of Ras itself, especially in the case of oncogenic mutation, as a drug target has been a challenge of such a high order that it has been referred to as the "Holy Grail" of cancer research or alternatively as an "un-druggable" target. Ras binds GTP with high affinity, with Kd values in the picomolar range, whereas GTP itself is present in micromolar quantities in the cell cytoplasm [6,17]. Another issue is that, rather than the standard paradigm in which an enzyme inhibitor may be developed as a drug, drugging the activated form of Ras requires that one find a compound that can restore function of a defective enzyme. Nevertheless, some inroads have been made from recent research efforts. In the case of the G12C mutation, inhibitors have been identified that take advantage of a previously unappreciated binding pocket and reactivity of a cysteine residue to form a covalent bond within the catalytic domain of activated K-Ras [18]. A related approach, leveraging the reactivity of the cysteine in a GDP analog, SML-8-73-1, was also reported [19]. A proof-of-concept study also identified a GTP analog that supplied RasG12V with the missing aromatic amino group, mimicking that of the native GAP, which suggested possible approaches to this difficult problem [20]. In the case of wild-type Ras activated by RTKs or other means, interfering with GTP loading (GEF binding) may represent another promising approach. Albeit with potency and selectivity concerns, this has been demonstrated using stabilized peptides and small molecules that interfere with the Ras–Sos interactions or nucleotide exchange [8,11,21–24].

Ras Sub-Cellular Localization

Localization of Ras at the plasma membrane is considered a critical aspect of its signaling, and thus has been an active area of drug discovery for many years. Full maturation of the Ras protein requires prenyl group addition at the CAAX motif, and this enzymatic process has been targeted with farnesyl transferase (FTase) and geranylgeranyltransferase (GGTase) inhibitors, but with limited success and significant toxicity [10,25,26]. Non-peptidic farnesyl transferase inhibitors (FTIs) were identified through high-throughput screening campaigns and optimized lead compounds were a dramatic success in pre-clinical models of H-Ras-induced tumors [27], but resulted in disappointing efficacy in the clinic [28,29], presumably because of the predominance of mutations in K-*ras*, which can use the compensatory GGTase to achieve prenyl group–mediated membrane insertion [10,11]. Nonetheless, FTIs have received new interest for the treatment of other conditions aside from cancer, as well as the small percentage of cancers harboring H-*ras* mutations [30].

Palmitoylation of K-Ras 4A, H-Ras and N-Ras isoforms, but not K-Ras4B, is required for proper plasma membrane association. This reversible process affects transport between intra-cellular and plasma membranes [31]. Inhibition by a small molecule disrupts Ras localization and tumor cell growth [32]. However, the ability of the most commonly mutated Ras isoform, Ras4B, to evade this mechanism as well as the large number of palmitoylation substrates raises concerns about specificity, which need to be addressed before this can be considered a fully validated drug target [31].

In addition to prenylation, methylation and proteolytic cleavage of amino-terminal residues are important steps in maturation of the Ras protein [26]. As a pre-requisite for methylation, proteolytic cleavage of the amino terminal amino acids is achieved by Ras converting CAAX endopeptidase (RCE). Thus, this enzyme has garnered interest as a drug target of small molecule inhibitors identified via high-throughput screening of libraries of small molecules [33] and natural products [34]. ICMT inhibitors have been developed through rational design [35–37] and could induce apoptosis or loss of transformation traits in tumor cells [38,39]. However, concerns about the potential for paradoxical tumorigenic effects, as well as toxicity due to the multitude of known and predicted cellular substrates of ICMT and RCE have dampened enthusiasm for this Ras targeting strategy [40,41].

Phosphodiesterase 6 delta subunit (PDEδ) is a prenyl-binding protein that serves to regulate the membrane localization of K-Ras and other prenylated proteins [42]. Deltarasin is a prototype compound that interferes with the K-Ras–PDEδ interaction, thus disrupting K-Ras plasma membrane localization, Ras-dependent signaling, and tumor growth [43,44]. A class of compounds thought to act by a similar mechanism are related to farnesylated salicylate (salirasib) that can disrupt membrane localization of Ras through competition with galectin proteins [45,46]. These have demonstrated activity in vitro and in pre-clinical mouse models of cancer. Consequently, salirasib is currently in clinical trials for several cancer types [47,48].

Ras Protein–Protein Interaction With Effectors

The protein–protein interaction between Ras and downstream effector proteins, primary among these are the various Raf isoforms, are an interesting, although challenging, target. Certain non-steroidal anti-inflammatory drugs such as sulindac sulfide and indomethacin have been reported to inhibit Ras-induced malignant transformation [49]. The ability of sulindac sulfide to decrease activated Ras binding to and activation of c-Raf, as well as activation of downstream signaling or transcription by directly binding to Ras in a non-covalent manner, provided the impetus for rational drug discovery projects [50,51]. Resulting compounds selectively inhibited focus formation of Ras-transformed fibroblasts and preferentially inhibited growth of Ras-transfected Madin-Darby canine kidney epithelial cells [49,50,52,53]. A series of compounds, typified by MCP1, which was identified in a high-throughput yeast 2 hybrid screen also disrupts the H-Ras-Raf-1 interaction [54]. A derivative, MCP110, inhibited Ras-mediated cell growth, signaling with moderate potency and inhibited colon tumor growth in mouse xenograft models alone and in combination with microtubule targeting agents [55]. In contrast to the active search for inhibitors of Ras-Raf interaction, fewer compounds have been reported to disrupt Ras–PI3K interactions [14]. However, mutational studies of the p110α catalytic subunit demonstrate the importance of its RBD in the maintenance of K-Ras-driven lung tumors, indicating that this protein–protein interaction represents a promising target [15] (Fig. 8.1). Recently, rigosertib, which was originally identified as a polo-like kinase 1 (PLK1) inhibitor, has been reported to interfere with the Ras-PI3K interaction (as well as with Ral-GDS and A-, B-, and c-Raf isoforms) by acting as a "Ras mimetic" [15a].

INDIRECTLY TARGETING RAS

Ras Effector Pathways

Several inhibitors of the Ras effector kinase, Raf, have received US Food and Drug Administration (FDA) approval for the treatment of several malignancies. The c-Raf inhibitor, sorafenib, is active for the treatment of renal and hepatocellular carcinomas. However, likely owing to its modest selectivity, its activity may be attributed to inhibition of other important kinases [56]. Regorafenib is another Raf multi-kinase inhibitor that received FDA approval for colon cancer treatment [57]. In contrast, inhibitors of mutant V600E B-Raf, vemurafenib and dabrafanib, are quite selective for the activated form of Raf and have achieved striking clinical responses [58]. Unfortunately, resistance almost

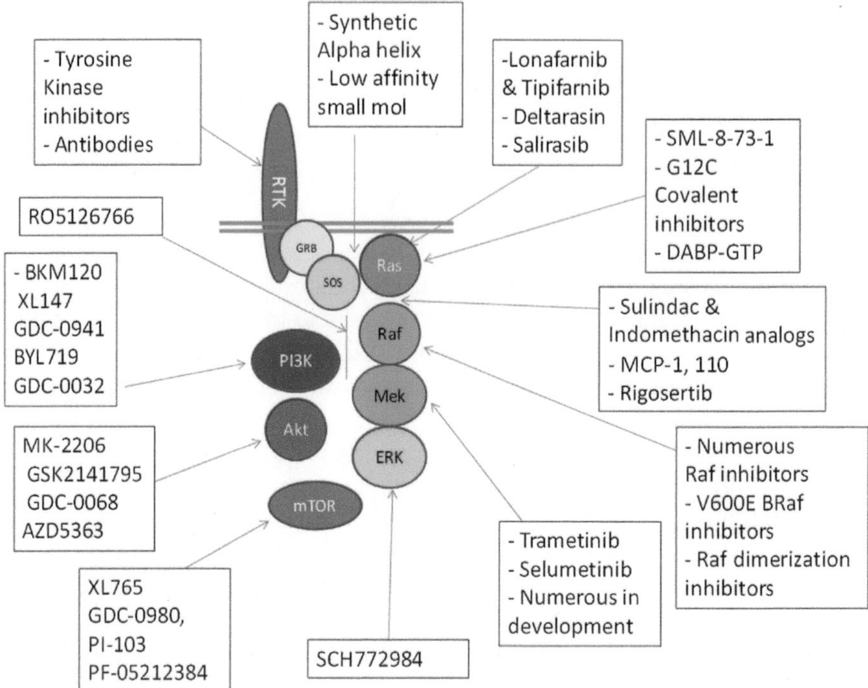

FIGURE 8.1

Inhibitors of canonical Ras activation and signaling. Inhibitors act at various (or multiple) locations within the canonical growth factor receptor/Ras/effector signaling pathway. Small molecule kinase inhibitors are approved for clinical use or in trials as described in the text. Experimental therapeutics that disrupt protein–protein interactions between Ras and Sos or Ras and Raf are in various stages of preclinical evaluation. Compounds that disrupt Ras membrane localization run the gamut of developmental stage, from a farnesyl transferase inhibitor that has been granted US Food and Drug Administration fast track status to investigational compounds such as salirasib and deltarasin. Direct Ras-binding inhibitors include small molecules that bind small surface pockets or covalent inhibitors that leverage a mutation-specific reactive amino acid.

invariably develops. In many cases, resistance is thought to be the result of a loss of feedback and paradoxical enhancement of homo- and hetero-dimer formation of the Raf isoforms or compensatory mutations in other survival pathways. Several new Raf inhibitors are being developed that may overcome these limitations by various means, including non-catalytic inhibitors [12].

Many of the MEK inhibitors that have been developed, such as trametinib or selumetinib, may suffer from similar problems with de-regulation of feedback loops; however, due to structural features of MEK, there is the potential for allosteric inhibitors to provide substantial selectivity advantages over ATP competitive mechanisms [13,59]. A novel dual inhibitor, RO5126766, binds MEK, thereby locking the inactive MEK and Raf into a non-productive complex, which may circumvent the issue of feedback loop inhibition and paradoxical ERK activation [59]. Likewise, new ERK inhibitors such as SCH772984, which has demonstrated in vivo anti-cancer activity in pre-clinical models, may represent another promising approach to avoid paradoxical effects resulting from feedback de-regulation [60].

Inhibitors of components of the PI3K/Akt/mTOR pathway have not been very successful as single agents, but may synergize with Raf/MEK/ERK pathway inhibitors, in addition to a wide array of cytotoxic agents, to block Ras-dependent tumor growth and survival. Considered to be a major effector pathway downstream of activated Ras, components of the PI3K/AKT/mTOR signaling cascade are also the subject of numerous drug discovery campaigns [61]. Some success has been achieved with allosteric inhibitors of the mTORC1 and 2, in the form of FDA-approved everolimus and temsirolomus for treatment of pancreatic, breast, and renal malignancies [62]. Several other mTORC1/2 inhibitors, including both allosteric and ATP competitive inhibitors are being evaluated in ongoing clinical trials [61]. Both isoform specific (BYL719 and GDC-0032) and pan-selective PI3K inhibitors (BKM120, XL147, and GDC-0941) are undergoing clinical trials for several indications, in many cases in hopes of addressing the issues of resistance to first-line therapies [63]. The outcome of current clinical trials of dual inhibitors of both PI3K and mTORC1 or 2 (eg, XL765, GDC-0980, PI-103, or PF-05212384) will address the hypothesis that these may circumvent feedback loops that are thought to limit the efficacy of the agents. In addition, several inhibitors of all three Akt isoforms (MK-2206, GSK2141795, GDC-0068, or AZD5363) are being evaluated in a range of tumor types [61].

The GTPases, Ral and Rac, are downstream of Ras and likely will present challenges as drug targets similar to those of Ras itself when the concentrations and affinity of nucleotides are considered. However, the infrequency of Ral A/B mutations in cancer suggests that avenues to target these GTPases are available such as interfering with the Ral GEF or enhancing RalGAP-mediated GTP hydrolysis, which are not feasible with respect to mutant Ras. A binding site identified in the structure of GDP-bound Ral is the target of a series of compounds (RBC6, 8, and 10) found by virtual screening. Analogous to compounds that interfere with Ras binding to Raf, these disrupt Ral binding to the Ral effector protein, RALBP1. RBC8 and a derivative, BQU57, demonstrated promising pre-clinical activity, significantly inhibiting lung tumor growth in a mouse xenograft model [64].

Autophagy is an important metabolic response to Ras-driven cellular proliferation, and it is also noteworthy that this process is regulated in part by Ral B [65], as discussed later. Downstream of RalB lies the TANK-binding kinase 1 (TBK1), an atypical IκB kinase, which is a target that has also been identified by phenotypic screening as a K-Ras synthetic lethal (RSL) gene [66,67]. Several inhibitors of TBK1 have been identified through high-throughput screening and rational drug design efforts by several groups that have demonstrated potent in vitro inhibitory activity and pre-clinical pharmacodynamic, anti-cancer, and anti-inflammatory activities [68]. Interestingly, an FDA-approved small molecule, amlexanox, has also been reported to inhibit TBK1 in vivo [69]. Rac, another small GTPase downstream of Ras [70], is activated by TIAM1 [70,71], which is activated by Ras and is important for Ras-mediated tumorigenesis [72]. Further downstream of Rac, the PAK family of kinases may represent a tractable target in Ras-driven cancers [73–75].

Ras-Activated Nutrient and Salvage Pathways

Metabolic changes in mutant Ras cells have long been recognized, and may represent an important difference from normal cells to provide insight for a fruitful class of targets. Rapidly growing tumors require an adequate supply of nutrients, thus the reason in part for angiogenesis and, in turn, anti-angiogenic therapeutics. At the cellular level, a number of adaptive changes occur, which enable tumor cells to sustain pathologic growth rates. Although many of these changes are not restricted to tumors with Ras mutations, they may be induced by Ras activation and in some cases required for survival of tumors harboring Ras mutations [76]. For example, the increased dependence on glucose metabolism and uptake (the Warburg effect) involves transporters and enzymes that provide an under-exploited class of targets such as lactate dehydrogenase or hexokinase 2 [77,78]. Altered glucose metabolism in turn may result in dependence on and alterations of metabolism of glutamine for raw materials for the anabolic pentose phosphate pathway and nicotinamide adenine dinucleotide phosphate (NADPH) for antioxidant defense by maintaining the intra-cellular glutathione pool [77]. These dependencies present several potential drug targets, for example, glutaminase, which is in fact the subject of several clinical trials [79,80].

At a cellular level, tumors need to internalize the fuel to support these metabolic changes, and the processes to provide tumor cells with needed nutrients differentiates them from normal cells, thus representing potential targets for cancer intervention. Autophagy, a starvation-associated salvage or quality control mechanism, enables tumor cells to recycle superfluous proteins and organelles. Ras-mediated metabolic changes can cause accumulation of defective mitochondria, which must be eliminated through autophagy. Thus, traffic of vesicles through this process, ultimately leading to lysosomal degradation, present a family of targets, which have been initially probed with the anti-parasitic drugs chloroquine and hydroxychloroquine [11,81]. Ras mutant tumors also increase fluid phase vesicle engulfment, a process called macro-pinocytosis. Here tumor cells utilize extracellular nutrient sources such as serum albumin, and inhibition of this process can selectively inhibit growth of tumor cells in vitro and in vivo [82]. This process may also be leveraged as a drug delivery mechanism as in the case of albumin-conjugated paclitaxel [78].

TARGETS AND INHIBITORS IDENTIFIED THROUGH PHENOTYPIC SCREENING

Aside from the approach of leveraging targets within the cast of well-known Ras effectors, phenotypic assays have been used to identify novel targets and exploit phenotypic "Achilles heels" of Ras-driven cancers. Activation of Ras, especially by mutation, has long been known to alter the phenotype of cells. Obviously, this includes neoplastic transformation itself, but in other ways that are both more subtle and less well understood. Nevertheless, these phenotypic changes may represent a fertile ground for identifying drug targets and developing novel treatments, targeting so-called non-oncogene addiction [97]. This is the basis for screening approaches that seek to identify RSL genes, small molecules, or natural products, which selectively inhibit growth or survival of tumor cells harboring Ras mutations (Table 8.1).

Table 8.1 Target Genes, Inhibitors, and Processes Outside of the Canonical Ras Signaling Pathways

Gene/ Target	Inhibitor(s)	Process/Phenotype	References
WT1	–	Senescence	[83]
–	ML210	Non-apoptotic cell death	[84]
VDAC	RSL3	Ferroptosis	[85]
VDAC	Erastin	Ferroptosis	[86]
PKCi	Oncrasin	Oxidative stress–induced apoptosis	[87]
ND	Lanperisone	Non-apoptotic oxidative stress	[88]
PLK1	Volasertib	Mitotic stress	[89]
APC/C	–	Mitotic stress	[90]
CDK4	PD0332991	Mitotic stress	[89]
CDC6	–	Mitotic stress	[91]
TPX2	–	Mitotic stress	[92]
BCL-XL	ABT-263, AT-101	Apoptosis	[93]
Survivin	YM155	Apoptosis	[92]
TAK1	5Z-7-oxo-zeaenol	Apoptosis	[94]
STK33	BRD-8899	Apoptosis	[95]
SNAIL2	–	Differentiation	[96]
TBK1	BX795	Apoptosis/NF-kB signaling	[66]
–	Hydroxychloroquine	Autophagy	[81]
–	EIPA	Macro-pinocytosis	[82]

APC/C, *Anaphase promoting complex/cyclosome;* BCL-XL, *B-cell lymphoma extra-large; CDC6, Cell division cycle 6;* CDK4, *cyclin-dependent kinase 4; EIPA, Ethylisopropylamiloride; ML210, [4-[bis(4-chlorophenyl)methyl]piperazin-1-yl]-(5-methyl-4-nitro-1,2-oxazol-3-yl)methanone; ND, Not determined;* NF-kB, *nuclear factor kB;* PKCi, *protein kinase C iota;* PLK1, *polo-like kinase 1;* RSL3, *Ras synthetic lethal 3; SNAIL2, Snail Family Zinc Finger 2;* STK33, *serine/threonine kinase;* TAK1, *transforming growth factor-β–activated kinase 1;* TBK1, *TANK-binding kinase 1;* TPX2, *targeting protein for Xklp2;* VDAC, *voltage-dependent anion channel;* WT1, *Wilms tumor 1.*

Indirectly Coupled Downstream Ras Effectors Identified Through Small Molecule Screening

Mitochondrial Voltage-Dependent Anion Channel

The most direct phenotypic approach to finding novel therapeutics with selective toxicity toward cells expressing activated Ras is through paired high-throughput screening to compare viability of a cell line (or lines) expressing mutant Ras to a cell line (or lines) lacking the mutation. One such project was conducted within the National Institutes of Health Molecular Libraries Screening Center program to identify compounds that were synthetically lethal to cells expressing oncogenic H-Ras. A lead compound, ML210, inhibited the growth of cells expressing mutant Ras with an IC_{50} of 71 nM, and was four-fold less active in cells lacking oncogenic Ras. Although the molecular target of ML210 is unknown, the compound was chemically optimized to eliminate reactive groups and improve pharmacologic properties [84]. Another high-throughput screening assay identified two compounds, RSL3 and RSL5, that induce non-apoptotic, MEK-dependent, oxidative cell death [85]. RSL5, like a previously identified RSL compound, erastin, binds a voltage-dependent anion channel [86]. An erastin analog, PRLX96936, demonstrated modest activity and manageable safety in a clinical trial, but its development status is not known [98].

Protein Kinase C iota

Yet another high-throughput screening assay identified a single compound, oncrasin, which was selectively active in vivo and in vitro against K-Ras mutant cell lines in a protein kinase C iota (PKC*i*)-dependent manner [87]. Subsequent studies have implicated other molecular mechanisms as well, but one oncrasin analog, NSC-743380, was reported to be highly potent and has demonstrated anti-tumor activity in a mouse model of K-Ras-driven renal cancer and is currently undergoing pre-clinical development [99,100].

Oxidative Stress

One of the hallmarks of cancer that has been recognized in recent years is a condition of oxidative stress [101]. Thus, it is not surprising that a synthetic lethal screen using embryonic fibroblasts derived from mice expressing the oncogenic K-Ras-G12D identified an FDA-approved compound (lanperisone) that induced non-apoptotic cell death by a mechanism involving oxidative stress [88]. Activated Ras alters the cell's ability to buffer oxidative stress, in part, because of alterations in metabolism described earlier. These changes represent a therapeutic target that potentially can be exploited by this novel class of Ras-selective compounds.

Targets Identified via Genetic Screening

Several intriguing targets have been identified by focused or genome-wide RNA interference (RNAi) screens using various screening strategies, technologies, and cellular model systems [90]. These targets represent potentially new avenues to inhibit Ras-driven cancers. The synthetic lethal approach has been used to identify RSL genes in cancer cells or cells engineered to express various mutant forms of Ras. Such synthetic lethal gene interactions may fall into the category of non-oncogene addiction [97], and are expected to be among the hallmarks of cancer [101].

Mitotic/Replicative Stress Genes

Mitotic stress, or chromosomal instability, is recognized as another hallmark of cancer that is associated with Ras mutations [101,102]. As one of the stress phenotypes of cancer, it represents a phenotype that may be rich in drug targets, which fits the description of non-oncogene addiction [97]. In addition to the well-established classes of DNA-damaging agents, topoisomerase inhibitors, and microtubule disruptors, numerous targets broadly described by mitotic stress are the subject of various drug discovery programs. Thus, it is not surprising that specific mitotic proteins have been identified by synthetic lethal screening (Table 8.1) [90]. In one screen, short hairpin RNAs (shRNAs) that selectively kill mutant Ras expressing colon tumor cells but not isogenic cells lacking mutant Ras were identified [103]. Among several functional categories of targets validated in a second isogenic cell pair were genes involved in mitosis, including Polo-like kinase 1 (PLK1) and anaphase promoting complex/cyclosome [89,90]. In fact, several PLK1 inhibitors have entered clinical trials [104]. In a separate study, CDK4 was reported to have a synthetic lethal interaction with K-Ras, resulting in senescence of a murine model of lung cancer, but not colon or pancreatic cancer cells. Anti-tumor activity was demonstrated in vivo with the CDK4 inhibitor PD0332991 (palbociclib) [89]. PD0332991 is FDA approved for the treatment of estrogen receptor-positive, HER2-negative metastatic breast cancer. In a similar vein, a target of CDK1, the CDC6 gene, was identified as an RSL in a small interfering RNA (siRNA) screen using isogenic colon cancers cells [91]. TPX2, a protein that binds the mitotic kinesin Eg5, was also identified as an RSL gene in a pooled siRNA screen of isogenic colon and lung cancer cells [92,105]. The theme of sensitivity to mitotic or replicative stress, although not novel, was reproduced in several screens, and thus the array of compounds in trials to target this process may be especially active against tumors with activated Ras [97].

Apoptosis-Related Genes

Several genes involved in regulation of apoptosis have been identified as RSLs through phenotypic screening of RNAi libraries of various focus and construction. For example, a synthetic lethal interaction was identified between K-Ras and the anti-apoptotic BH3 family gene, BCL-XL [93]. BCL-XL and other anti-apoptotic proteins in the Bcl-2 family are the target of drugs which have been identified through high-throughput screening or created through rational design. Bcl-2 family inhibitors or BH3 mimetics, such as ABT-263 and AT-101, are in clinical trials for laryngeal, prostate, and non–small cell lung, and other cancers [106]. Another anti-apoptotic protein, survivin, was identified as an RSL in a pooled siRNA screen of isogenic colon and lung cancer cells [92]. A small molecule survivin inhibitor, YM155, is undergoing clinical evaluation for a number of malignancies, but the results have been modest, and there are target specificity concerns about this compound [107]. Transforming growth factor-β-activated kinase 1 (TAK1) represents yet another apoptosis-related gene, which was identified as an RSL gene in a panel of K-Ras-dependent colon tumor cell lines [94,108]. Pharmacologic inhibition of TAK1 by a small molecule referred to as 5Z-7-oxo-zeaenol recapitulated this synthetic lethal interaction in vitro [94,109].

Other Targets

A variety of targets identified in such screens makes it challenging to group into broad functional classes, yet represent novel and potentially fruitful targets. The Serine/Threonine kinase, STK33, was identified as the predominant hit in an shRNA screen although small molecule inhibitors of this kinase have failed to confirm the necessity of STK33 kinase activity for survival of K-Ras-transformed cells [95,110]. Wilms tumor 1, the tumor suppressor identified in rare pediatric renal tumors, although perhaps not technically an RSL gene, was found to selectively induce senescence in K-Ras-expressing lung cancer cells [83]. SNAIL2, a zinc finger transcriptional repressor involved in epithelial-to-mesenchymal transition in Ras-dependent cells, was identified in a 2500 shRNA screen involving isogenic colon tumor cells [96]. As described previously, TBK1 was also identified as an RSL gene in an shRNA screen of >1000 kinase and phosphatase genes in 19 cell lines [66,68]. Despite the fact that several novel targets have been identified using the genetic synthetic lethal approach, the modest degree of overlap between top hits highlights concerns over the validity of such targets.

CONCLUSIONS

At present, there is a wide gap between the list of currently available FDA-approved drugs that target Ras and the goal of having drugs that will not only be active against tumors with activated Ras but also will preferentially inhibit Ras-driven tumors. Although targets downstream of Ras within the canonical signaling pathways are represented in the FDA-approved list of targeted cancer therapeutics, currently available inhibitors do not achieve such goals. These include Raf and Mek inhibitors within the mitogen-activated protein kinase (MAPK) pathway and mTOR inhibitors in the PI3K/Akt/mTOR pathway. Unfortunately, the efficacy and durability of response to the available drugs targeting these canonical pathways has been, with a few notable exceptions, disappointing. In addition to novel compounds targeting these proteins, compounds are also in development for targeting other components of the MAPK and PI3K/Akt/mTOR pathways. These include directly targeting the Ras catalytic domain or protein–protein interactions between Ras and Son of Sevenless (SOS) or Raf. In addition, a large number of potential drug targets for treating Ras-driven cancers have been identified through phenotypic approaches. At present, compounds that act on several of these targets are undergoing clinical trials, but to be sure, much work remains to find a drug for the currently un-druggable Ras-driven tumor.

List of Acronyms and Abbreviations

BCL-XL B-cell lymphoma extra-large
CDK4 Cyclin-dependent kinase 4
CRC Colorectal cancer
ERK Extracellular signal-regulated kinase
FTase Farnesyl transferase
GDP Guanosine diphosphate
GEF Guanine nucleotide exchange factor
GGTase Geranylgeranyltransferase
Grb2 Growth factor receptor-bound protein
GTP Guanosine triphosphate
HTS High-throughput screen
ICMT Isoprenylcysteine carboxyl methyltransferase
MAPK Mitogen-activated protein kinase
MEK Mitogen/extracellular signal-regulated kinase
mTOR Mechanistic target of rapamycin
NADPH Nicotinamide adenine dinucleotide phosphate
NF1 Neurofibromin
PAK p21 Activated kinase
PDEδ Phosphodiesterase 6 delta subunit
PDK Phosphoinositide-dependent kinase
PI3K Phosphoinositide 3-kinase
PKCi Protein kinase C iota
PLK1 Polo-like kinase 1
Raf Rapidly accelerated fibrosarcoma
Ral Ras like
Ras Rat sarcoma oncogene
RasGAP P120-Ras GTPase activating protein
RBD Ras-binding domain
RCE Ras converting CAAX endopeptidase
RNAi RNA interference
RSL Ras synthetic lethal
RTKs Receptor tyrosine kinases
shRNA Short hairpin RNA
siRNA Small interfering RNA
SOS Son of Sevenless

STK33 Serine/threonine kinase 33
TAK1 Transforming growth factor-β-activated kinase 1
TBK1 TANK-binding kinase 1
TIAM1 T-cell lymphoma invasion and metastasis 1
TPX2 Targeting protein for Xklp2
VDAC Voltage-dependent anion channel
WT1 Wilms tumor 1

References

[1] Fernandez-Medarde A, Santos E. Ras in cancer and developmental diseases. Genes Cancer 2011;2:344–58.

[2] Der CJ, Krontiris TG, Cooper GM. Transforming genes of human bladder and lung carcinoma cell lines are homologous to the ras genes of Harvey and Kirsten sarcoma viruses. Proc Natl Acad Sci USA 1982;79:3637–40.

[3] Parada LF, Tabin CJ, Shih C, Weinberg RA. Human EJ bladder carcinoma oncogene is homologue of Harvey sarcoma virus ras gene. Nature 1982;297:474–8.

[4] Massarelli E, Varella-Garcia M, Tang X, Xavier AC, Ozburn NC, Liu DD, et al. II KRAS mutation is an important predictor of resistance to therapy with epidermal growth factor receptor tyrosine kinase inhibitors in non-small-cell lung cancer. Clin Cancer Res 2007;13:2890–6. United States.

[5] Gysin S, Salt M, Young A, McCormick F. Therapeutic strategies for targeting ras proteins. Genes Cancer 2011;2:359–72.

[6] Takashima A, Faller DV. Targeting the RAS oncogene. Expert Opin Ther Targets 2013;17 :507–31.

[7] Jackman DM, Miller VA, Cioffredi LA, Yeap BY, Janne PA, Riely GJ, et al. Impact of epidermal growth factor receptor and KRAS mutations on clinical outcomes in previously untreated non-small cell lung cancer patients: results of an online tumor registry of clinical trials. Clin Cancer Res 2009;15:5267–73. United States.

[8] Bos JL, Rehmann H, Wittinghofer A. GEFs and GAPs: critical elements in the control of small G proteins. Cell 2007;129:865–77. United States.

[9] Fiordalisi JJ, Johnson 2nd RL, Weinbaum CA, Sakabe K, Chen Z, Casey PJ, et al. High affinity for farnesyltransferase and alternative prenylation contribute individually to K-Ras4B resistance to farnesyltransferase inhibitors. J Biol Chem 2003;278:41718–27. United States.

[10] Zverina EA, Lamphear CL, Wright EN, Fierke CA. Recent advances in protein prenyltransferases: substrate identification, regulation, and disease interventions. Curr Opin Chem Biol 2012;16:544–52.

[11] Cox AD, Fesik SW, Kimmelman AC, Luo J, Der CJ. Drugging the undruggable RAS: mission possible? Nat Rev Drug Discov 2014;13:828–51.

[12] Holderfield M, Deuker MM, McCormick F, McMahon M. Targeting RAF kinases for cancer therapy: BRAF-mutated melanoma and beyond. Nat Rev Cancer 2014;14:455–67.

[13] Deschenes-Simard X, Kottakis F, Meloche S, Ferbeyre G. ERKs in cancer: friends or foes? Cancer Res 2014;74:412–9.

[14] Ramjaun AR, Downward J. Ras and phosphoinositide 3-kinase: partners in development and tumorigenesis. Cell Cycle 2007;6:2902–5. United States.

[15] Castellano E, Sheridan C, Thin MZ, Nye E, Spencer-Dene B, Diefenbacher ME, et al. Requirement for interaction of PI3-kinase p110alpha with RAS in lung tumor maintenance. Cancer Cell 2013;24:617–30.

[15a] Athuluri-Divakar SK, Vasquez-Del Carpio R, Dutta K, Baker SJ, Cosenza SC, Basu I, et al. A small molecule RAS-mimetic disrupts RAS association with effector proteins to block signaling. Cell 2016;165:643–55.

[16] Kashatus DF. Ral GTPases in tumorigenesis: emerging from the shadows. Exp Cell Res 2013;319:2337–42.

[17] Finkel T, Der CJ, Cooper GM. Activation of ras genes in human tumors does not affect localization, modification, or nucleotide binding properties of p21. Cell 1984;37:151–8.

[18] Ostrem JM, Peters U, Sos ML, Wells JA, Shokat KM. K-Ras(G12C) inhibitors allosterically control GTP affinity and effector interactions. Nature 2013;503:548–51.

[19] Lim SM, Westover KD, Ficarro SB, Harrison RA, Choi HG, Pacold ME, et al. Therapeutic targeting of oncogenic K-Ras by a covalent catalytic site inhibitor. Angew Chem Int Ed Engl 2014;53:199–204.

[20] Ahmadian MR, Zor T, Vogt D, Kabsch W, Selinger Z, Wittinghofer A, et al. Guanosine triphosphatase stimulation of oncogenic Ras mutants. Proc Natl Acad Sci USA 1999;96:7065–70.

[21] Palmioli A, Sacco E, Abraham S, Thomas CJ, Di Domizio A, De Gioia L, et al. First experimental identification of Ras-inhibitor binding interface using a water-soluble Ras ligand. Bioorg Med Chem Lett 2009;19:4217–22. England.

[22] Patgiri A, Yadav KK, Arora PS, Bar-Sagi D. An orthosteric inhibitor of the Ras-Sos interaction. Nat Chem Biol 2011;7:585–7.

[23] Sun Q, Burke JP, Phan J, Burns MC, Olejniczak ET, Waterson AG, et al. Discovery of small molecules that bind to K-Ras and inhibit Sos-mediated activation. Angew Chem Int Ed Engl 2012;51:6140–3.

[24] Maurer T, Garrenton LS, Oh A, Pitts K, Anderson DJ, Skelton NJ, et al. Small-molecule ligands bind to a distinct pocket in Ras and inhibit SOS-mediated nucleotide exchange activity. Proc Natl Acad Sci USA 2012;109:5299–304.

[25] Lobell RB, Omer CA, Abrams MT, Bhimnathwala HG, Brucker MJ, Buser CA, et al. Evaluation of farnesyl: protein transferase and geranylgeranyl: protein transferase inhibitor combinations in preclinical models. Cancer Res 2001;61:8758–68.

[26] Berndt N, Hamilton AD, Sebti SM. Targeting protein prenylation for cancer therapy. Nat Rev Cancer 2011;11:775–91.

[27] Liu M, Bryant MS, Chen J, Lee S, Yaremko B, Lipari P, et al. Antitumor activity of SCH 66336, an orally bioavailable tricyclic inhibitor of farnesyl protein transferase, in human tumor xenograft models and wap-ras transgenic mice. Cancer Res 1998;58:4947–56.

[28] Van Cutsem E, van de Velde H, Karasek P, Oettle H, Vervenne WL, Szawlowski A, et al. Phase III trial of gemcitabine plus tipifarnib compared with gemcitabine plus placebo in advanced pancreatic cancer. J Clin Oncol 2004;22:1430–8. United States.

[29] Rao S, Cunningham D, de Gramont A, Scheithauer W, Smakal M, Humblet Y, et al. Phase III double-blind placebo-controlled study of farnesyl transferase inhibitor R115777 in patients with refractory advanced colorectal cancer. J Clin Oncol 2004;22:3950–7. United States.

[30] Majmudar JD, Hodges-Loaiza HB, Hahne K, Donelson JL, Song J, Shrestha L, et al. Amide-modified prenylcysteine based ICMT inhibitors: structure-activity relationships, kinetic analysis and cellular characterization. Bioorg Med Chem 2011;20:283–95.

[31] Resh MD. Targeting protein lipidation in disease. Trends Mol Med 2012;18:206–14.

[32] Ducker CE, Griffel LK, Smith RA, Keller SN, Zhuang Y, Xia Z, et al. Discovery and characterization of inhibitors of human palmitoyl acyltransferases. Mol Cancer Ther 2006;5:1647–59. United States.

[33] Manandhar SP, Hildebrandt ER, Schmidt WK. Small-molecule inhibitors of the Rce1p CaaX protease. J Biomol Screen 2007;12:983–93. United States.

[34] Bharate SB, Singh B, Vishwakarma RA. Modulation of k-Ras signaling by natural products. Curr Med Chem 2012;19:2273–91.

[35] Ramanujulu PM, Yang T, Yap SQ, Wong FC, Casey PJ, Wang M, et al. Functionalized indoleamines as potent, drug-like inhibitors of isoprenylcysteine carboxyl methyltransferase (ICMT). Eur J Med Chem 2013;63:378–86.

[36] Bergman JA, Hahne K, Song J, Hrycyna CA, Gibbs RA. S-farnesyl-thiopropionic acid (FTPA) triazoles as potent inhibitors of isoprenylcysteine carboxyl methyltransferase. ACS Med Chem Lett 2012;3:15–9.

[37] Judd WR, Slattum PM, Hoang KC, Bhoite L, Valppu L, Alberts G, et al. Discovery and SAR of methylated tetrahydropyranyl derivatives as inhibitors of isoprenylcysteine carboxyl methyltransferase (ICMT). J Med Chem 2011;54:5031–47.

[38] Winter-Vann AM, Baron RA, Wong W, dela Cruz J, York JD, Gooden DM, et al. A small-molecule inhibitor of isoprenylcysteine carboxyl methyltransferase with antitumor activity in cancer cells. Proc Natl Acad Sci USA 2005;102:4336–41. United States.

[39] Wang M, Hossain MS, Tan W, Coolman B, Zhou J, Liu S, et al. Inhibition of isoprenylcysteine carboxylmethyltransferase induces autophagic-dependent apoptosis and impairs tumor growth. Oncogene 2010;29:4959–70.

[40] Wahlstrom AM, Cutts BA, Karlsson C, Andersson KM, Liu M, Sjogren AK, et al. Rce1 deficiency accelerates the development of K-RAS-induced myeloproliferative disease. Blood 2007;109:763–8. United States.

[41] Court H, Amoyel M, Hackman M, Lee KE, Xu R, Miller G, et al. Isoprenylcysteine carboxylmethyltransferase deficiency exacerbates KRAS-driven pancreatic neoplasia via notch suppression. J Clin Invest 2013;123:4681–94.

[42] Chandra A, Grecco HE, Pisupati V, Perera D, Cassidy L, Skoulidis F, et al. The GDI-like solubilizing factor PDEdelta sustains the spatial organization and signalling of Ras family proteins. Nat Cell Biol 2011;14:148–58. England.

[43] Zimmermann G, Papke B, Ismail S, Vartak N, Chandra A, Hoffmann M, et al. Small molecule inhibition of the KRAS-PDEdelta interaction impairs oncogenic KRAS signalling. Nature 2013;497:638–42.

[44] Zimmermann G, Schultz-Fademrecht C, Kuchler P, Murarka S, Ismail S, Triola G, et al. Structure guided design and kinetic analysis of highly potent benzimidazole inhibitors targeting the PDEdelta prenyl binding site. J Med Chem 2014;57:5435–48.

[45] Levy R, Grafi-Cohen M, Kraiem Z, Kloog Y. Galectin-3 promotes chronic activation of K-Ras and differentiation block in malignant thyroid carcinomas. Mol Cancer Ther 2010;9 :2208–19. United States: Aacr.

[46] Rotblat B, Ehrlich M, Haklai R, Kloog Y. The Ras inhibitor farnesylthiosalicylic acid (Salirasib) disrupts the spatiotemporal localization of active Ras: a potential treatment for cancer. Methods Enzymol 2008;439:467–89. United States.

[47] Mackenzie GG, Bartels LE, Xie G, Papayannis I, Alston N, Vrankova K, et al. A novel Ras inhibitor (MDC-1016) reduces human pancreatic tumor growth in mice. Neoplasia 2013;15:1184–95.

[48] Laheru D, Shah P, Rajeshkumar NV, McAllister F, Taylor G, Goldsweig H, et al. Integrated preclinical and clinical development of S-trans, trans-farnesylthiosalicylic acid (FTS, Salirasib) in pancreatic cancer. Invest New Drugs 2012;30:2391–9.

[49] Gala M, Sun R, Yang VW. Inhibition of cell transformation by sulindac sulfide is confined to specific oncogenic pathways. Cancer Lett 2002;175:89–94. Ireland.

[50] Herrmann C, Block C, Geisen C, Haas K, Weber C, Winde G, et al. Sulindac sulfide inhibits Ras signaling. Oncogene 1998;17:1769–76.

[51] Pan MR, Chang HC, Hung WC. Non-steroidal anti-inflammatory drugs suppress the ERK signaling pathway via block of Ras/c-Raf interaction and activation of MAP kinase phosphatases. Cell Signal 2008;20:1134–41. England.

[52] Karaguni IM, Glusenkamp KH, Langerak A, Geisen C, Ullrich V, Winde G, et al. New indene-derivatives with anti-proliferative properties. Bioorg Med Chem Lett 2002;12:709–13. England.

[53] Waldmann H, Karaguni IM, Carpintero M, Gourzoulidou E, Herrmann C, Brockmann C, et al. Sulindac-derived Ras pathway inhibitors target the Ras-Raf interaction and downstream effectors in the Ras pathway. Angew Chem Int Ed Engl 2004;43:454–8.

[54] Kato-Stankiewicz J, Hakimi I, Zhi G, Zhang J, Serebriiskii I, Guo L, et al. Inhibitors of Ras/Raf-1 interaction identified by two-hybrid screening revert Ras-dependent transformation phenotypes in human cancer cells. Proc Natl Acad Sci USA 2002;99:14398–403. United States.

[55] Gonzalez-Perez V, Reiner DJ, Alan JK, Mitchell C, Edwards LJ, Khazak V, et al. Genetic and functional characterization of putative Ras/Raf interaction inhibitors in *C. elegans* and mammalian cells. J Mol Signal 2010;5:2.

[56] Wilhelm SM, Carter C, Tang L, Wilkie D, McNabola A, Rong H, et al. BAY 43-9006 exhibits broad spectrum oral antitumor activity and targets the RAF/MEK/ERK pathway and receptor tyrosine kinases involved in tumor progression and angiogenesis. Cancer Res 2004;64:7099–109.

[57] Wilhelm SM, Dumas J, Adnane L, Lynch M, Carter CA, Schutz G, et al. Regorafenib (BAY 73-4506): a new oral multikinase inhibitor of angiogenic, stromal and oncogenic receptor tyrosine kinases with potent preclinical antitumor activity. Int J Cancer 2011;129:245–55.

[58] Basile KJ, Le K, Hartsough EJ, Aplin AE. Inhibition of mutant BRAF splice variant signaling by next-generation, selective RAF inhibitors. Pigment Cell Melanoma Res 2014;27:479–84.

[59] Zhao Y, Adjei AA. The clinical development of MEK inhibitors. Nat Rev Clin Oncol 2014;11:385–400.

[60] Morris EJ, Jha S, Restaino CR, Dayananth P, Zhu H, Cooper A, et al. Discovery of a novel ERK inhibitor with activity in models of acquired resistance to BRAF and MEK inhibitors. Cancer Discov 2013;3:742–50.

[61] Chia S, Gandhi S, Joy AA, Edwards S, Gorr M, Hopkins S, et al. Novel agents and associated toxicities of inhibitors of the pi3k/Akt/mtor pathway for the treatment of breast cancer. Curr Oncol 2015;22:33–48.

[62] Hubbard PA, Moody CL, Murali R. Allosteric modulation of Ras and the PI3K/AKT/mTOR pathway: emerging therapeutic opportunities. Front Physiol 2015;5:478.

[63] Burris 3rd HA. Overcoming acquired resistance to anticancer therapy: focus on the PI3K/AKT/mTOR pathway. Cancer Chemother Pharmacol 2013;71:829–42.

[64] Yan C, Liu D, Li L, Wempe MF, Guin S, Khanna M, et al. Discovery and characterization of small molecules that target the GTPase Ral. Nature 2014;515:443–7.

[65] Bodemann BO, Orvedahl A, Cheng T, Ram RR, Ou YH, Formstecher E, et al. A. RalB and the exocyst mediate the cellular starvation response by direct activation of autophagosome assembly. Cell 2011;144:253–67.

[66] Barbie DA, Tamayo P, Boehm JS, Kim SY, Moody SE, Dunn IF, et al. Systematic RNA interference reveals that oncogenic KRAS-driven cancers require TBK1. Nature 2009;462:108–12.

[67] Chien Y, Kim S, Bumeister R, Loo YM, Kwon SW, Johnson CL, et al. RalB GTPase-mediated activation of the IkappaB family kinase TBK1 couples innate immune signaling to tumor cell survival. Cell 2006;127:157–70. United States.

[68] Yu T, Yang Y, Yin Q, Hong S, Son YJ, Kim JH, et al. TBK1 inhibitors: a review of patent literature (2011–2014. Expert Opin Ther Pat 2015;1–12.

[69] Reilly SM, Chiang SH, Decker SJ, Chang L, Uhm M, Larsen MJ, et al. An inhibitor of the protein kinases TBK1 and IKK-varepsilon improves obesity-related metabolic dysfunctions in mice. Nat Med 2013;19:313–21.

[70] Wennerberg K, Rossman KL, Der CJ. The Ras superfamily at a glance. J Cell Sci 2005;118 :843–6. England.

[71] Michiels F, Habets GG, Stam JC, van der Kammen RA, Collard JG. A role for Rac in Tiam1-induced membrane ruffling and invasion. Nature 1995;375:338–40.

[72] Malliri A, van der Kammen RA, Clark K, van der Valk M, Michiels F, Collard JG. Mice deficient in the Rac activator Tiam1 are resistant to Ras-induced skin tumours. Nature 2002;417: 867–71. England.

[73] Baker NM, Yee Chow H, Chernoff J, Der CJ. Molecular pathways: targeting RAC-p21-activated serine-threonine kinase signaling in RAS-driven cancers. Clin Cancer Res 2014;20:4740–6.

[74] Rudolph J, Crawford JJ, Hoeflich KP, Chernoff J. p21-activated kinase inhibitors. Enzymes 2013;34:157–80.

[75] Yi C, Maksimoska J, Marmorstein R, Kissil JL. Development of small-molecule inhibitors of the group I p21-activated kinases, emerging therapeutic targets in cancer. Biochem Pharmacol 2010;80:683–9.

[76] Ying H, Kimmelman AC, Lyssiotis CA, Hua S, Chu GC, Fletcher-Sananikone E, et al. Oncogenic Kras maintains pancreatic tumors through regulation of anabolic glucose metabolism. Cell 2012;149:656–70.

[77] Bryan N, Raisch KP. Identification of a mitochondrial-binding site on the amino-terminal end of hexokinase II. Biosci Rep 2015;35.

[78] White E. Exploiting the bad eating habits of Ras-driven cancers. Genes Dev 2013;27: 2065–71.

[79] Son J, Lyssiotis CA, Ying H, Wang X, Hua S, Ligorio M, et al. Glutamine supports pancreatic cancer growth through a KRAS-regulated metabolic pathway. Nature 2013;496:101–5.

[80] Gross MI, Demo SD, Dennison JB, Chen L, Chernov-Rogan T, Goyal B, et al. Antitumor activity of the glutaminase inhibitor CB-839 in triple-negative breast cancer. Mol Cancer Ther 2014;13:890–901.

[81] Mancias JD, Kimmelman AC. Targeting autophagy addiction in cancer. Oncotarget 2011;2:1302–6.

[82] Commisso C, Davidson SM, Soydaner-Azeloglu RG, Parker SJ, Kamphorst JJ, Hackett S, et al. Macropinocytosis of protein is an amino acid supply route in Ras-transformed cells. Nature 2013;497:633–7.

[83] Vicent S, Chen R, Sayles LC, Lin C, Walker RG, Gillespie AK, et al. Wilms tumor 1 (WT1) regulates KRAS-driven oncogenesis and senescence in mouse and human models. J Clin Invest 2010;120:3940–52.

[84] Bittker JA, Weiwer M, Shimada K, Yang WS, MacPherson L, Dandapani S, et al. Screen for ras-selective lethal compounds and VDAC ligands - probe 1. Probe Reports from the NIH molecular libraries program. Bethesda MD: National Center for Biotechnology Information; 2011.

[85] Yang WS, Stockwell BR. Synthetic lethal screening identifies compounds activating iron-dependent, nonapoptotic cell death in oncogenic-RAS-harboring cancer cells. Chem Biol 2008;15:234–45. England.

[86] Dolma S, Lessnick SL, Hahn WC, Stockwell BR. Identification of genotype-selective antitumor agents using synthetic lethal chemical screening in engineered human tumor cells. Cancer Cell 2003;3:285–96. United States.

[87] Guo W, Wu S, Liu J, Fang B. Identification of a small molecule with synthetic lethality for K-ras and protein kinase C iota. Cancer Res 2008;68:7403–8. United States.

[88] Shaw AT, Winslow MM, Magendantz M, Ouyang C, Dowdle J, Subramanian A, et al. Selective killing of K-ras mutant cancer cells by small molecule inducers of oxidative stress. Proc Natl Acad Sci USA 2011;108:8773–8.

[89] Puyol M, Martin A, Dubus P, Mulero F, Pizcueta P, Khan G, et al. A synthetic lethal interaction between K-Ras oncogenes and Cdk4 unveils a therapeutic strategy for non-small cell lung carcinoma. Cancer Cell 2010;18:63–73.

[90] Yu B, Luo J. Synthetic lethal genetic screens in Ras mutant cancers. In: Enzymes. 2013. p. 201–19. 2013/01/01 ed., vol. 34 Pt. B.

[91] Steckel M, Molina-Arcas M, Weigelt B, Marani M, Warne PH, Kuznetsov H, et al. Determination of synthetic lethal interactions in KRAS oncogene-dependent cancer cells reveals novel therapeutic targeting strategies. Cell Res 2012;22:1227–45.

[92] Sarthy AV, Morgan-Lappe SE, Zakula D, Vernetti L, Schurdak M, Packer JC, et al. Survivin depletion preferentially reduces the survival of activated K-Ras-transformed cells. Mol Cancer Ther 2007;6:269–76.

[93] Corcoran RB, Cheng KA, Hata AN, Faber AC, Ebi H, Coffee EM, et al. Synthetic lethal interaction of combined BCL-XL and MEK inhibition promotes tumor regressions in KRAS mutant cancer models. Cancer Cell 2013;23:121–8.

[94] Singh A, Sweeney MF, Yu M, Burger A, Greninger P, Benes C, et al. TAK1 inhibition promotes apoptosis in KRAS-dependent colon cancers. Cell 2012;148:639–50.

[95] Scholl C, Frohling S, Dunn IF, Schinzel AC, Barbie DA, Kim SY, et al. Synthetic lethal interaction between oncogenic KRAS dependency and STK33 suppression in human cancer cells. Cell 2009;137:821–34.

[96] Wang Y, Ngo VN, Marani M, Yang Y, Wright G, Staudt LM, et al. Critical role for transcriptional repressor Snail2 in transformation by oncogenic RAS in colorectal carcinoma cells. Oncogene 2010;29:4658–70.

[97] Luo J, Solimini NL, Elledge SJ. Principles of cancer therapy: oncogene and non-oncogene addiction. Cell 2009;136:823–37. United States.

[98] Voorhees PM, Schlossman RL, Gasparetto CJ, Berdeja JG, Morris J, Jacobstein DA, et al. An open-label, dose escalation, multi-center phase 1 study of PRLX 93936, an agent synthetically active against the activated ras pathway, in the treatment of relapsed or relapsed and refractory multiple myeloma. In: 56th ASH annual meeting and exposition, San Francisco, CA. 2014.

[99] Guo W, Wu S, Wang L, Wei X, Liu X, Wang J, et al. Antitumor activity of a novel oncrasin analogue is mediated by JNK activation and STAT3 inhibition. PLoS One 2011;6:e28487. United States.

[100] Wu S, Wang L, Huang X, Cao M, Hu J, Li H, et al. Prodrug oncrasin-266 improves the stability, pharmacokinetics, and safety of NSC-743380. Bioorg Med Chem 2014;22:5234–40.

[101] Hanahan D, Weinberg RA. Hallmarks of cancer: the next generation. Cell 2011;144:646–74.

[102] Dikovskaya D, Cole JJ, Mason SM, Nixon C, Karim SA, McGarry L, et al. Mitotic stress is an integral part of the oncogene-induced senescence program that promotes multinucleation and cell cycle arrest. Cell Rep 2015;12:1483–96.

[103] Luo J, Emanuele MJ, Li D, Creighton CJ, Schlabach MR, Westbrook TF, et al. A genome-wide RNAi screen identifies multiple synthetic lethal interactions with the Ras oncogene. Cell 2009;137:835–48.

[104] Yim H. Current clinical trials with polo-like kinase 1 inhibitors in solid tumors. Anticancer Drugs 2013;24:999–1006.

[105] Balchand SK, Mann BJ, Titus J, Ross JL, Wadsworth P. TPX2 inhibits Eg5 by interactions with both motor and microtubule. J Biol Chem 2015;290:17367–79.

[106] Vela L, Marzo I. Bcl-2 family of proteins as drug targets for cancer chemotherapy: the long way of BH3 mimetics from bench to bedside. Curr Opin Pharmacol 2015;23:74–81.

[107] Rauch A, Hennig D, Schafer C, Wirth M, Marx C, Heinzel T, et al. Survivin and YM155: how faithful is the liaison? Biochim Biophys Acta 2014;1845:202–20.

[108] Mihaly SR, Ninomiya-Tsuji J, Morioka S. TAK1 control of cell death. Cell Death Differ 2014;21:1667–76.

[109] Rawlins P, Mander T, Sadeghi R, Hill S, Gammon G, Foxwell B, et al. Inhibition of endotoxin-induced TNF-alpha production in macrophages by 5Z-7-oxo-zeaenol and other fungal resorcylic acid lactones. Int J Immunopharmacol 1999;21:799–814.

[110] Luo T, Masson K, Jaffe JD, Silkworth W, Ross NT, Scherer CA, et al. STK33 kinase inhibitor BRD-8899 has no effect on KRAS-dependent cancer cell viability. Proc Natl Acad Sci USA 2012;109:2860–5.

CHAPTER 9

GTP-Competitive Inhibitors of RAS Family Members

J.C. Hunter[1], N.S. Gray[2], K.D. Westover[1]

[1]*The University of Texas Southwestern Medical Center at Dallas, Dallas, TX, United States;*
[2]*Dana Farber Cancer Institute, Boston, MA, United States*

CONTENTS

INTRODUCTION

RAS proteins have been pursued as therapeutic cancer targets for almost as long as their protein sequences have been known [1–4]. Two of the initial key observations about RAS were that it binds guanosine nucleosides [2,5] and that it signals in a GTP-dependent manner [6,7]. Although initially there was interest in developing GTP-competitive RAS inhibitors, the idea was quickly dismissed given the observation that GTP and GDP are tightly held by RAS with estimated affinities in the low nanomolar to sub-nanomolar range [5,8]. At the time there were no clear examples of rational drug design resulting in highly potent competitive inhibitors that were chemically distinct from natural ligands, similar to the kinase inhibitors of today, to serve as a counterargument. The conclusion was that a synthetic small molecule inhibitor with adequate potency to compete with GTP would never be achieved. Attention therefore shifted away from the active site.

Casey et al. demonstrated that RAS is farnesylated [9], and this modification was also shown to be essential for oncogenic transformation of cells [10]. Major efforts were, therefore, devoted to inhibition of the RAS farnesylating enzyme, farnesyl transferase (FT). Pre-clinical evaluation of FT inhibitors showed impressive activity in tumor cells lines and animal models, achieving significant anti-tumoral response rates [11–16]. However, clinical trials of the leading candidate compounds tipfarnib and lonafarnib were disappointing achieving minimal tumor responses despite measurable FT inhibition in the majority of patients [17,18].

The apparent failure of FT inhibitors shifted attention to other proteins that participate in RAS signaling pathways, both upstream and downstream of RAS. From these efforts advanced inhibitors of epidermal growth factor receptor (EGFR), RAF, phosphoinositide 3-kinase (PI3K), mitogen-activated protein

Conquering RAS. http://dx.doi.org/10.1016/B978-0-12-803505-4.00009-6

kinase (MEK), extracellular signal-regulated kinase (ERK), protein kinase B (AKT), and mammalian target of rapamycin (mTOR) have now been developed and many are now approved for use in clinical settings. However, none of these inhibitors have proved an adequate solution to address RAS-driven cancers [19]. A recurring theme for why these compounds fail in the oncogenic RAS context is biological complexity. One of the most striking examples is BRAF. BRAF functions immediately downstream of RAS to propagate pro-growth signals and therefore potent BRAF inhibitors were expected to shut down the pathway. Paradoxically, the inhibitors impaired feedback inhibitory activity of BRAF leading to up-regulation of the pathway. Clinically, this was manifest by emergence of skin cancers in patients receiving BRAF inhibitors [20,21]. As another example, MEK inhibitors have also shown inhibitory activity in the MAPK pathway but in doing so activate PI3K [22].

The complexity surrounding RAS biology and the corollary challenges with designing therapeutics to address pathologic RAS signaling have another dimension that is revealed by a deepening understanding that not all RAS mutations function similarly to stimulate cancer. Hints of this emerge from the observation that not only is the distribution of oncogenic mutations found in the closely related RAS isoforms, NRAS, HRAS, and KRAS, non-uniform across diseases, but also the specific mutations for specific RAS isoforms cluster within particular diseases [23]. KRAS, the most commonly mutated RAS isoform in cancer (85% of mutated RAS observed in tumor specimens), provides a good example. KRAS G12D is the most common mutation in pancreatic and colorectal cancer (approximately 20,000 of 32,000 cases annually for pancreas and 25,000 of 60,000 cases annually for colorectal), whereas KRAS G12C is overwhelmingly the most common in lung cancer (23,000 of 45,000 cases annually) [24]. In the case of KRAS G12C, the mutation is caused by changes in the genome after tobacco smoke exposure. Chemicals in tobacco smoke form DNA adducts resulting in genome-wide G→T transversions, and within the KRAS locus this yields the activating KRAS G12C mutation. This likely explains why lung cancer, which is associated with tobacco smoke exposure, has a high incidence of KRAS G12C mutated tumors. However, there is also evidence that biological selection for specific mutations occurs in certain contexts [25]. Recent data from our group provide evidence that differential structural and biochemical features may play a role in how this selection occurs [26]. In summary, all of this implies that different RAS mutations likely favor different signaling pathways which depend on context, and understanding which pathways specific mutant isoforms favor may be important for designing therapies tailored to specific KRAS mutant isoforms [26–30].

Given the complexity surrounding RAS biology and the lack of success in addressing RAS-driven tumors with indirect inhibitors, we and others have returned to exploring the possibility of making direct inhibitors of KRAS. Two major efforts have garnered particular attention: (1) a covalent approach that targets a transient pocket next to switch II opposite the GTP-binding site [31] and (2) a covalent approach for targeting the GTP-binding site [32,33]. Here we present a theoretical basis for targeting the GTP-binding site and proof-of-concept data.

RATIONALE FOR TARGETING THE RAS ACTIVE SITE

Guanine nucleotide (GN) binding regulatory proteins depend on binding to GTP for biological activity [34]. The group led by Cooper demonstrated that RAS interacts directly with RAF kinase in a GTP-dependent manner [6]. The group led by Wittinghofer later quantitated the interaction [35] and then showed the structural basis for this phenomenon [36]. In simplest terms they showed that GTP-bound (activated) RAS passes signals based on a specific conformation characterized by "closure" of two dynamic structural elements called switch 1 and switch 2 around GTP. The activated conformation is controlled by several key interactions including those between the residues T35 and G60, and the gamma phosphate of GTP (Fig. 9.1A). This is distinct from the GDP-bound form in which switch 2 is "open" based primarily on loss of interactions between the guanine nucleoside and switch 2. It is worth noting that in the hundreds of X-ray and nuclear magnetic resonance structures of RAS superfamily proteins available through the protein database (PDB), structural variability is primarily evident in the switches while the remainder of the protein changes little (Fig. 9.1B). These and other observations have resulted in the general consensus that information is conveyed by RAS to other proteins via the conformation of the switches.

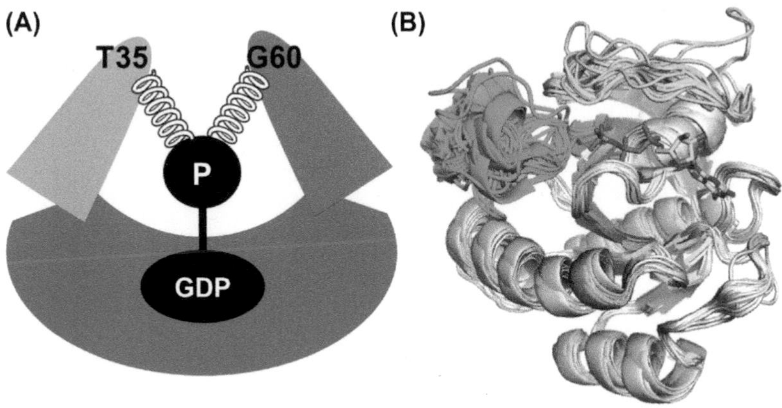

FIGURE 9.1

(A) Activated RAS architecture. Switch 1 (yellow) and switch 2 (red) are held in a closed conformation when RAS (blue) is GTP bound (left). However, when the gamma phosphate is lost it returns to an open conformation. (B) Global alignment of 50 RAS superfamily structures. Switch 1 (defined as residues 26–40) is in yellow, switch 2 (defined as residues 57–70) is in green.

The transition from GDP to GTP happens in normal physiology when growth factors stimulate extracellular receptors to induce nucleotide exchange with the help of guanosine nucleotide exchange factors (GEFs), such as Son of sevenless (Sos), which hold RAS in an extended, fully open conformation allowing GDP to dissociate and GTP to bind [37]. Conversion back to GDP happens via GTP hydrolysis, which occurs slowly within RAS alone, but can be stimulated by GTPase-activating proteins (GAPs) [38]. The GTP-bound, closed RAS conformation is competent to interact with RAS "effectors," such as RAF kinases and PI3K, which propagate and amplify signals from RAS [39]. Cancer-causing mutations in RAS or other members of the pathway tip the balance in favor of a higher probability that RAS is GTP bound instead of GDP bound, producing signals that result in the cancer state. Notably all regulators (GEFs and GAPs) and effectors of RAS that have been characterized structurally interact with switch 1 and/or switch 2 in some manner (Fig. 9.2).

One reason to be optimistic about targeting the GTP-binding site as a perturbation strategy is that interactions between RAS and cancer-related effectors specifically require that the switches be in the closed conformation to interact. Therefore, small molecules likely do not need to satisfy structural requirements other than pushing RAS away from the active, closed conformation to be effective. In other words, there is no requirement to achieve a specific "inactive" conformation, only to avoid an activated conformation. In summary, we believe that targeting the GTP-binding site for therapeutic purposes has an excellent chance of disrupting cancer-related KRAS signaling because the contents of the GTP-binding site control KRAS signaling activity.

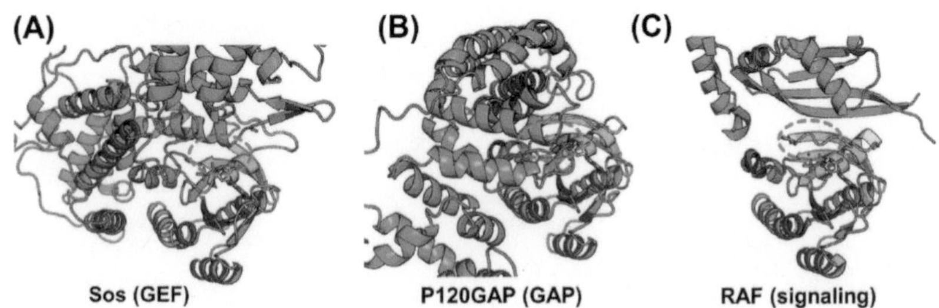

FIGURE 9.2

Location of effector, GAP and GEF binding on RAS. X-ray crystal structures for RAS in complex with (A) Sos (PDB 1BKD), (B) p120GAP (PDB 1WQ1), and (C) RAF (PDB 3DDC) are shown. Each constitutes a distinct class of RAS-interacting protein (Raf, signaling; p120GAP, nucleotide hydrolysis; SOS, nucleotide exchange) and each contacts RAS on a different portion of the RAS surface. For all complexes RAS has been superimposed for comparable orientation between panels. RAS is cyan except for switch 1 (yellow) and switch 2 (red) and the interacting protein is in green. GDP is shown as sticks for orientation and is highlighted with a *pink dotted line*.

A BRIEF HISTORY OF COVALENT INHIBITORS

We hypothesized that a covalent inhibitor may be able to overcome the challenges that initially deterred efforts to target the RAS active site. Covalent modification of biological macromolecules is used extensively in Nature to regulate or alter biological systems. This is true not only within cells, which use ubiquitination, acetylation, phosphorylation, and a host of other modifications to effect cellular processes, but also of interactions between cells as in the covalent modification of steroid hormone receptors by various hormones [40] or between different organisms as exemplified by thioester bonds found within complement proteins of the immune system [41] and in proteins on the pili of gram-positive bacteria that allow them to adhere to host cells [42]. All of these examples are notable for elegant mechanisms that confine the covalent reactivity to a specific, defined space, preventing many off-target effects.

Use of covalent molecules for therapeutic purposes also has a long history. Some of the oldest examples include aspirin, of which natural forms have been used for over 2000 years, dating back to ancient Egyptian records [43]. Other notable widely used examples include penicillin, proton pump inhibitors, and clopidogrel (Plavix). By some estimates up to one-third of all approved medications work by covalent mechanisms [44]. As experience with structure-assisted drug design has grown, multiple new examples of targeted covalent compounds have emerged. A common method for developing covalent inhibitors is to identify a potent reversible chemical scaffold and modify this to include an electrophilic warhead based on the binding mode of the scaffold and the location of the targeted amino acid, usually a cysteine [45]. Cysteine is a prime target for selective modification based on the strongly nucleophilic side chain sulfhydryl. In addition, cysteines do not occur at high frequencies in the proteome providing opportunities to maintain selectivity [46]. US Food and Drug Administration–approved examples of covalent compounds now include ibrutinib, which targets Bruton's tyrosine kinase (BTK), and afatinib, which targets EGFR [47,48]. Notably in phase III trials ibrutinib showed impressive activity for chronic lymphocytic leukemia (CLL) by improving not only progression-free survival but also overall survival [49]. Multiple additional covalent agents are now in the pipeline including targeted agents to BMX non-receptor tyrosine kinase (BMX), c-Jun N-terminal kinase (JNK), erb-b2 receptor tyrosine kinase 3 (HER3), cyclin-dependent kinase 8 (CDK8), and TGF-beta activated kinase 1 (TAK1) [27,50–52].

SIMULATIONS OF GTP-COMPETITIVE RAS INHIBITORS

To test the idea that the GTP-binding site of KRAS might be targetable using an irreversible inhibitor we performed simulations using enzyme kinetics software. At their core, simulations are composed of relatively simple calculations that take into account the concentrations of components (enzyme, ligands, and substrates) and rate constants describing how quickly binding events or chemical conversions occur. However, when multiple transformations are linked together in a multi-component reaction scheme, the mathematics become more complicated and computers provide valuable assistance. An elementary scheme for the interaction of KRAS with a GTP competitive inhibitor is shown in Fig. 9.3, where E = RAS, I = inhibitor, EI = a transient complex of inhibitor and RAS, EI* = a covalent complex between KRAS and inhibitor, EGTP = a transient complex between RAS and GTP, K_M = the dissociation constant for GTP, K_i = the inhibition constant for inhibitor, and K_{inact} = the rate constant for covalent inactivation of KRAS. Simulation conditions are as follows: KRAS = 400 nM [53], GTP = 200 μM [54–64], Inhibitor = 50 μM, K_M = 500 pM [65–67], and was done using Gepasi [68].

As expected, these simulations show that reversible inhibitors would require binding constants in the low picomolar range to effectively compete with GTP (likely impossible to achieve). Predictions for the picomolar affinity requirement are not shown, but we provide results showing that a reversible inhibitor with a 10 nM K_i is not expected to displace GTP from KRAS (orange curve in Fig. 9.3). However, by introducing the inactivation term (a non-zero K_{inact}), simulations show that compounds with likely achievable binding constants (in the nanomolar range) would provide inhibitory activity over a range of likely achievable K_{inact} values (Fig. 9.3B and C).

AN OPPORTUNITY TO COVALENTLY TARGET THE KRAS ACTIVE SITE

We noted that the KRAS G12C mutation is the most common KRAS mutation in lung cancer, present in 23,000 new cases of lung cancer and 5000 cases of other cancer types annually [24]. Cysteine 12 (C12) is solvent-accessible and adjacent to the active site, near the usual position of the gamma-phosphate of

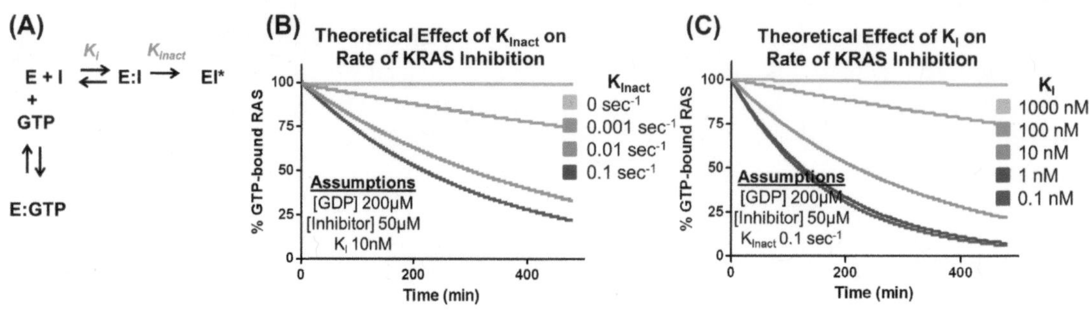

FIGURE 9.3 K_i/K_{inact}.
(A) Two parameters, K_i and K_{inact}, describe the kinetics of irreversible inhibition. (B,C) Modeling of irreversible inhibition of G12C KRAS predicts a K_{inact} of 0.1 s^{-1} and K_i of 10 nM will yield 50% inhibition of KRAS G12C in ~4 h.

the native GTP. To investigate if a covalent compound could replace the natural GTP or GDP ligands bound to KRAS G12C and inactivate the protein, we designed and synthesized a covalent tool compound based on the structure of GDP, SML-8-73-1 [32,33]. SML-8-73-1 contains an electrophilic warhead extending from the beta-phosphate, which undergoes Michael addition to C12, forming a stable thioether linkage (Fig. 9.4A and B). We analyzed SML-8-73-1 for the ability to form a covalent bond with purified KRAS G12C using mass spectrometry (MS) and found it could quantitatively label KRAS G12C (Fig. 9.4C–F). An X-ray co-crystal structure also showed SML-8-73-1 to be covalently bound to KRAS G12C (Fig. 9.4B). It is worth noting that in all of these experimental systems, GDP-bound KRAS protein was the starting material and required SML-8-73-1 to displace GDP to form a covalent bond.

To further characterize the interactions between SML-8-73-1 and KRAS, we measured K_{inact}/K_i, a preferred enzymological parameter for assessing covalent probes as mentioned previously [69]. To obtain these numbers we used the method described by Copeland [69], which requires a series of reaction time courses run over a range of compound concentrations (Fig. 9.5). Reaction rates are plotted versus compound concentrations and fit to the curve parameterized according to the equation in Fig. 9.5B. Using the fit, estimates for K_{inact} and K_i can be extracted. SML-8-73-1 displayed excellent potency with a K_i of 9 nM and a K_{inact} of 0.84 min^{-1}. According to our simulations these values should enable an inhibitor to covalently label >50% of the

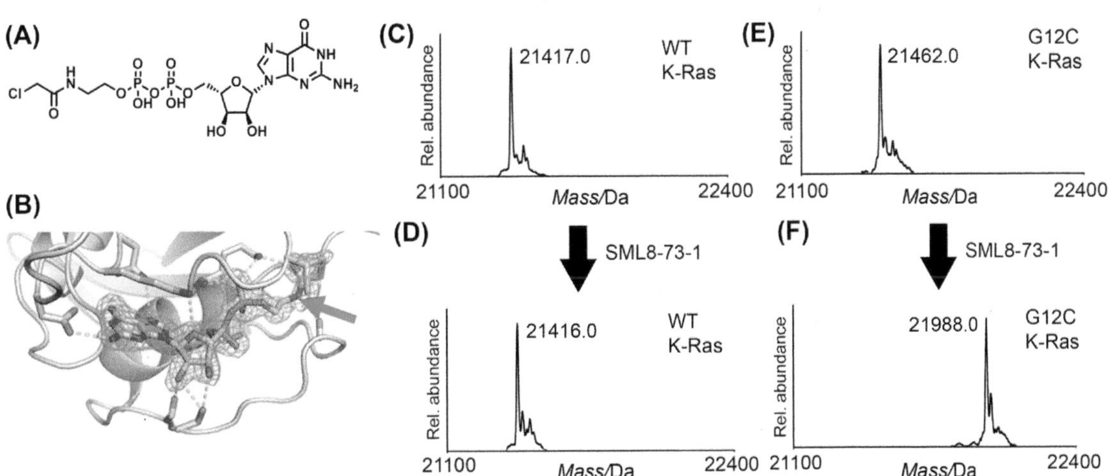

FIGURE 9.4

(A) Chemical structure of the SML-8-73-1 compound. (B) Electron density corresponding to an X-ray crystal structure of SML-8-73-1 bound to C12 (indicated with *red arrow*) of KRAS G12C. (C–F) SML-8-7-31 reacts quantitatively with G12C K-Ras, but does not label WT K-Ras. De-convoluted mass spectra obtained for K-Ras G12C (E,F) and WT K-Ras (C,D) before (C,E) and after (D,F) incubation with SML-8-7-31.

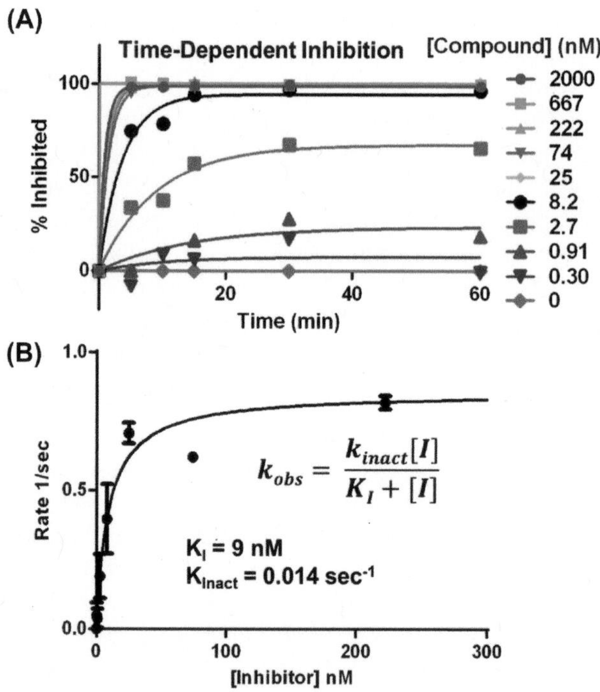

FIGURE 9.5 K_{inact}/K_i.

(A) Determination of K_i and K_{inact} for SML-8-73-1. A series of time courses at different concentrations of inhibitors (B) can be re-plotted as K_{obs} versus [I] The resulting curve can be fit to the equation shown and K_{inact}/K_i can thereby be obtained.

enzyme within 4 h, significantly less than the 12-h half-life that has been reported for KRAS within the cells [70].

Kinetic GDP Displacement Assay

In addition to the K_{inact}/K_i assessment, we characterized the kinetics of GDP and GTP displacement from the active site of KRAS G12C by SML-8-73-1. We developed a robust biochemical assay that uses purified, GDP-loaded KRAS G12C and a cysteine-reactive compound, 7-diethylamino-3-(4-maleimidophenyl)-4-methylcoumarin, or CPM, which fluoresces after reacting with a thiol group (Fig. 9.6A). CPM efficiently reacts with KRAS G12C to yield a ~10-fold higher fluorescent signal than seen with wild-type GN bound KRAS [32,33]. Because we start by using GDP-loaded protein this assay provides a composite metric for the ability of compounds to compete with GTP and GDP for the active site. Our analysis showed rapid displacement of GDP by SML-8-73-1 from the active site of KRAS G12C (Fig. 9.6B, blue curve). As expected, addition of 1 mM GTP and GDP to the reaction slowed the displacement, but given enough time inactivation proceeded to completion (Fig. 9.6B, red curve).

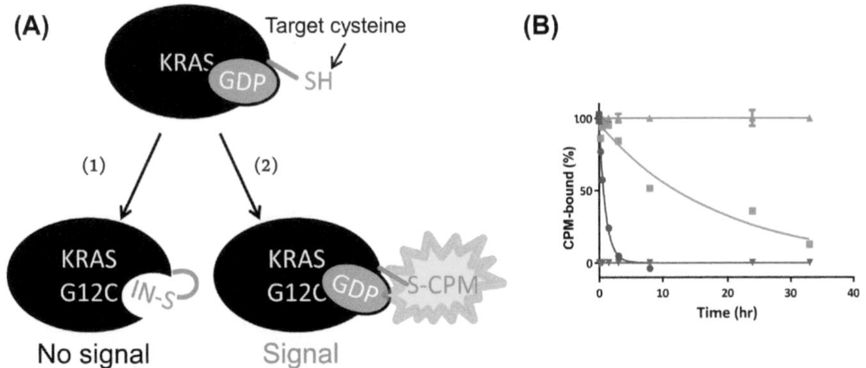

FIGURE 9.6 Kinetic GDP displacement assay.
(A) Covalent assay principle. If an inhibitor (IN) binds to C12 (pathway 1), then CPM cannot bind and no signal is detected. If no IN binds (pathway 2), the cysteine is available for CPM. (B) KRAS G12C sample data. KRAS G12C protein (green) was exposed to SML in the presence of 1 mM concentrations of GTP and GDP (red) or absence (blue). WT KRAS (purple) is a negative control.

This provides an important demonstration of the ability of SML-8-73-1 to displace GDP from the active site of KRAS G12C.

Disruption of Interactions With Ras-Binding Domain

Activated KRAS interacts directly with RAF kinase to amplify and propagate pro-cancer signals. The GTP-bound, closed conformation of RAS is required for this interaction to occur. Conversely, for inactivation of KRAS signaling, KRAS must be in an open conformation with switch 2 extended. We evaluated for the open versus closed conformation of SML-bound KRAS in several ways. We first used hydrogen exchange (HX) MS. HX MS provides a means of assessing protein conformational dynamics by measuring the rate of hydrogen exchange of backbone amide hydrogens [71]. We performed HX MS on K-Ras G12C bound to several different ligands including a non-hydrolyzable GTP mimic (GMPPNP), GDP, or SML-8-73-1. Hydrogen exchange into each complex was compared by monitoring the deuteration of over 40 peptides produced by HX labeling followed by proteolytic digestion. The peptides comprising portions of the nucleotide-binding pocket (Fig. 9.7) showed significant variability in labeling signatures. When K-Ras G12C was bound to covalent inhibitor SML-8-73-1, the deuterium incorporation in residues 7–20 and 114–120 resembled the GDP-bound protein (Fig. 9.7) suggesting that the compound likely stabilizes an inactive form of K-Ras G12C.

We later confirmed this by solving a 1.9 Å co-crystal structure of KRAS G12C bound to SML 8-73-1. In this structure, switch 2 is clearly in the open, inactive conformation (Fig. 9.8) suggesting that KRAS G12C:SML-8-73-1 will not effectively interact with RAF. To confirm that SML inactivates KRAS G12C we

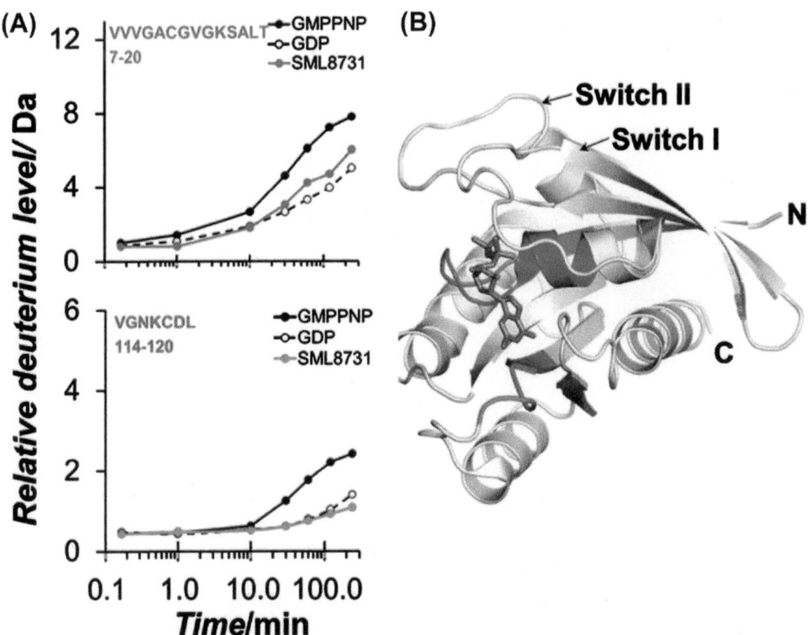

FIGURE 9.7 Hydrogen exchange/mass spectrometry confirms that SML-bound G12C KRAS is in the inactive conformation.
(A) Deuterium incorporation at the indicated peptide positions was measured by mass spectrometry for GMPPNP (black), GDP (white), or SML-8-73-1 (blue or red) bound KRAS. (B) The location of each peptide measured in (A, blue and red) is displayed on the X-ray structure of HRAS (PDBID 4Q21) [92]. *Reprinted from the original publication Lim SM, Westover KD, Ficarro SB, Harrison RA, Choi HG, Pacold ME, et al. Therapeutic targeting of oncogenic k-ras by a covalent catalytic site inhibitor. Angew Chem 2014;53(1):199–204 with permission from John Wiley & Sons.*

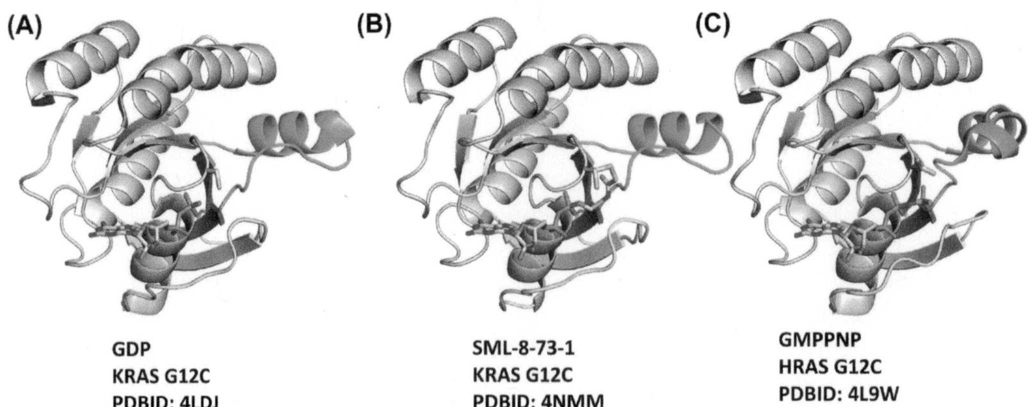

FIGURE 9.8 Ligand-dependent structures of K-Ras.
(A) G12C K-Ras bound to GDP (B) SML-8-73-1 and G12C HRAS bound to GMPPNP (C, 4L9W) are shown with switch 1 colored yellow, and switch 2 colored green. When bound to SML-8-73-1, G12C K-Ras assumes a conformation nearly identical to GDP-bound G12C K-Ras with both switch regions in the inactive conformation (RMSD ~0.17 Å). The GMPPNP bound form shows switch 2 is "closed."

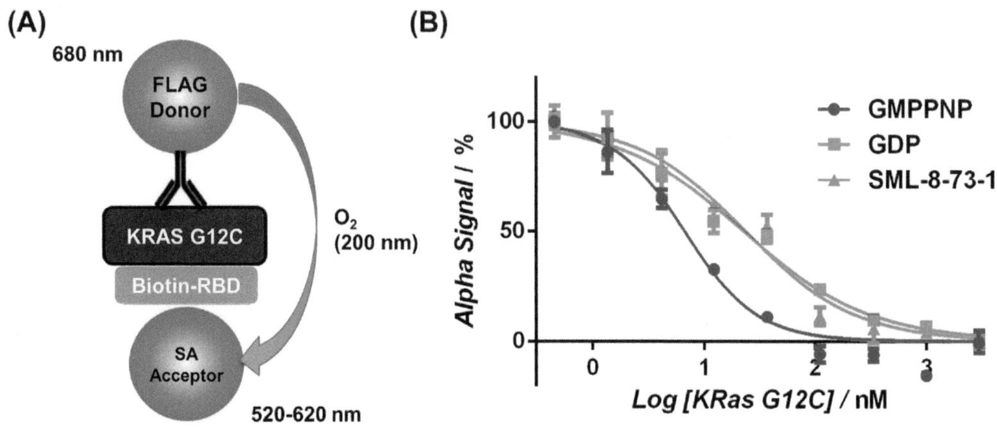

FIGURE 9.9 Ras:RAF interaction assay.

(A) Assay principle: FLAG-tagged KRASG12C loaded with a GTP analogue allowed to form a complex with biotin–Ras-binding domain (RBD). Complexes are then competed apart using untagged KRAS G12C loaded with test compounds. (B) Sample data showing that SML disrupts KRAS:RBD interaction.

developed a RAS:RAF interaction assay to measure the strength of interaction between KRAS G12C and the Ras-binding domain (RBD) of RAF kinase. The assay showed that SML effectively reduces the affinity of KRAS G12C for Raf RBD to the level of GDP-bound KRAS (Fig. 9.9B).

Selectivity of SML-Class Compounds

A major concern in developing GTP-competitive compounds targeted to KRAS is that they will have cross-reactivity with other GTP-binding proteins because the amino acid sequence of the GTP-binding site is well conserved across the RAS superfamily. To asses if SML does in fact cross-react with other RAS family members we employed a proteomics approach to survey which GTP-binding proteins display a high affinity for SML. This approach is based on a compound profiling technology originally developed to study kinase inhibitors [72]. In this approach, lysates derived from KRAS G12C homozygous cancer cells such as MIA PaCa-2 are treated with SML followed by a GTP-desthiobiotin probe. This probe covalently targets a lysine that is conserved within the RAS superfamily so that it non-specifically biotinylates over 100 GTP-dependent enzymes. Biotinylated proteins are isolated using streptavidin beads and quantified using MS. Selectivity is assessed based on the ability of our test compounds to specifically protect KRAS from biotinylation compared with other GTP-dependent enzymes. This approach allowed us to detect approximately ~90 GTP-dependent enzymes. The data for SML are shown in Fig. 9.10 and demonstrate that other than KRAS G12C (red bar), SML shows reactivity with only three other GTPases, GUF1, EFTUD1, and ARL3 (orange bars). These data are notable because they provide an example where addition of a covalent

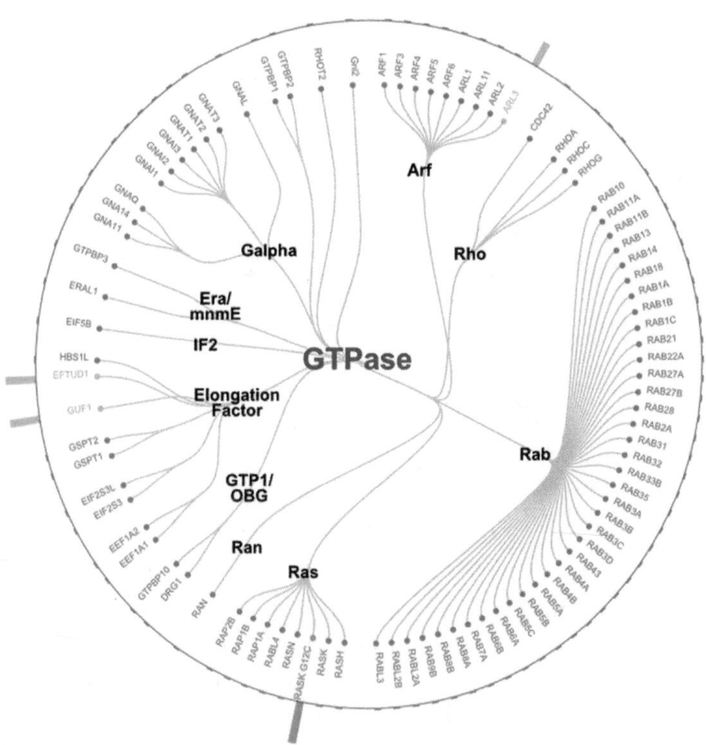

FIGURE 9.10 ActivX profiling.
Selectivity of SML-8-73-1 in G12C KRAS-expressing cells at 100 μM as determined by GTP-desthiobiotin competition experiment. GTPases detected but not inhibited by SML are listed in green. Targets inhibited by SML-8-73-1 are in orange and red (KRAS is most inhibited). Degree of inhibition is indicated by bar height.

substituent is the basis for covalent inhibitor selectivity. It is also notable for being the first example of comprehensive profiling of a direct-acting GTP inhibitor against a large panel of GTPases.

LIMITATIONS OF SML-CLASS COMPOUNDS

Studying the biological effects of KRAS G12C inhibition using SML-8-73-1 has been challenging because this compound contains a polar bisphosphate pharmacophore which does not allow the compound to penetrate cellular membranes. We attempted to overcome this by using a caging strategy wherein charged ions are modified to mask a charged group and thereby allow for passive cellular uptake [73]. The caging moiety was designed so it would be removed intra-cellularly by enzymatic cleavage. Unfortunately, we found that caged bisphosphates, where both phosphates are modified, are unstable as a consequence of hydrolysis of the phosphate ester [33]. We now realize that additional innovation will be required to translate these proof-of-concept studies into pre-clinical candidate compounds. This will likely require re-engineering the bisphosphate portion of the molecule and perhaps discovery of new inhibitor scaffold chemotypes, as has been accomplished with kinases. Interestingly, there is one report of a GTP-competitive small molecule of a RAS family member that is chemically unrelated to guanosine. The compound, CID-1067700, was discovered by high-throughput screening using RAB7a (RAS protein from Rat Brain) as a model for the RAS superfamily [74,75]. Although the experience with GTP-competitive inhibitors of the RAS superfamily is limited, there are reasons for optimism.

APPLICABILITY OF CYSTEINE TARGETING OF THE RAS SUPERFAMILY

It may be possible to apply the same covalent inhibitor strategy to other members of the RAS superfamily. The family is composed of 160 small globular GDP/GTP-binding proteins with a remarkably high degree of structural and sequence conservation. The family is sub-classified into five groups comprising the Ras, Rho/Rac, Rab, Ran, Arf, and families based on sequence similarity. As a group, RAS family GTPases regulate an extensive range of cellular processes including cell proliferation, division, migration, vesicle transport, exocytosis, and cytoskeletal remodeling, among others [76–79]. Misregulation of small GTPase activity has been implicated in a number of diseases and may be attractive drug targets in various conditions including cancer, neurodegenerative, and vascular disease among others [80–85].

As a group, however, most small GTPases exhibit the same obstacles that have discouraged attempts to directly target the RAS active site, namely high-affinity for GDP/GTP combined with a high concentration of endogenous nucleotide and no obvious alternative binding pockets [86–88]. Another substantial hurdle is the necessity to design an inhibitor with a high degree of selectivity given the large number of small GTPase family members and other GDP/GTP-binding proteins within the cell. To minimize off-target effects, small molecules targeting the nucleotide-binding pocket small GTPases must be exquisitely selective in binding to the intended target protein. As we have demonstrated, both of these challenges may be overcome through covalent small molecules, which can be designed with extremely high potency and selectivity through reaction with precisely located cysteine residues within the active site of the intended target protein.

To predict the general utility of targeting RAS superfamily proteins by targeting cysteine residues near the GDP/GTP-binding pocket, we analyzed the relative frequency of cysteine in the residues surrounding the active site of all 160 small GTPase family members. Among the four regions near the nucleotide pocket

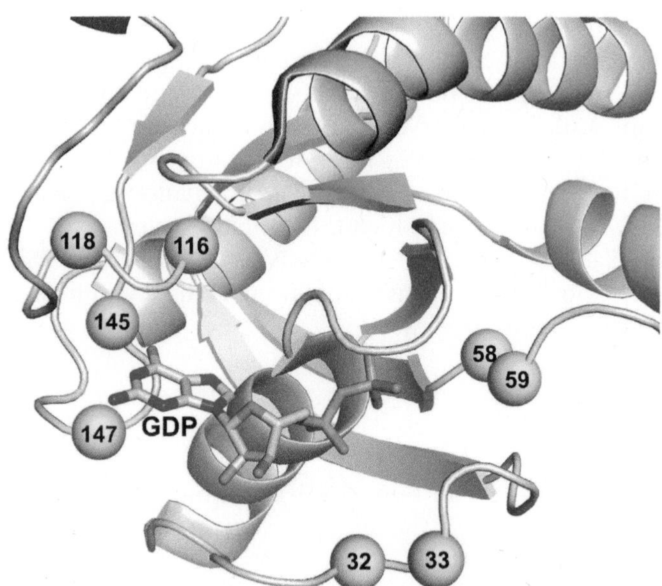

FIGURE 9.11 Cysteine residues occur in multiple locations near the GDP/GTP-binding site in various RAS superfamily members.

The primary sequence of all 160 RAS family members was aligned using Clustal omega server to determine the location and relative frequency of cysteine residues at locations near the GDP/GTP-binding site. Residues within 5 Å of the nucleotide were analyzed and highlighted in yellow if a cysteine is present in one or more RAS family members. See Table 9.1 for details. Figure was prepared using Pymol Software.

(residues 28–34, 58–59, 116–120, 145–147) cysteine is relatively rare, present in only 2% of all sites across the family (56 occurrences of 2721 total residues). Cysteine occurs most often at position 118 and is present at this location in 18% of small GTPases. At positions 145, 116, 32, and 147, cysteine occurs in only 7, 4, 3, or 1% of small GTPases, respectively, and is found in only one family member at positions 33, 58, and 59 (Fig. 9.11 and Table 9.1). Among these cysteine-containing small GTPases are notable promising therapeutic targets including K-RAS, N-RAS, E-RAS, H-RAS, RAB, and Rap1 [89–91]. Overall this relative paucity of cysteine residues near the guanine-binding pocket suggests that if cysteine reactive covalent compounds could be designed they would likely exhibit a high degree of selectivity toward their desired target and may become an attractive approach for modulating this class of signaling enzymes.

CONCLUSION

Our early experience with covalent inhibitors targeting the GTP-binding site of KRAS together with the recent clinical success of covalent kinase inhibitors argue

Table 9.1 Cysteine Residues in GTPases

Amino Acid Position							
32	**33**	**58**	**59**	**116**	**118**	**145**	**147**
Gem	NKIRas1	RabL5	FLJ22655	RhoBTB1	Di-Ras1	Arf1	Rerg
Rab14				RhoBTB2	Di-Ras2	Arf3	Ris/RasL12
Rab39A				RhoD	E-Ras	Arf4	
RasL10B				Rif	H-Ras.1	Arf5	
				Rnd1	H-Ras.2	Arf6	
				Rnd2	K-Ras4A	Arl4	
				Rnd3	K-Ras4B	Arl5A.1	
					NKIRas2	Arl5A.2	
					N-Ras	Arl7	
					Rab10	Arl8	
					Rab13	FLJ22595	
					Rab1A		
					Rab22A		
					Rab22B		
					Rab32		
					Rab33A		
					Rab33B		
					Rab38		
					Rab39A		
					Rab39B		
					Rab3A		
					Rab3B		
					Rab3C		
					Rab3D		
					Rab7L1		
					Rab8A		
					Rab8B		
					Rap1A		
					Rap1B		

All 160 RAS superfamily members were aligned using the Clustal Omega Sequence Alignment Server. GTPases with cysteine residues within 5 Å of the nucleotide-binding pocket were identified as potentially targetable through cysteine-reactive, small molecule active site inhibitors.

that cysteine-targeted GTPase inhibitors/activators will eventually yield novel tool compounds and possibly therapeutics for inhibiting a broad range of signaling pathways regulated by the RAS superfamily of GTPases. Clearly there are many hurdles before this can be achieved but the biochemical tools to move this concept forward have now been developed and the field is ripe for this advancement.

References

[1] Der CJ, Krontiris TG, Cooper GM. Transforming genes of human bladder and lung carcinoma cell lines are homologous to the ras genes of Harvey and Kirsten sarcoma viruses. Proc Natl Acad Sci USA 1982;79(11):3637–40.

[2] Scolnick EM, Papageorge AG, Shih TY. Guanine nucleotide-binding activity as an assay for src protein of rat-derived murine sarcoma viruses. Proc Natl Acad Sci USA 1979;76(10):5355–9.

[3] Kirsten WH, Mayer LA. Morphologic responses to a murine erythroblastosis virus. J Natl Cancer Inst 1967;39(2):311–35.

[4] Young HA, Shih TY, Scolnick EM, Rasheed S, Gardner MB. Different rat-derived transforming retroviruses code for an immunologically related intracellular phosphoprotein. Proc Natl Acad Sci USA 1979;76(7):3523–7.

[5] Manne V, Yamazaki S, Kung H-F. Guanosine nucleotide binding by highly purified Ha-ras-encoded p21 protein produced in *Escherichia coli*. Proc Natl Acad Sci USA 1984;81(22):6953–7.

[6] Vojtek AB, Hollenberg SM, Cooper JA. Mammalian Ras interacts directly with the serine/threonine kinase Raf. Cell 1993;74(1):205–14.

[7] Block C, Janknecht R, Herrmann C, Nassar N, Wittinghofer A. Quantitative structure-activity analysis correlating Ras/Raf interaction in vitro to Raf activation in vivo. Nat Struct Biol 1996;3(3):244–51.

[8] John J, Sohmen R, Feuerstein J, Linke R, Wittinghofer A, Goody RS. Kinetics of interaction of nucleotides with nucleotide-free H-ras p21. Biochemistry 1990;29(25):6058–65.

[9] Casey PJ, Solski PA, Der CJ, Buss JE. P21ras is modified by a farnesyl isoprenoid. Proc Natl Acad Sci USA 1989;86(21):8323–7.

[10] Willumsen BM, Norris K, Papageorge AG, Hubbert NL, Lowy DR. Harvey murine sarcoma virus p21 ras protein: biological and biochemical significance of the cysteine nearest the carboxy terminus. EMBO J 1984;3(11):2581–5.

[11] End DW, Smets G, Todd AV, Applegate TL, Fuery CJ, Angibaud P, et al. Characterization of the antitumor effects of the selective farnesyl protein transferase inhibitor R115777 in vivo and in vitro. Cancer Res 2001;61(1):131–7.

[12] Kohl NE, Mosser SD, deSolms SJ, Giuliani EA, Pompliano DL, Graham SL, et al. Selective inhibition of ras-dependent transformation by a farnesyltransferase inhibitor. Science 1993;260(5116):1934–7.

[13] Kohl NE, Omer CA, Conner MW, Anthony NJ, Davide JP, Desolms SJ, et al. Inhibition of farnesyltransferase induces regression of mammary and salivary carcinomas in ras transgenic mice. Nat Med 1995;1(8):792–7.

[14] Kohl NE, Wilson FR, Mosser SD, Giuliani E, deSolms SJ, Conner MW, et al. Protein farnesyltransferase inhibitors block the growth of ras-dependent tumors in nude mice. Proc Natl Acad Sci USA 1994;91(19):9141–5.

[15] Nagasu T, Yoshimatsu K, Rowell C, Lewis MD, Garcia AM. Inhibition of human tumor xenograft growth by treatment with the farnesyl transferase inhibitor B956. Cancer Res 1995;55(22):5310–4.

[16] Mangues R, Corral T, Kohl NE, Symmans WF, Lu S, Malumbres M, et al. Antitumor effect of a farnesyl protein transferase inhibitor in mammary and lymphoid tumors overexpressing N-ras in transgenic mice. Cancer Res 1998;58(6):1253–9.

[17] Adjei AA, Mauer A, Bruzek L, Marks RS, Hillman S, Geyer S, et al. Phase II study of the farnesyl transferase inhibitor R115777 in patients with advanced non-small-cell lung cancer. J Clin Oncol 2003;21(9):1760–6.

[18] Doll RJ, Kirschmeier P, Bishop WR. Farnesyltransferase inhibitors as anticancer agents: critical crossroads. Curr Opin Drug Discov Dev 2004;7(4):478–86.

[19] Singh H, Longo DL, Chabner BA. Improving prospects for targeting RAS. J Clin Oncol 2015.

[20] Poulikakos PI, Zhang C, Bollag G, Shokat KM, Rosen N. RAF inhibitors transactivate RAF dimers and ERK signalling in cells with wild-type BRAF. Nature 2010;464(7287):427–30.

[21] Flaherty KT, Puzanov I, Kim KB, Ribas A, McArthur GA, Sosman JA, et al. Inhibition of mutated, activated BRAF in metastatic melanoma. N Engl J Med 2010;363(9):809–19.

[22] Turke AB, Song Y, Costa C, Cook R, Arteaga CL, Asara JM, et al. MEK inhibition leads to PI3K/AKT activation by relieving a negative feedback on ERBB receptors. Cancer Res 2012;72(13):3228–37.

[23] Prior IA, Lewis PD, Mattos C. A comprehensive survey of Ras mutations in cancer. Cancer Res 2012;72(10):2457–67.

[24] Forbes SA, Bindal N, Bamford S, Cole C, Kok CY, Beare D, et al. COSMIC: mining complete cancer genomes in the Catalogue of Somatic Mutations in Cancer. Nucleic Acids Res 2011;39(Database issue):D945–50.

[25] Westcott PM, Halliwill KD, To MD, Rashid M, Rust AG, Keane TM, et al. The mutational landscapes of genetic and chemical models of Kras-driven lung cancer. Nature 2015;517(7535):489–92.

[26] Hunter JC, Manandhar A, Carrasco MA, Gurbani D, Gondi S, Westover KD. Biochemical and structural analysis of common cancer-associated KRAS mutations. Mol Cancer Res 2015;13(9):1325–35.

[27] Shepherd FA, Domerg C, Hainaut P, Janne PA, Pignon JP, Graziano S, et al. Pooled analysis of the prognostic and predictive effects of KRAS mutation status and KRAS mutation subtype in early-stage resected non-small-cell lung cancer in four trials of adjuvant chemotherapy. J Clin Oncol 2013;31(17):2173–81.

[28] Kim ES, Herbst RS, Wistuba II, Lee JJ, Blumenschein Jr GR, Tsao A, et al. The BATTLE trial: personalizing therapy for lung cancer. Cancer Discov 2011;1(1):44–53.

[29] Karachaliou N, Mayo C, Costa C, Magri I, Gimenez-Capitan A, Molina-Vila MA, et al. KRAS mutations in lung cancer. Clin Lung Cancer 2013;14(3):205–14.

[30] Garassino MC, Marabese M, Rusconi P, Rulli E, Martelli O, Farina G, et al. Different types of K-Ras mutations could affect drug sensitivity and tumour behaviour in non-small-cell lung cancer. Ann Oncol 2011;22(1):235–7.

[31] Ostrem JM, Peters U, Sos ML, Wells JA, Shokat KM. K-Ras(G12C) inhibitors allosterically control GTP affinity and effector interactions. Nature 2013;503(7477):548–51.

[32] Hunter JC, Gurbani D, Ficarro SB, Carrasco MA, Lim SM, Choi HG, et al. In situ selectivity profiling and crystal structure of SML-8-73-1, an active site inhibitor of oncogenic K-Ras G12C. Proc Natl Acad Sci USA 2014;111.

[33] Lim SM, Westover KD, Ficarro SB, Harrison RA, Choi HG, Pacold ME, et al. Therapeutic targeting of oncogenic k-ras by a covalent catalytic site inhibitor. Angew Chem 2014;53(1):199–204.

[34] Spiegel AM. Signal transduction by guanine nucleotide binding proteins. Mol Cell Endocrinol 1987;49(1):1–16.

[35] Herrmann C, Martin GA, Wittinghofer A. Quantitative-analysis of the complex between P21(Ras) and the ras-binding domain of the human Raf-1 protein-kinase. J Biol Chem 1995;270(7):2901–5.

[36] Nassar N, Horn G, Herrmann C, Block C, Janknecht R, Wittinghofer A. Ras/Rap effector specificity determined by charge reversal. Nat Struct Biol 1996;3(8):723–9.

[37] Boriack-Sjodin PA, Margarit SM, Bar-Sagi D, Kuriyan J. The structural basis of the activation of Ras by Sos. Nature 1998;394(6691):337–43.

[38] Settleman J, Albright CF, Foster LC, Weinberg RA. Association between GTPase activators for Rho and Ras families. Nature 1992;359(6391):153–4.

[39] Pacold ME, Suire S, Perisic O, Lara-Gonzalez S, Davis CT, Walker EH, et al. Crystal structure and functional analysis of Ras binding to its effector phosphoinositide 3-kinase gamma. Cell 2000;103(6):931–43.

[40] Takahashi N, Breitman TR. Covalent modification of proteins by ligands of steroid hormone receptors. Proc Natl Acad Sci USA 1992;89(22):10807–11.

[41] Law S, Dodds AW. The internal thioester and the covalent binding properties of the complement proteins C3 and C4. Protein Sci 1997;6(2):263–74.

[42] Walden M, Edwardt JM, Dziewulska AM, Bergmann R, Saalbach G, Kan SY, et al. An internal thioester in a pathogen surface protein mediates covalent host binding. Elife 2015;4:e06638.

[43] Jeffreys D. Aspirin: the extraordinary story of a wonder drug. Bloomsbury Publishing; 2010.

[44] Robertson JG. Mechanistic basis of enzyme-targeted drugs. Biochemistry 2005;44(15): 5561–71.

[45] Zhang J, Yang PL, Gray NS. Targeting cancer with small molecule kinase inhibitors. Nat Rev Cancer 2009;9(1):28–39.

[46] Chalker JM, Bernardes GJL, Lin YA, Davis BG. Chemical modification of proteins at cysteine: opportunities in chemistry and biology. Chem Asian J 2009;4(5):630–40.

[47] Liu Q, Sabnis Y, Zhao Z, Zhang T, Buhrlage SJ, Jones LH, et al. Developing irreversible inhibitors of the protein kinase cysteinome. Chem Biol 2013;20(2):146–59.

[48] Singh J, Petter RC, Baillie TA, Whitty A. The resurgence of covalent drugs. Nat Rev Drug Discov 2011;10(4):307–17.

[49] Byrd JC, Brown JR, O'Brien S, Barientos JC, Kay NE, Reddy NM, et al. Ibrutinib versus Ofatumumab in previously treated chronic lymphoid leukemia. N Engl J Med 2014;371(3): 213–23.

[50] Zhou W, Ercan D, Chen L, Yun CH, Li D, Capelletti M, et al. Novel mutant-selective EGFR kinase inhibitors against EGFR T790M. Nature 2009;462(7276):1070–4.

[51] Zhang T, Inesta-Vaquera F, Niepel M, Zhang J, Ficarro SB, Machleidt T, et al. Discovery of potent and selective covalent inhibitors of JNK. Chem Biol 2012;19(1):140–54.

[52] Hur W, Velentza A, Kim S, Flatauer L, Jiang X, Valente D, et al. Clinical stage EGFR inhibitors irreversibly alkylate Bmx kinase. Bioorg Med Chem Lett 2008;18(22):5916–9.

[53] Fujioka A, Terai K, Itoh RE, Aoki K, Nakamura T, Kuroda S, et al. Dynamics of the Ras/ERK MAPK cascade as monitored by fluorescent probes. J Biol Chem 2006;281(13):8917–26.

[54] Werner A, Siems W, Schmidt H, Rapoport I, Gerber G, Toguzov RT, et al. Determination of nucleotides, nucleosides and nucleobases in cells of different complexity by reversed-phase and ion-pair high-performance liquid chromatography. J Chromatogr B 1987;421:257–65.

[55] Fürst W, Hallström S. Simultaneous determination of myocardial nucleotides, nucleosides, purine bases and creatine phosphate by ion-pair high-performance liquid chromatography. J Chromatogr B 1992;578(1):39–44.

[56] Pilz R, Willis R, Boss G. The influence of ribose 5-phosphate availability on purine synthesis of cultured human lymphoblasts and mitogen-stimulated lymphocytes. J Biol Chem 1984;259(5):2927–35.

[57] Hauschka PV. Analysis of nucleotide pools in animal cells. Methods Cell Biol 1973;7: 361–462.

[58] Snyder FF, Cruikshank MK, Seegmiller JE. A comparison of purine metabolism and nucleotide pools in normal and hypoxanthine-guanine phosphoribosyltransferase-deficient neuroblastoma cells. Biochim Biophys Acta 1978;543(4):556–69.

[59] Jackson RC, Boritzki TJ, Morris HP, Weber G. Purine and pyrimidine ribonucleotide contents of rat liver and hepatoma 3924A and the effect of ischemia. Life Sci 1976;19(10):1531–6.

[60] Weber G, Lui MS, Jayaram HN, Pillwein K, Natsumeda Y, Faderan MA, et al. Regulation of purine and pyrimidine metabolism by insulin and by resistance to tiazofurin. Adv Enzyme Regul 1985;23:81–99.

[61] Keppler DO, Pausch J, Decker K. Selective uridine triphosphate deficiency induced by D-galactosamine in liver and reversed by pyrimidine nucleotide precursors effect on ribonucleic acid synthesis. J Biol Chem 1974;249(1):211–6.

[62] de Korte D, Haverkort WA, van Gennip AH, Roos D. Nucleotide profiles of normal human blood cells determined by high-performance liquid chromatography. Anal Biochem 1985;147(1):197–209.

[63] Wright DG. A role for guanine ribonucleotides in the regulation of myeloid cell maturation. Blood 1987;69(1):334–7.

[64] Pilz RB, Huvar I, Scheele JS, Van den Berghe G, Boss GR. A decrease in the intracellular guanosine 5′-triphosphate concentration is necessary for granulocytic differentiation of HL-60 cells, but growth cessation and differentiation are not associated with a change in the activation state of Ras, the transforming principle of HL-60 cells. Cell Growth Differ 1997;8(1):53–9.

[65] Feuerstein J, Goody RS, Wittinghofer A. Preparation and characterization of nucleotide-free and metal ion-free p21"apoprotein". J Biol Chem 1987;262(18):8455–8.

[66] Finkel T, Der CJ, Cooper GM. Activation of ras genes in human tumors does not affect localization, modification, or nucleotide binding properties of p21. Cell 1984;37(1):151–8.

[67] Hattori S, Ulsh L, Halliday K, Shih T. Biochemical properties of a highly purified v-rasH p21 protein overproduced in *Escherichia coli* and inhibition of its activities by a monoclonal antibody. Mol Cell Biol 1985;5(6):1449–55.

[68] Mendes P. GEPASI: a software package for modelling the dynamics, steady states and control of biochemical and other systems. Comput Appl Biosci 1993;9(5):563–71.

[69] Copeland RA. Evaluation of enzyme inhibitors in drug discovery a guide for medicinal chemists and pharmacologists. Hoboken (NJ): J. Wiley; 2005.

[70] Shukla S, Allam US, Ahsan A, Chen G, Krishnamurthy PM, Marsh K, et al. KRAS protein stability is regulated through SMURF2: UBCH5 complex-mediated β-TrCP1 degradation. Neoplasia (New York, NY) 2014;16(2):115–28.

[71] Marcsisin SR, Engen JR. Hydrogen exchange mass spectrometry: what is it and what can it tell us? Anal Bioanal Chem 2010;397(3):967–72.

[72] Patricelli MP, Szardenings AK, Liyanage M, Nomanbhoy TK, Wu M, Weissig H, et al. Functional interrogation of the kinome using nucleotide acyl phosphates. Biochemistry 2007;46(2):350–8.

[73] Adams SR, Tsien RY. Controlling cell chemistry with caged compounds. Annu Rev Physiol 1993;55(1):755–84.

[74] Agola JO, Sivalingam D, Cimino DF, Simons PC, Buranda T, Sklar LA, et al. Quantitative bead-based flow cytometry for assaying Rab7 GTPase interaction with the Rab-interacting lysosomal protein (RILP) effector protein. Methods Mol Biol 2015:331–54.

[75] Agola JO, Hong L, Surviladze Z, Ursu O, Waller A, Strouse JJ, et al. A competitive nucleotide binding inhibitor: in vitro characterization of Rab7 GTPase inhibition. ACS Chem Biol 2012;7(6):1095–108.

[76] Goitre L, Trapani E, Trabalzini L, Retta SF. The Ras superfamily of small GTPases: the unlocked secrets. Methods Mol Biol 2014;1120:1–18.

[77] Cox AD, Der CJ. Ras history: the saga continues. Small GTPases 2010;1(1):2–27.

[78] Rougerie P, Delon J. Rho GTPases: masters of T lymphocyte migration and activation. Immunol Lett 2012;142(1–2):1–13.

[79] Stankiewicz TR, Linseman DA. Rho family GTPases: key players in neuronal development, neuronal survival, and neurodegeneration. Front Cell Neurosci 2014;8:314.

[80] Kelleher FC, McArthur GA. Targeting NRAS in melanoma. Cancer J 2012;18(2):132–6.

[81] Cooper JM, Bodemann BO, White MA. The RalGEF/Ral pathway: evaluating an intervention opportunity for Ras cancers. Enzymes 2013;34(Pt. B):137–56.

[82] Shirakawa R, Horiuchi H. Ral GTPases: crucial mediators of exocytosis and tumourigenesis. J Biochem 2015;157(5):285–99.

[83] Alan JK, Lundquist EA. Mutationally activated Rho GTPases in cancer. Small GTPases 2013;4(3):159–63.

[84] Tang SC, Chen YC. Novel therapeutic targets for pancreatic cancer. World J Gastroenterol 2014;20(31):10825–44.

[85] Lin KB, Tan P, Freeman SA, Lam M, McNagny KM, Gold MR. The Rap GTPases regulate the migration, invasiveness and in vivo dissemination of B-cell lymphomas. Oncogene 2010;29(4):608–15.

[86] Zhang B, Zhang Y, Wang Z, Zheng Y. The role of Mg^{2+} cofactor in the guanine nucleotide exchange and GTP hydrolysis reactions of Rho family GTP-binding proteins. J Biol Chem 2000;275(33):25299–307.

[87] Zhao J, Wang WN, Tan YC, Zheng Y, Wang ZX. Effect of Mg(2+) on the kinetics of guanine nucleotide binding and hydrolysis by Cdc42. Biochem Biophys Res Commun 2002;297(3):653–8.

[88] Frech M, Schlichting I, Wittinghofer A, Chardin P. Guanine nucleotide binding properties of the mammalian RalA protein produced in *Escherichia coli*. J Biol Chem 1990;265(11):6353–9.

[89] Wheeler DB, Zoncu R, Root DE, Sabatini DM, Sawyers CL. Identification of an oncogenic RAB protein. Science 2015;350.

[90] Fleming JB, Shen GL, Holloway SE, Davis M, Brekken RA. Molecular consequences of silencing mutant K-ras in pancreatic cancer cells: justification for K-ras-directed therapy. Mol Cancer Res 2005;3(7):413–23.

[91] Lin KB, Freeman SA, Gold MR. Rap GTPase-mediated adhesion and migration: a target for limiting the dissemination of B-cell lymphomas? Cell Adh Migr 2010;4(3):327–32.

[92] Milburn MV, Tong L, deVos AM, Brunger A, Yamaizumi Z, Nishimura S, et al. Molecular switch for signal transduction: structural differences between active and inactive forms of protooncogenic ras proteins. Science 1990;247(4945):939–45.

CHAPTER 10

Next-Generation Strategies to Target RAF

D.D. Stuart

Novartis Institutes for Biomedical Research, Cambridge, MA, United States

INTRODUCTION

Tumor cells proliferate, invade, metastasize, and survive through a combination of altered signal transduction cues brought about by genetic and epigenetic mechanisms. No single oncogene has a greater impact on these phenotypes than mutant RAS. These oncogenic properties coupled with its high frequency in human tumors make mutant RAS an excellent target for cancer therapy. As described elsewhere in this book, targeting RAS directly has proved to be a significant challenge and therefore the downstream kinases RAF proto-oncogene serine/threonine-protein kinase (RAF), mitogen-activated protein kinase (MEK), and extracellular signal-regulated kinase (ERK) have been the focus of therapeutic intervention. From their inception, sorafenib and PD184352 were the very first RAF and MEK inhibitors, respectively, to enter human clinical trials with the goal of treating tumors with activated or mutant RAS [1,2]. Almost two decades have passed since these molecules were created and still no small molecule targeting this pathway has been approved for treating RAS mutant tumors. Multiple factors are responsible for the lack of clinical efficacy of these early inhibitors but the single biggest factor is arguably a lack of thorough understanding of the complexity of RAS-RAF-MEK-ERK signaling. Targeting the RAF kinases may still represent our best opportunity for treating RAS mutant tumors; however, a deeper understanding of RAF kinase biology and signaling must be applied in drug discovery efforts.

The RAF family of serine/threonine kinases consists of three isoforms: ARAF, BRAF, and CRAF (RAF1), which transmit signals for proliferation, differentiation, and survival from the RAS small GTPases (KRAS, NRAS, and HRAS). RAF kinases are named after the retroviral oncogene *v-raf*, which was shown to induce rapidly accelerated fibrosarcoma [3]. The mammalian homolog, CRAF, was subsequently identified as were the two other isoforms A- and BRAF (reviewed in Ref. [4]). RAF kinases are the first kinase nodes in the

CONTENTS

Conquering RAS. http://dx.doi.org/10.1016/B978-0-12-803505-4.00010-2

RAS-RAF-MEK-ERK signaling pathway. In normal tissues, activated RAS recruits RAF to the plasma membrane resulting in activation and phosphorylation of the dual-specificity kinases MEK1/2, which in turn phosphorylate and activate ERK1/2 [mitogen-activated protein kinase (MAPK)], the ultimate kinase in the pathway. For an excellent and very thorough review of the steps in RAF kinase activation and signaling, refer to the review written by Lavoie and Therrien [4]. This chapter will focus only on the aspects of RAF kinase biology that are relevant to the development of inhibitors for therapeutic intervention.

Although the ATP pocket in RAF has been the only druggable site demonstrated to date, the entire protein as well as the differences/similarities between the isoforms should be considered in the development of next-generation inhibitors (Fig. 10.1). Existing ATP-competitive inhibitors have taught us that allosteric regulation beyond the ATP pocket plays a significant role in the regulation of RAF kinase activity. Furthermore, the other domains of RAF could represent novel points of therapeutic intervention. RAF kinases share a high degree of structural homology and can be broken down into two major domains: the N-terminal regulatory domain and the C-terminal kinase domain. The N-terminal regulatory domain can be further subdivided into the RAS-binding domain (RBD) and cysteine-rich domain (CRD), which together make up conserved region 1 (CR1) and is critical for interaction with RAS, as well as a serine/threonine-rich domain (CR2), which contains regulatory phosphorylation sites (eg, Ser259) and interacts with 14-3-3. The C-terminal domain includes the ATP-binding domain, which contains all of the catalytic activity. With a high degree of structural homology, especially in the kinase domain, there is very little opportunity for the creation of selective inhibitors. In fact, vemurafenib and dabrafenib are often referred to as BRAF inhibitors, a label that is

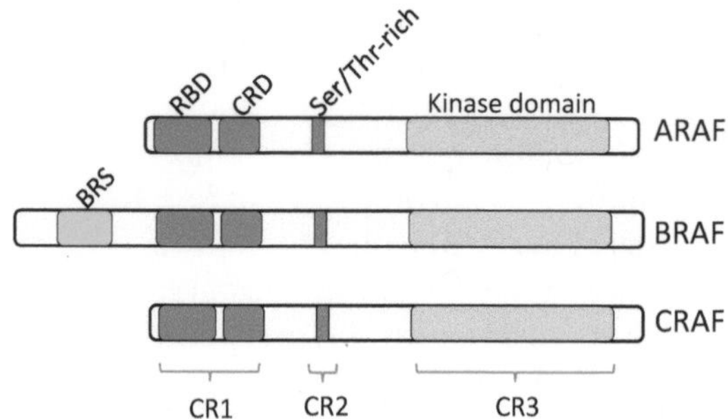

FIGURE 10.1

Primary structures and domains of RAF kinases. RAF kinases can be divided into three domains CR1, CR2, and CR3. The Ras-binding domain (RBD), cysteine-rich domain (CRD), serine/threonine-rich sequences, and kinase domains are highlighted. BRAF contains an extension at the N-terminus called the BRAF-specific sequence. *Adapted from Lavoie H, Therrien M. Regulation of RAF protein kinases in ERK signalling. Nat Rev Mol Cell Biol 2015;16(5):281–98.*

misleading. These molecules are capable of binding and inhibiting purified CRAF, BRAF, and BRAF^{V600E} in biochemical assays; it is only in the cellular context that they selectively inhibit BRAF^{V600E}. As discussed in the following section, this is because BRAF^{V600E} usually signals as a monomer, whereas CRAF and wild-type BRAF signal as heterodimers (or homo-dimers) and these inhibitors fail to potently inhibit both protomers of a RAF dimer (more details in subsequent sections). A more accurate description would be to refer to vemurafenib, dabrafenib, and encorafenib as RAF monomer inhibitors [5].

RAF MUTATIONS AND ONCOGENESIS

The most frequent and best characterized RAF mutation occurs in BRAF at codon 600 and results in an amino acid change from valine to glutamic acid (V600E), less frequently to lysine (V600K), or aspartic acid (V600D) and these mutations result in constitutive, RAS-independent activation of BRAF [6]. Following a thorough enzymatic characterization of BRAF mutations, Wan et al. grouped them into high, intermediate, and low activity relative to wild-type BRAF [7]. They postulated that the high-activity mutants G469A, E586K, V600E/D/K/R, K601E function by directly phosphorylating MEK1/2, whereas the impaired-activity mutants G466E/V, D594V, and G596R function by trans-activating CRAF (all codon numbers corrected to reflect correct start codon for BRAF). Although this was not the first inference of the importance of RAF transactivation, this was clear evidence that non-V600 BRAF mutants might signal differently than wild-type or BRAF^{V600} mutants. Convincing evidence for a non-catalytic role of BRAF in trans-activating CRAF came from a genetically engineered mouse model in which kinase dead BRAF (D594A) was induced in melanocytes along with mutant KRAS (G12D), resulting in tumors in 100% of the animals, whereas expression of either mutant gene alone did not result in tumorigenesis [8].

The role of RAF dimerization attracted significant attention with the observation that vemurafenib and other RAF inhibitors in this class such as dabrafenib and encorafenib paradoxically activate RAF in cells expressing wild-type BRAF [8–10]. Although the models proposed in each of these manuscripts differ slightly, they are consistent in demonstrating a role for activated RAS and trans-activation of one RAF protomer by a drug-bound protomer. The pharmacological activity of these inhibitors in cells expressing wild-type BRAF especially with mutant RAS explains the underlying mechanism responsible for cutaneous lesions observed in patients and clearly illustrate why inhibitors in this class will be ineffective against RAS mutant tumors.

A comprehensive characterization of a wide range of clinically observed BRAF mutations was described by the Rosen laboratory [5]. They used an engineered cell-based model in which SKBR3 breast cancer cells, which have amplified *ERBB2*, are transfected to express a wide range of BRAF mutants

(V600E/D/K/R/M, K601E/N/T, L597Q/V, G469A/VR, G464V/E) individually and with or without an additional mutation that prevents dimerization (BRAF^{R509H}) or interaction with RAS (BRAF^{R188L}). By measuring phospho-MEK and phospho-ERK in these cells in the presence or absence of lapatinib, each mutant could be characterized for its RAS dependence and the necessity for dimerization. To summarize this extensive body of work, none of these mutants required RAS for kinase activation, only the V600 mutations functioned as a monomer and was sensitive to vemurafenib, all of the other mutations were able to form homo-dimers in a RAS-independent manner. However, wild-type BRAF and all mutants required RAS for hetero-dimerization with CRAF.

In contrast to BRAF, ARAF and CRAF are rarely found to be mutated in human tumors. In a single case report, a compound ARAF mutation (ARAF^{F351L}; Q347_A348del) was found in tumor cells from a patient with Langerhans cell histiocytosis [11]. A low frequency (3/564 samples) of mutations in ARAF codon 214 were also identified in lung adenocarcinoma [12] and one report suggests up to 11% of intra-hepatic cholangiocarcinomas may have mutations in ARAF, although the functional consequences of the mutations are not clear [13]. CRAF mutations have been described mostly in rare genetic disorders, the "RASopothies" (reviewed in Ref. [14]); however, in prostate and melanoma tumor samples, CRAF fusions have been described which delete the N-terminal regulatory domain and join the C-terminal kinase domain with a range of partners; however, these are quite rare (4 of 450 prostate and 1 of 310 melanoma tumors) [15].

Although CRAF does not appear to be a target for oncogenic mutations, a critical and specific role for CRAF in tumorigenesis has been described using genetically engineered mouse models of KRAS mutant lung cancer. In two independent studies, CRAF knock-out was shown to prevent KRAS mutant lung tumorigenesis. Blasco and colleagues demonstrated that when *C-Raf*$^{lox/lox}$ was deleted in the lungs of adult mice through intra-tracheal administration of adenovirus-Cre recombinase, induction of lung tumorigenesis by K-Ras^{G12V} was completely prevented [16]. In contrast, these tumors were not dependent on B-Raf, or either Mek1, Mek2, Erk1, or Erk2. Similar results were obtained by Karreth and colleagues who used an inducible K-Ras^{G12D} lung tumor model and demonstrated a similar dependency on C-Raf but not B-Raf [17]. The absolute dependency for C-Raf in both models was further highlighted by the observation that the only tumors that arose in the *C-Raf*$^{lox/lox}$ mice were "escapers" in which *C-Raf* excision was incomplete and C-Raf protein was expressed.

It is clear from clinical and pre-clinical genetic studies that RAF kinases can act as critical mediators of oncogenic RAS signaling (CRAF) or as mutant oncogenic drivers themselves (eg, BRAF^{V600E}). Although clinical proof of concept for RAF inhibition has only been demonstrated in the BRAF^{V600} mutant setting, RAF kinases are the most proximal signaling node to mutant RAS, which has so far remained a significant challenge to drug. Therefore, RAF kinases remain critical nodes for therapeutic intervention in both BRAF mutant and RAS mutant tumors. However, as described in subsequent sections, different types of RAF inhibitors will need to be developed in RAS mutant versus BRAF mutant tumors.

RAF INHIBITORS: OLD LESSONS RE-LEARNED

Significant efforts to develop RAF kinase inhibitors began in the late 1990s following the rationale of targeting CRAF downstream of mutant RAS and activated growth factor receptor signaling. Sorafenib (described later) was the first RAF inhibitor to enter human clinical trials prior to the discovery of oncogenic BRAF mutations. However, the earliest efforts to target RAF taught valuable lessons that were not fully realized until years later: Hall-Jackson and colleagues recognized the paradoxical activation of RAF by ATP-competitive small molecule inhibitors [18,19]. They demonstrated that SB203580, previously identified as an inhibitor of c-Jun N-terminal kinase (JNK), p38γ, and p38δ, also inhibits the kinase activity of purified CRAF in biochemical assays. However, following treatment of cells with the compound, the kinase activity paradoxically increased in CRAF immuno-precipitated from cell lysates. Similar results were observed with ZM-336372, an ATP competitive CRAF inhibitor discovered from biochemical cascade screen [19]. In the biochemical assay using partially purified recombinant proteins, ZM-336372 demonstrated the expected sigmoidal dose–response with $IC_{50}=70\,nM$; however, in cells the compound failed to suppress phospho mitogen activated protein kinase (pMEK) or phospho extracellular signal-regulatory kinase (pERK) but instead activated these targets following drug wash-out. The authors attributed the increased activity to the modulation of RAF activity through feedback and, although the RAS-RAF-MEK-ERK pathway has been shown to be highly regulated by negative feedback loops [20,21], the results could also be explained by the transactivation of RAF dimers by ATP-competitive RAF inhibitors, as discovered subsequently [8–10].

The disconnect observed between the effect of putative RAF inhibitors in biochemical assays using purified or partially purified recombinant protein and the activity observed in cells extends beyond the paradoxical effect. Biochemically potent inhibitors often fail to demonstrate evidence of pathway inhibition in cells, an observation that was also made in the early attempts to develop CRAF inhibitors [22,23]. Differences in ATP concentrations likely account for some but not all of the shift between biochemical and cellular potency. The simplest explanation is that biochemical assays using purified CRAF (often kinase domain) do not accurately reflect the state of activated CRAF in a cell. Most importantly, the role of dimerization and transactivation of dimer partners in biochemical assays has not been demonstrated, and RAS is clearly not the driver of activation.

Sorafenib (BAY 43-9006) was the first small molecule RAF kinase inhibitor to enter human clinical trials. Although it has been much maligned as a very weak RAF inhibitor, it should be recognized that sorafenib was developed before the discovery of oncogenic BRAF mutations in 2002 [6]. Sorafenib was optimized in cell-based assays using HCT116 and MiaPaCa cells, both of which express mutant KRAS and furthermore, there was limited optimization against off-target kinases [1]. Given the complexity in RAF kinase signaling and the differences based on genetic context (eg, in cells with RAS versus BRAF mutations), it may not be surprising that sorafenib is not active in $BRAF^{V600E}$ tumors. This highlights the need for optimization of inhibitors in appropriate cellular systems. RAF265 was the next RAF inhibitor to enter human clinical trials in 2006, and, although its potency was optimized in cells expressing $BRAF^{V600E}$, it also inhibited multiple receptor tyrosine kinases (RTKs) which likely led to dose-limiting toxicities and lacked robust efficacy in melanoma that subsequent, more specific inhibitors demonstrated [24].

The first evidence of the true potential for a potent, selective RAF inhibitor in BRAF mutant tumors came from pre-clinical tool compounds SB590885 and PLX4720. King and colleagues used the phrase "genetic therapeutic index" to describe the pharmacological profile of SB590885, a potent and selective RAF inhibitor that lacked the off-target effects of the previously described molecules. SB590885 was highly selective for BRAF and CRAF in biochemical assays, showing single-digit nanomolar IC_{50}s and at least 1000-fold selectivity against a panel of 46 other kinases. In cell-based assays, SB590885 inhibited pERK in tumor cell lines expressing $BRAF^{V600E}$, but paradoxically increased pERK levels in cells expressing wild-type RAF and this led to selective inhibition of proliferation in cell lines expressing $BRAF^{V600E}$ [25]. In a human tumor xenograft mouse model expressing BRAFV600E, SB590885 inhibited tumor growth but lacked the exposure necessary to drive tumor regression. This work represents a key turning point in the development of RAF kinase inhibitors as it demonstrated that a potent and selective RAF inhibitor would be significantly more effective against tumors expressing $BRAF^{V600E}$. This critical feature was further demonstrated by the work of Tsai and colleagues in their characterization of PLX4720, a close chemical analog of vemurafenib (PLX4032) [26]. Similar to SB590885, PLX4720 was highly selective for BRAF and $BRAF^{V600E}$ in biochemical assays and demonstrated selective inhibition of pERK and proliferation in cells expressing $BRAF^{V600E}$. PLX4720 had superior pharmacokinetic properties compared with SB590885 and demonstrated tumor regression in a $BRAF^{V600E}$ xenograft model. The authors also demonstrated that PLX4720 completely lacked activity in a wild-type BRAF model. Thus, the "genetic therapeutic index" was extended to in vivo models and this set the stage for the first report of clinical efficacy of the selective RAF inhibitor, vemurafenib, in the following year [27].

Vemurafenib demonstrated significant efficacy in patients with $BRAF^{V600E}$ melanoma, with 81% of patients experiencing partial or complete tumor response and an acceptable safety profile in a phase I clinical trial [28]. The response rates (RRs) reported for vemurafenib and other RAF inhibitors in this class (dabrafenib and encorafenib) are much higher than observed for MEK inhibitors, such as trametinib (22% RR). The superior efficacy of RAF inhibitors is most likely explained by their ability to achieve superior RAF-MEK-ERK pathway suppression in $BRAF^{V600E}$ tumors due to improved tolerability compared with MEK inhibitors. The improved tolerability comes from the fact that RAF inhibitors do not inhibit the RAF-MEK-ERK pathway in normal tissues, as occurs with MEK inhibitors [29]. In fact, the common toxicities observed between vemurafenib, dabrafenib, and encorafenib are skin rash, hyper-keratosis, keratoacanthomas, and cutaneous squamous carcinomas, all of which occur partially or entirely due to activation of the RAF-MEK-ERK pathway [28,30–32]. Support for this mechanism comes from the observation that these toxicities may be decreased when RAF inhibitors are combined with a MEK inhibitor [33–35]. Thus, the inability of RAF inhibitors to inhibit the RAF-MEK-ERK pathway in cells expressing wild-type BRAF, including normal tissues, is the key to their improved therapeutic index. However, their lack of inhibition of RAS-driven signaling is also a key escape mechanism in resistance. The majority of patients have a relapse on single-agent vemurafenib or dabrafenib by ~6 months and RAF-MEK-ERK pathway reactivation appears in approximately half of these patients [36,37]. The key to resistance in these tumors appears to be through activation of drug-resistant RAF dimers (reviewed in Ref. [38]). For a thorough review of RAF inhibitors currently in clinical trials targeting $BRAF^{V600mut}$ tumors, refer to Ref. [39].

TARGETING RAF IN RAS MUTANT TUMORS

With the approval of vemurafenib and dabrafenib in the treatment of $BRAFV600^{mut}$ melanoma, and others in various stages of clinical development, the focus on the next generation of RAF inhibitors has been shifted to treating RAS mutant tumors. As described later, there are several RAF inhibitors in various stages of pre-clinical and early clinical development. In undertaking this approach, it is worthwhile reviewing why $BRAF^{V600mut}$ melanomas are sensitive to RAF inhibitors and considering the differences in targeting RAF in the context of activated mutant RAS.

Signaling through the RTK-RAS-RAF-MEK-ERK pathway is tightly controlled through negative feedback (Fig. 10.2). For example, activated ERK can phosphorylate the N-terminal domains of CRAF and BRAF, preventing engagement with RAS and subsequent RAF activation [40,41]. Activated ERK also dampens

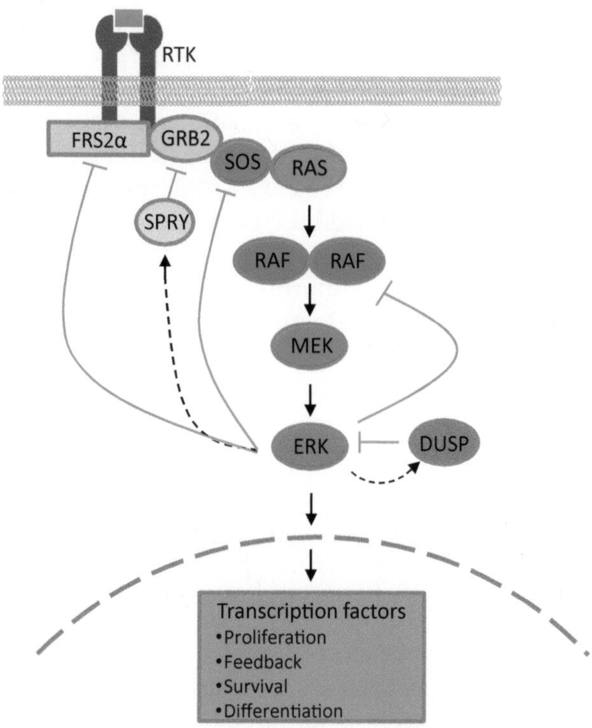

FIGURE 10.2

RAF kinase signaling in the context activated RAS. When RAS is activated by growth factor stimulation of receptor tyrosine kinases (RTKs), it recruits RAF dimers to the plasma membrane where they are activated through phosphorylation and relief of auto-inhibition by the N-terminal regulatory domain and phosphorylate MEK, which in turn phosphorylates and activates ERK. Although activated ERK is primarily responsible for phosphorylating cytoplasmic and nuclear substrates promoting survival, proliferation, and differentiation, it also participates in mediating negative feedback through direct and indirect mechanisms. Direct mechanisms of negative feedback include ERK phosphorylation of multiple sites on RAF, SOS, RTKs, and FRS2 that serve to inhibit downstream activation. Indirect mechanisms include transcriptional activation of negative regulators such as the dual-specificity phosphatases (DUSP) and sprouty proteins (SPRY). *Reviewed in and adapted from Lito P, Rosen N, Solit DB. Tumor adaptation and resistance to RAF inhibitors. Nat Med 2013;19(11):1401–9.*

signals from RTKs through direct ERK-mediated phosphorylation of the scaffolding/adapter protein FRS2α, which prevents the assembly of signaling complexes downstream of activated RTKs [42,43]. FRS2α also appears to be a key node in SPRY1/2-mediated negative feedback. SPRY2 expression is positively regulated by RAF-MEK-ERK pathway signaling and serves to negatively regulate RTK signaling to RAS by binding GRB2 and preventing its interaction with SOS through interaction in complex with FRS2α [44]. However, $BRAF^{V600E}$ is constitutively active, requiring no stimulation from upstream RAS and therefore renders it insensitive to these upstream mechanisms of regulation [6,21]. It is the engagement of these negative feedback loops that make BRAF^{V600E} melanoma tumor cells initially sensitive to RAF inhibitors: treatment does not result in immediate feedback-mediated pathway re-activation because the negative feedback in these cells has maintained low levels of activated RAS. However, sustained ERK suppression leads to loss of negative feedback, gradual increase in RAS-GTP, and induction of inhibitor-resistant RAF dimers [20].

In proposing RAF as a therapeutic target in RAS mutant tumors, it is necessary to consider not only how the target itself has changed, monomeric BRAF^{V600E} versus homo-/hetero-dimeric CRAF/BRAF, but also how the pathway is differentially regulated through feedback in the setting of mutant RAS versus BRAF^{V600E}. Vemurafenib is clearly unable to inhibit signaling at therapeutically relevant concentrations in RAS mutant tumors due to its inability to inhibit RAF dimers. Therefore alternative chemical matter, possibly with different binding modes, will be required to inhibit RAF in the context of mutant RAS. However, even if an effective inhibitor of RAF dimers is discovered, regulation of the pathway through feedback could compromise the ability to potently suppress ERK output. CRAF, BRAF, and wild-type RAS alleles are subject to the negative feedback mechanisms described earlier and loss of this negative feedback following suppression of phospho-ERK could serve to reactive the pathway, for example, by recruiting additional RAF dimers.

As described in previous sections, our current understanding of RAF biology has accelerated in recent years in part by the development of specific inhibitors. The role of dimerization and allosteric transactivation is just beginning to shed light on the efficacy and toxicity of specific inhibitors in the context of mutant and wild-type RAF. However, hypothesis-driven drug discovery efforts are still hampered by a lack of robust in vitro tools. For example, recombinant RAF proteins used in biochemical kinase assays are often restricted to the kinase domain and lack the regulatory N-terminal domain or the other pathway components necessary for RAF activation in a cell. Such assays are obviously unable to identify novel mechanisms of inhibition, such as disruption of the RAS–RAF interaction, and may also lack the ability to differentiate inhibitors based on differences in allosteric modulation. The clearest example is the observation that molecules such as vemurafenib potently inhibit wild-type RAF isoforms in biochemical assays and yet display opposing effects in cells expressing BRAF^{V600E}(inhibition) compared with BRAF^{wt} (activation). This is in spite of the fact that dimerization appears to be necessary for wild-type B/CRAF but not BRAF^{V600E} kinase activity in these assays [45]. More elaborate biochemical assay formats such as RAS-BRAF-MEK have been developed, but their biological relevance and ability to differentiate RAF inhibitors is unclear at this time.

Leveraging Cell-Based Assays

For the reasons outlined earlier, near-term strategies to develop next-generation RAF inhibitors will have to rely heavily on cell-based assays, with hit-finding and lead-optimization campaigns using multiple cell lines. This is the strategy applied in the development of the so-called paradox-breaker RAF inhibitors described by Plexxikon. The goal of this program was to develop selective RAF inhibitors that potently inhibit pathway signaling and proliferation in cell lines expressing of $BRAF^{V600E}$, but do not paradoxically activate the pathway at similar concentrations in cells expressing wild-type BRAF. Such a molecule would theoretically lack the hyperplasia, hyperkeratosis, and skin lesions observed with vemurafenib but maintain efficacy against tumor cells expressing $BRAF^{V600E}$. The lead optimization effort for this program, as described, was to select for molecules that potently inhibit phospho-MEK/ERK in A375 and Colo829 melanoma cells (both $BRAF^{V600E}$) but fail to activate these markers at similar concentrations in a panel of $BRAF^{wt}$ cell lines: B9 (HRAS mutant) squamous cell carcinoma, IPC-298(NRAS mutant) melanoma, and HCT116 (KRAS mutant) colorectal cancer cells [46]. The clinical candidate PLX8394 and an analog PLX7904 were shown to effectively inhibit signaling and proliferation in $BRAF^{V600E}$ melanoma cell lines, including lines with acquired resistance to vemurafenib or the pre-clinical analog PLX4720, which have acquired an NRAS mutation or expression of a $BRAF^{V600E}$ splice variant that preferentially dimerizes [47,48].

Structural analyses of the interaction between BRAF and paradox breakers have attempted to explain the bio-physical underpinnings for the lack of RAF activation by these compounds [46,49]. Although PLX8394 and PLX7904 are structurally very similar to vemurafenib and PLX4720, co-crystal structures indicate that one key difference is that paradox breakers interact with a Leu505 on the C-terminal end of the αC-helix. Leu505 is one of four residues that form the hydrophobic or regulator spine on kinases and the αC-helix forms key contacts at the RAF dimer interface [50]. Further support for this hypothesis comes from the observation that the paradoxical activation of dabrafenib was dramatically reduced by engineering it to include the PLX7902 tail [46]. Arora and colleagues used in silico molecular simulations to propose that differences in hydrogen bond interactions with gatekeeper residue of BRAF (T529) induces conformational changes in the αC-helix and the activation loop leading to differential effects between paradox breakers and activators [49]. A major shortcoming to both studies is that only a limited number of RAF inhibitors were analyzed and more thorough analysis with additional inhibitors and biological validation is warranted to further support the hypothesis. Whether paradox-breakers such as PLX8394 effectively treat $BRAF^{V600mut}$ tumors in the absence of side-effects from paradoxical activation awaits clinical proof of concept.

As discussed earlier, the differences in cellular activity of RAF inhibitors in $BRAF^{V600E}$ versus $BRAF^{wt}$ cells is not revealed using biochemical assays with purified RAF kinase domains. In describing the pharmacological profile of a novel RAF inhibitor, LY3009120, Peng et al. used a cell-based method to determine binding of the inhibitor to different RAF isoforms. LY3009120 was shown to bind CRAF, $BRAF^{V600E}$, and ARAF with similar affinity, whereas vemurafenib had a >1000-fold shift in binding affinity between $BRAF^{V600E}$ and CRAF in A375 cells [51]. The authors went on to demonstrate that LY3009120 lacks the paradoxical activation in RAS mutant cells, even though CRAF-BRAF dimerization was induced to the same degree as by vemurafenib; however, LY3009120 effectively inhibited signaling. Furthermore, LY3009120 exhibited some evidence of anti-proliferative activity in RAS mutant tumor cells in vitro and in vivo. This compound is undergoing a phase I clinical trial in patients with solid tumors (ClinicalTrials.gov).

Yao et al. characterized the inability of inhibitors like vemurafenib to inhibit RAF dimers as negative co-operativity [5]. In a thorough characterization of the cellular activity of a comprehensive list of clinically observed BRAF mutants against a range of RAF inhibitors, Yao et al. determined that atypical BRAF mutants (ie, non-V600) preferentially dimerize in a RAS-independent manner and are resistant to most inhibitors. The authors used an innovative approach to identify inhibitors with improved activity against RAF dimers by leveraging the long off-rate of the RAF inhibitor LGX818 (encorafenib). Treatment with a high concentration of LGX818 led to complete suppression of pERK levels in SK-MEL-30 ($NRAS^{Q61K}$) cells, but 1 h following wash-out, pERK was increased above baseline. This hyper-activation is presumably due to retention of LGX818 at the first protomer leading to transactivation of the second un-occupied protomer. For most inhibitors, including vemurafenib and dabrafenib, much higher concentrations were required to inhibit signaling compared with cells expressing $BRAF^{V600E}$ (A375). The IC_{50}s for inhibition in each cell line were compared, with the ratio being indicative of the relative affinity of a compound for RAF dimers versus $BRAF^{V600E}$ monomers. Although encorafenib, vemurafenib, and dabrafenib all demonstrated 50- to >135-fold shifts in IC_{50}, the type II inhibitor BGB659 (compound 27 from Gould et al. [52]) demonstrated no shift, suggesting that it is equipotent at inhibiting RAF dimers as monomeric $BRAF^{V600E}$. The authors then went on to demonstrate that BGB659 is an effective inhibitor of RAF-MEK-ERK signaling in cells expressing mutant NRAS and a wide range of atypical BRAF mutations, including p61-$BRAF^{V600E}$ in vemurafenib-resistant melanoma cells in vitro and in vivo [5]. This study highlights both the complexity of RAF signaling as well as the degree of allosteric regulation at play in RAF dimerization. The observation that ATP-competitive inhibitors can impact the affinity for the second site of an unbound RAF protomer and that different inhibitors have different affinities for that site have significant implications for design of next-generation RAF inhibitors.

The efficacy of RAF inhibitors that are potent against RAF dimers might be improved in combination with MEK inhibitors in RAS mutant tumors [53]. Lambda and colleagues determined that CRAF knock-down was synthetically lethal with MEK inhibition in RAS mutant tumors [54]. Pharmacological proof of concept was demonstrated with RAF265 and the MEK inhibitor selumetinib in KRAS mutant colorectal cancer with the combination providing more potent suppression of proliferation and induction of apoptosis in vitro compared with either single agent alone. The authors attributed the combination efficacy to the ability of RAF inhibition to suppress RAS-mediated RAF activation that occurs with MEK inhibitor-mediated relief of negative feedback in RAS mutant tumors. This mechanism was further elucidated and expanded by Lito and colleagues who demonstrated that the combination of MEK inhibition and CRAF knock-down was more efficacious than MEK inhibition alone in KRAS mutant tumor cells [55]. They further demonstrated that allosteric MEK inhibitors are less able to inhibit MEK when activated by CRAF versus $BRAF^{V600E}$ and that some MEK inhibitors are able to suppress CRAF-mediated MEK activation better than others (selumetinib, PD0325901) due to decreased affinity of inhibitor-bound MEK to CRAF (trametinib) or the prevention of CRAF-mediated MEK phosphorylation (CH5126766). These studies support further evaluation of combinations of MEK and RAF inhibitors in RAS mutant tumors.

Novel Modes of Inhibition

The examples described earlier are illustrative of the opportunity that exists to select for specific pharmacological properties of ATP-competitive RAF inhibitors and reveal the underlying allosteric regulation in RAF kinase signaling, including dimerization. Preventing dimerization may also be a relevant mechanism of inhibition worth pursuing and this was investigated by Freeman et al. [56]. The dimer interface of BRAF (amino acids 503–521) was used to design a peptide that, when expressed in COS cells, inhibited CRAF-BRAF dimerization and decreased pMEK. Furthermore, a cell-permeable TAT-tagged version of the peptide was shown to inhibit CRAF-BRAF dimerization, signaling, and proliferation in RAS^{mut} tumor cell lines. This study provides the first pharmacological proof of concept that inhibition of RAF dimerization is a potential therapeutic; however, additional work will be required to optimize the pharmacological properties of the peptides, or determine if the RAF dimerization interface could be targeted by small molecules.

Another RAF regulatory mechanism that has been proposed as a point of pharmacological intervention is the CRAF:14-3-3 interaction [57]. The importance of 14-3-3 binding to CRAF activity is multi-faceted, with both positive and negative regulatory effects. Binding of 14-3-3 to the C-terminus of CRAF through interaction with Ser621 appears to be required for activation [58,59]. In contrast, in the presence of 14-3-3 binding to the CR2 domain (pSer259 and pSer233), CRAF recruitment to the plasma membrane, binding to RAS, and subsequent activation is prevented [60]. The observation that germline mutations in this region of CRAF disrupt 14-3-3 binding and are associated with Noonan and Leopard syndromes, support the role of 14-3-3 binding in this region as a regulatory mechanism of RAF activation [60,61]. Cotylenin A, a natural product that was originally discovered as a phytotoxin [62], was shown to induce differentiation and inhibit proliferation of tumor cells [63,64]. In plants, cotylenin A modulates the activated complex between the plant plasma membrane H^+-ATPase and 14-3-3, causing loss of water; however, Molzan et al. demonstrated that cotylenin A interacts with a CRAF phosphopeptide coinciding with the 14-3-3 binding site and stabilized a peptide/14-3-3 complex [57]. Although the therapeutic utility of cotylenin A has not been demonstrated, these studies provide evidence that stabilization of inhibitory 14-3-3 interactions with CRAF is possible with a small molecule.

SUMMARY

Over the past decade the complexity of RAF kinase signaling has been unraveled through the use of molecular biology and genetics and the generation of therapeutic agents. The first X-ray crystal structure of the BRAF kinase domain may not have been elucidated without the presence of a potent ligand, sorefenib bound to the ATP site [7]. The X-ray crystal structure of BRAF has provided insight into the structure–activity relationship of inhibitors, and structural insights into basic RAF biology, including dimerization. The lack of $BRAF^{V600E}$ dimerization is key to the pharmacological activity of clinically approved RAF inhibitors which demonstrate selective inhibitory activity against $BRAF^{V600E}$ mutant tumors.

The biological insights into RAF biology can be leveraged to create the next generation of RAF inhibitors. The marketing approval of RAF inhibitors in $BRAF^{V600mut}$ melanoma provides proof of concept that targeting RAF can be more efficacious than targeting MEK due to a wider therapeutic index. Can RAF inhibitors be developed which also differentiate in wild-type BRAF tumors, or tumors with atypical BRAF mutations? Or can CRAF-selective inhibitors be developed in KRAS mutant tumors where genetically engineered mouse models point to CRAF as a unique and critical node in mutant KRAS signaling? With the challenges faced in targeting RAS directly with small molecule inhibitors, such a RAF inhibitor may represent the best option for treating RAS mutant tumors as single agents or in combination with upstream (RTK) or downstream (MEK, ERK) inhibitors. Given the critical unmet medical need in patients with RAS mutant tumors, there will be continued focus on understanding the biology and therapeutic potential of targeting RAF and scientific discoveries will go hand in hand with therapeutic advances.

References

[1] Lyons J, Wilhelm S, Hibner B, Bollag G. Discovery of a novel Raf kinase inhibitor. Endocr Relat Cancer 2001;8(3):219–25.

[2] Sebolt-Leopold JS, Dudley DT, Herrera R, Becelaere KV, Wiland A, Gowan RC, et al. Blockade of the MAP kinase pathway suppresses growth of colon tumors in vivo. Nat Med 1999;5(7):810–6.

[3] Rapp UR, Goldsborough MD, Mark GE, Bonner TI, Groffen J, Reynolds FH, et al. Structure and biological activity of v-raf, a unique oncogene transduced by a retrovirus. Proc Natl Acad Sci USA 1983;80(14):4218–22.

[4] Lavoie H, Therrien M. Regulation of RAF protein kinases in ERK signalling. Nat Rev Mol Cell Biol 2015;16(5):281–98.

[5] Yao Z, Torres NM, Tao A, Gao Y, Luo L, Li Q, et al. BRAF mutants evade ERK-dependent feedback by different mechanisms that determine their sensitivity to pharmacologic inhibition. Cancer Cell 2015;28(3):370–83. http://dx.doi.org/10.1016/j.ccell.2015.08.001.

[6] Davies H, Bignell GR, Cox C, Stephens P, Edkins S, Clegg S, et al. Mutations of the BRAF gene in human cancer. Nature 2002;417(6892):949–54.

[7] Wan PT, Garnett MJ, Roe SM, Lee S, Niculescu-Duvaz D, Good VM, et al. Mechanism of activation of the RAF-ERK signaling pathway by oncogenic mutations of B-RAF. Cell 2004;116:855–67.

[8] Heidorn SJ, Milagre C, Whittaker S, Nourry A, Niculescu-Duvas I, Dhomen N, et al. Kinase-dead BRAF and oncogenic RAS cooperate to drive tumor progression through CRAF. Cell 2010;140(2):209–21.

[9] Poulikakos P, Zhang C, Bollag G, Shokat K, Rosen N. RAF inhibitors transactivate RAF dimers and ERK signalling in cells with wild-type BRAF. Nature 2010;464(7287):427–30.

[10] Hatzivassiliou G, Song K, Yen I, Brandhuber BJ, Anderson DJ, Alvarado R, et al. RAF inhibitors prime wild-type RAF to activate the MAPK pathway and enhance growth. Nature 2010;464(7287):431–5.

[11] Nelson DS, Quispel W, Badalian-Very G, van Halteren AG, van den Bos C, Bovee JV, et al. Somatic activating ARAF mutations in Langerhans cell histiocytosis. Blood 2014;123(20):3152–5.

[12] Imielinski M, Greulich H, Kaplan B, Araujo L, Amann J, Horn L, et al. Oncogenic and sorafenib-sensitive ARAF mutations in lung adenocarcinoma. J Clin Invest 2014;124(4):1582–6.

[13] Sia D, Losic B, Moeini A, Cabellos L, Hao K, Revill K, et al. Massive parallel sequencing uncovers actionable FGFR2-PPHLN1 fusion and ARAF mutations in intrahepatic cholangiocarcinoma. Nat Commun 2015;6:6087.

[14] Tidyman WE, Rauen KA. The RASopathies: developmental syndromes of Ras/MAPK pathway dysregulation. Curr Opin Genet Dev 2009;19(3):230–6.

[15] Palanisamy N, Ateeq B, Kalyana-Sundaram S, Pflueger D, Ramnarayanan K, Shankar S, et al. Rearrangements of the RAF kinase pathway in prostate cancer, gastric cancer and melanoma. Nat Med 2010;16(7):793–8.

[16] Blasco RB, Francoz S, Santamaria D, Canamero M, Dubus P, Charron J, et al. c-Raf, but not B-Raf, is essential for development of K-Ras oncogene-driven non-small cell lung carcinoma. Cancer Cell 2011;19(5):652–63.

[17] Karreth FA, Frese KK, DeNicola GM, Baccarini M, Tuveson DA. C-Raf is required for the initiation of lung cancer by K-Ras(G12D). Cancer Discov 2011;1(2):128–36.

[18] Hall-Jackson CA, Goedert M, Hedge P, Cohen P. Effect of SB 203580 on the activity of c-Raf in vitro and in vivo. Oncogene 1999;18(12):2047–54.

[19] Hall-Jackson CA, Eyers PA, Cohen P, Goedert M, Boyle FT, Hewitt N, et al. Paradoxical activation of Raf by a novel Raf inhibitor. Chem Biol 1999;6(8):559–68.

[20] Lito P, Pratilas C, Joseph E, Tadi M, Halilovic E, Zubrowski M, et al. Relief of profound feedback inhibition of mitogenic signaling by raf inhibitors attenuates their activity in BRAFV600E melanomas. Cancer Cell 2012;22(5):668–82.

[21] Pratilas CA, Taylor BS, Ye Q, Viale A, Sander C, Solit DB, et al. V600EBRAF is associated with disabled feedback inhibition of RAF-MEK signaling and elevated transcriptional output of the pathway. Proc Natl Acad Sci USA 2009;106(11):4519–24.

[22] Lackey K, Cory M, Davis R, Frye SV, Harris PA, Hunter RN, et al. The discovery of potent cRaf1 kinase inhibitors. Bioorg Med Chem Lett 2000;10(3):223–6.

[23] Wood E, Crosby RM, Dickerson S, Frye SV, Griffin R, Hunter R, et al. A prodrug approach to the design of cRaf1 kinase inhibitors with improved cellular activity. Anti-Cancer Drug Des 2001;16(1):1–6.

[24] Sharfman WH, Hodi FS, Lawrence DP, Flaherty KT, Amaravadi RK, Kim KB, et al. Results from the first-in-human (FIH) phase I study of the oral RAF inhibitor RAF265 administered daily to patients with advanced cutaneous melanoma. ASCO Meet Abstr 2011;29(Suppl. 15):8508.

[25] King AJ, Patrick DR, Batorsky RS, Ho ML, Do HT, Zhang SY, et al. Demonstration of a genetic therapeutic index for tumors expressing oncogenic BRAF by the kinase inhibitor SB-590885. Cancer Res 2006;66(23):11100–5.

[26] Tsai J, Lee J, Wang W, Zhang J, Cho H, Mamo S, et al. Discovery of a selective inhibitor of oncogenic B-Raf kinase with potent antimelanoma activity. Proc Natl Acad Sci USA 2008;105:3041–6.

[27] Flaherty K, Puzanov I, Sosman J, Kim K, Ribas A, McArthur G, et al. Phase I study of PLX4032: proof of concept for V600E BRAF mutation as a therapeutic target in human cancer. ASCO Meet Abstr 2009;27(15S):9000.

[28] Flaherty KT, Puzanov I, Kim KB, Ribas A, McArthur GA, Sosman JA, et al. Inhibition of mutated, activated BRAF in metastatic melanoma. NEJM 2010;363(9):809–19.

[29] Gadiot J, Hooijkaas AI, Deken MA, Blank CU. Synchronous BRAF(V600E) and MEK inhibition leads to superior control of murine melanoma by limiting MEK inhibitor induced skin toxicity. OncoTargets Ther 2013;6:1649–58.

[30] Falchook GS, Long GV, Kurzrock R, Kim KB, Arkenau TH, Brown MP, et al. Dabrafenib in patients with melanoma, untreated brain metastases, and other solid tumours: a phase 1 dose-escalation trial. Lancet 2012;379(9829):1893–901.

[31] Dummer R, Robert C, Nyakas M, McArthur GA, Kudchadkar RR, Gomez-Roca C, et al. Initial results from a phase I, open-label, dose escalation study of the oral BRAF inhibitor LGX818 in patients with BRAF V600 mutant advanced or metastatic melanoma. ASCO Meet Abstr 2013;31(Suppl. 15):9028.

[32] Su F, Viros A, Milagre C, Trunzer K, Bollag G, Spleiss O, et al. Ras mutations in cutaneous squamous-cell carcinomas in patients treated with BRAF inhibitors. N Engl J Med 2012;366(3):207–15.

[33] Flaherty KT, Infante JR, Daud A, Gonzalez R, Kefford RF, Sosman J, et al. Combined BRAF and MEK inhibition in melanoma with BRAF V600 mutations. N Engl J Med 2012;367(18): 1694–703.

[34] Kefford R, Miller WH, Tan DS-W, Sullivan RJ, Long G, Dienstmann R, et al. Preliminary results from a phase Ib/II, open-label, dose-escalation study of the oral BRAF inhibitor LGX818 in combination with the oral MEK1/2 inhibitor MEK162 in BRAF V600-dependent advanced solid tumors. ASCO Meet Abstr 2013;31(Suppl. 5):9029.

[35] Ribas A, Gonzalez R, Pavlick A, Hamid O, Gajewski TF, Daud A, et al. Combination of vemurafenib and cobimetinib in patients with advanced BRAF(V600)-mutated melanoma: a phase 1b study. Lancet Oncol 2014;15(9):954–65.

[36] Trunzer K, Pavlick AC, Schuchter L, Gonzalez R, McArthur GA, Hutson TE, et al. Pharmacodynamic effects and mechanisms of resistance to vemurafenib in patients with metastatic melanoma. J Clin Oncol 2013;31(14):1767–74.

[37] Shi H, Hugo W, Kong X, Hong A, Koya RC, Moriceau G, et al. Acquired resistance and clonal evolution in melanoma during BRAF inhibitor therapy. Cancer Discov 2013;4(1):80–93. http://dx.doi.org/10.1158/2159-8290.CD-13-0642.

[38] Lito P, Rosen N, Solit DB. Tumor adaptation and resistance to RAF inhibitors. Nat Med 2013;19(11):1401–9.

[39] Hertzman Johansson C, Egyhazi Brage S. BRAF inhibitors in cancer therapy. Pharmacol Ther 2014;142(2):176–82.

[40] Dougherty M, Muller J, Ritt D, Zhou M, Zhou X, Copeland T, et al. Regulation of Raf-1 by direct feedback phosphorylation. Mol Cell 2005;17:215–24.

[41] Ritt DA, Monson DM, Specht SI, Morrison DK. Impact of feedback phosphorylation and raf heterodimerization on normal and mutant B-Raf signaling. Mol Cell Biol 2010;30(3):806–19.

[42] Lax I, Wong A, Lamothe B, Lee A, Frost A, Hawes J, et al. The docking protein FRS2alpha controls a MAP kinase-mediated negative feedback mechanism for signaling by FGF receptors. Mol Cell 2002;10(4):709–19.

[43] Sato T, Gotoh N. The FRS2 family of docking/scaffolding adaptor proteins as therapeutic targets of cancer treatment. Expert Opin Ther Targets 2009;13(6):689–700.

[44] Hanafusa H, Torii S, Yasunaga T, Nishida E. Sprouty1 and Sprouty2 provide a control mechanism for the Ras/MAPK signalling pathway. Nat Cell Biol 2002;4(11):850–8.

[45] Holderfield M, Merritt H, Chan J, Wallroth M, Tandeske L, Zhai H, et al. Raf inhibitors activate the MAPK pathway by relieving inhibitory autophosphorylation. Cancer Cell 2013;23:594–602.

[46] Zhang C, Spevak W, Zhang Y, Burton EA, Ma Y, Habets G, et al. RAF inhibitors that evade paradoxical MAPK pathway activation. Nature 2015;526(7574):583–6. http://dx.doi.org/10.1038/nature14982.

[47] Basile KJ, Le K, Hartsough EJ, Aplin AE. Inhibition of mutant BRAF splice variant signaling by next-generation, selective RAF inhibitors. Pigment Cell Melanoma Res 2014;27(3):479–84.

[48] Le K, Blomain ES, Rodeck U, Aplin AE. Selective RAF inhibitor impairs ERK1/2 phosphorylation and growth in mutant NRAS, vemurafenib-resistant melanoma cells. Pigment Cell Melanoma Res 2013;26(4):509–17.

[49] Arora R, Di Michele M, Stes E, Vandermarliere E, Martens L, Gevaert K, et al. Structural investigation of B-Raf paradox breaker and inducer inhibitors. J Med Chem 2015;58(4):1818–31.

[50] Taylor SS, Kornev AP. Protein kinases: evolution of dynamic regulatory proteins. Trends Biochem Sci 2011;36(2):65–77.

[51] Peng SB, Henry JR, Kaufman MD, Lu WP, Smith BD, Vogeti S, et al. Inhibition of raf isoforms and active dimers by LY3009120 leads to anti-tumor activities in RAS or BRAF mutant cancers. Cancer Cell 2015;28(3):384–98. http://dx.doi.org/10.1016/j.ccell.2015.08.002.

[52] Gould AE, Adams R, Adhikari S, Aertgeerts K, Afroze R, Blackburn C, et al. Design and optimization of potent and orally bioavailable tetrahydronaphthalene Raf inhibitors. J Med Chem 2011;54(6):1836–46.

[53] Atefi M, Titz B, Avramis E, Ng C, Wong DJ, Lassen A, et al. Combination of pan-RAF and MEK inhibitors in NRAS mutant melanoma. Mol Cancer 2015;14:27.

[54] Lamba S, Russo M, Sun C, Lazzari L, Cancelliere C, Grernrum W, et al. RAF suppression synergizes with MEK inhibition in KRAS mutant cancer cells. Cell Rep 2014;8(5):1475–83.

[55] Lito P, Saborowski A, Yue J, Solomon M, Joseph E, Gadal S, et al. Disruption of CRAF-mediated MEK activation is required for effective MEK inhibition in KRAS mutant tumors. Cancer Cell 2014;25(5):697–710.

[56] Freeman AK, Ritt DA, Morrison DK. Effects of Raf dimerization and its inhibition on normal and disease-associated Raf signaling. Mol Cell 2013;49(4):751–8.

[57] Molzan M, Kasper S, Roglin L, Skwarczynska M, Sassa T, Inoue T, et al. Stabilization of physical RAF/14-3-3 interaction by cot Lenin A as treatment strategy for RAS mutant cancers. ACS Chem Biol 2013;8(9):1869–75.

[58] Tzivion G, Luo Z, Avruch J. A dimeric 14-3-3 protein is an essential cofactor for Raf kinase activity. Nature 1998;394(6688):88–92.

[59] Yip-Schneider MT, Miao W, Lin A, Barnard DS, Tzivion G, Marshall MS. Regulation of the Raf-1 kinase domain by phosphorylation and 14-3-3 association. Biochem J 2000;351(Pt 1):151–9.

[60] Light Y, Paterson H, Marais R. 14-3-3 antagonizes Ras-mediated Raf-1 recruitment to the plasma membrane to maintain signaling fidelity. Mol Cell Biol 2002;22(14):4984–96.

[61] Molzan M, Schumacher B, Ottmann C, Baljuls A, Polzien L, Weyand M, et al. Impaired binding of 14-3-3 to C-RAF in Noonan syndrome suggests new approaches in diseases with increased Ras signaling. Mol Cell Biol 2010;30(19):4698–711.

[62] Sassa T, Tojyo T, Munakata K. Isolation of a new plant growth substance with cytokinin-like activity. Nature 1970;227(5256):379.

[63] Asahi K, Honma Y, Hazeki K, Sassa T, Kubohara Y, Sakurai A, et al. Cotylenin A, a plant-growth regulator, induces the differentiation in murine and human myeloid leukemia cells. Biochem Biophys Res Commun 1997;238(3):758–63.

[64] Honma Y, Ishii Y, Yamamoto-Yamaguchi Y, Sassa T, Asahi K, Cotylenin A. A differentiation-inducing agent, and IFN-alpha cooperatively induce apoptosis and have an antitumor effect on human non-small cell lung carcinoma cells in nude mice. Cancer Res 2003;63(13):3659–66.

CHAPTER 11

Targeting Metabolic Vulnerabilities in *RAS*-Mutant Cells

A.D. Rao[1,2], G.A. McArthur[1,2]

[1]*Peter MacCallum Cancer Centre, East Melbourne, VIC, Australia;* [2]*University of Melbourne, Parkville, VIC, Australia*

CONTENTS

INTRODUCTION

RAS is a frequently mutated oncogene, for which there is currently no standard targeted therapy available, and novel approaches for targeting RAS are a key priority in current cancer research. Approaches being investigated range from direct targeting of mutant RAS to efforts focused on targeting key downstream effector pathways that in turn affect the hallmarks of cancer. There is an emerging understanding of the role of the *RAS* oncogene in the metabolic re-programming of cancer cells [1,2]. This chapter aims to review the role of oncogenic *RAS* in tumor metabolism, thereby revealing potential metabolic vulnerabilities in RAS-mutant cells that could be exploited as therapeutic targets.

AN INTRODUCTION TO CANCER METABOLISM

Cancer Metabolism: An Emerging Hallmark of Cancer

In the recent update of the "Hallmarks of Cancer" [3], "reprogramming energy metabolism" was named as an emerging hallmark of cancer due to the critical role of altered energy metabolism in supporting the biosynthetic requirements of aberrant cell proliferation in tumors. Although a detailed description of the field of cancer metabolism is beyond the scope of this chapter, the key aspects are summarized in the following sections. Interested readers are also directed to comprehensive reviews of the field [2,4,5].

The field of cancer metabolism research dates back decades to Otto Warburg's [6] observations that even in the presence of oxygen, cancer cells seem to re-program their energy metabolism to depend on glycolysis, a phenomenon known as aerobic glycolysis or the "Warburg effect." Although originally thought to be a result of impaired mitochondrial function, emerging evidence suggests that cancer cells also use oxidative phosphorylation (OXPHOS) and

Conquering RAS. http://dx.doi.org/10.1016/B978-0-12-803505-4.00011-4

in fact have functional mitochondria [7]. A proposed explanation for this paradox is the role of glucose degradation in providing intermediates for essential biosynthetic pathways, such as citrate and glycerol for lipid synthesis or ribose for nucleotides [8]. Thus metabolic re-programming in cancer cells can affect catabolic and anabolic metabolism to both meet energy requirements and provide the building blocks necessary for rapid tumor cell proliferation.

The Role of Oncogenes in Metabolic Re-programming

There is an evolving appreciation that oncogene-driven re-programming of cancer cell metabolism is a central mechanism underlying the altered metabolism that is characteristic of cancer cells [2]. Although the focus of this chapter is on the role of *RAS*, alterations in other oncogenes and signaling pathways have been implicated in re-programming metabolism in cancer cells—key examples include the phosphoinositide 3-kinase (PI3K)/AKT/mammalian target of rapamycin (mTOR) C1 pathway, myelocytomatosis oncogene cellular homolog (MYC), hypoxia-inducible factor 1 alpha (HIF1α), and v-raf murine sarcoma viral oncogene homolog B1 (BRAF).

For example, AKT1 has a role in increasing glycolysis through phosphorylation of glycolytic enzymes, such as hexokinase, and glucose uptake by increasing the expression of glucose transporters [9–11]. Increased glycolysis can also be the result of prolonged AKT signaling to forkhead box O (FOXO) transcription factors that are key transcriptional regulators of glycolytic metabolism [12]. Downstream of AKT, mTORC1 also has a role in metabolism, particularly integrating protein and lipid synthesis with growth signals [13]. mTORC also has important indirect effects on glycolysis through its interaction with HIF1α [14,15]. In addition to being activated by mTORC, the transcription factor HIF1α can also be activated by mutations in tumor suppressors (eg, von Hippel-Lindau [16] or succinate dehydrogenase [17]) with a resultant increase in glycolysis through effects on glucose transporters and glycolytic enzymes [18,19]. In the context of hypoxia, stabilization of HIF1α results in increased glycolysis and decreased OXPHOS [20]. An important effect of HIF1α is on pyruvate dehydrogenase kinase (PDK1), which inhibits pyruvate dehydrogenase (PDH) activity, affecting mitochondrial metabolism [21,22]. Interestingly, HIF1α is also thought to cooperate with MYC resulting in a further enhancement of aerobic glycolysis [23]. Glycolytic enzymes [24] and lactate dehydrogenase A [25] are targets of MYC, highlighting MYC's critical role in glucose metabolism. MYC also has an important role in glutaminolysis, affecting both glutamine transporters and glutaminase 1 [26], further demonstrating its role in cancer metabolism.

The role of mutant *BRAF* in re-programming cancer metabolism has been prominent in the literature. Recent work has identified mechanisms by which *BRAF* regulates glycolysis in melanoma in the context of BRAF inhibitor therapy, involving a transcriptional network centered on HIF1α, MYC, and thioredoxin-interacting protein (TXNIP) [27], also demonstrating that changes in glucose metabolism play a role in resistance to targeted therapies, specifically in the context of treatment with a BRAF inhibitor. Another study has demonstrated metabolic re-programming that results in increased OXPHOS as an early adaptation following BRAF inhibition [28]. These examples illustrate the potential therapeutic implications of obtaining an understanding of the impact of oncogenic alterations on cancer metabolism.

As in the context of *BRAF*-mutant tumors, in *RAS*-mutant tumors there is the potential for co-operation with other important key oncogenes and signaling pathways that are known to alter cancer metabolism. In the case of *RAS*, seminal work regarding transformation of embryonic fibroblasts demonstrated strong co-operation between both oncogenic RAS and MYC to achieve tumorigenic conversion [29]. Similarly, PI3K also has a role in RAS-induced transformation as one of the effector pathways of RAS [30]. As discussed earlier, oncogenic MYC and elements of the PI3K pathway (particularly AKT and mTOR) have important roles in altered cancer cell metabolism, thus highlighting the interplay between various oncogenic abnormalities that might co-operate with RAS to re-program metabolism.

Given our evolving understanding of the importance of metabolic changes in the development of tumors, and in particular the role that oncogenes play in determining this metabolic re-programming, a detailed understanding of the role of *RAS* in cancer metabolism has the potential to reveal vulnerabilities amenable to therapeutic intervention.

METABOLISM IN RAS-MUTANT CELLS

RAS and the Warburg Effect

A seminal study in cancer metabolism demonstrated that the *RAS* oncogenes increase glycolytic metabolism and glucose uptake in cancer cells [31]. In this study, isogenic colorectal cancer cell lines containing mutated or wild-type *KRAS* were subjected to transcriptomic analyses, which revealed up-regulation of the *GLUT1* gene, which encodes a key glucose transporter, in the *KRAS*-mutant cells. These cells were shown to have an increase in glucose uptake and lactate production, consistent with increased glycolysis. When cells were cultured in a low-glucose environment, cells with a *KRAS* mutation demonstrated a growth advantage. Interestingly, in these cells, the glycolysis inhibitor 3-bromopyruvate (which inhibits hexokinase) was shown to preferentially suppress cell growth.

The concept that *RAS* oncogenes induce metabolic re-programming has been further substantiated in models of mouse fibroblasts [32] transformed with *KRAS* and human carcinoma cell lines harboring *KRAS* mutations [33]. These studies, using a combination of transcriptional profiling, ^{13}C Carbon flux analysis, and chemical and biological perturbations, illustrated a phenotype of increased glycolysis, decreased OXPHOS, and increased glutamine use. There is also evidence to suggest that in fibroblasts transformed by *HRAS*, HIF1α has a role in the regulation of glucose uptake via glucose transporter (GLUT) 1 [34]. Furthermore, *RAS*-transformation in mouse fibroblasts or human bronchial epithelial cells has been demonstrated to be under the metabolic control of 6-phosphofructo-2-kinase [35].

In a murine model of $KRAS^{G12D}$ mutant pancreatic cancer, extinction of $KRAS^{G12D}$ resulted in decreased glucose uptake and lactate production, as well as down-regulation of GLUT1 and rate-limiting enzymes for glycolysis such as hexokinase 1 and 2 [36]. In this study, the role of downstream effector pathways of RAS on glycolysis was also dissected using pharmacologic inhibitors and short hairpin ribonucleic acid (shRNA). Use of an MEK (mitogen activated protein kinase/extracellular signal-related kinase kinase) inhibitor confirmed the importance of the mitogen-activated protein kinase (MAPK) pathway and showed a reduction in the expression of glycolytic genes such as *GLUT1* and *HK1*. However, the use of either an mTORC1 inhibitor or a pan PI3K inhibitor did not result in significant changes in glucose metabolism in this model. Investigation of key potential transcriptional regulators of glucose metabolism, through the shRNA knockdown of MYC and HIF1α, demonstrated a significant role for MYC but not HIF1α in altering metabolism in these *RAS*-mutant cells.

Taken together, these findings illustrate through a variety of models the important role of the *RAS* oncogene in re-programming cancer cells to utilize aerobic glycolysis and induce the Warburg effect.

Beyond Warburg: Other Aspects of Glucose and Glutamine Metabolism in RAS-Mutant Cells

In addition to alterations in glycolysis, there is an emerging understanding of the role of mutant *RAS* in controlling other aspects of metabolism, particularly mitochondrial metabolism, glutamine metabolism and redox homeostasis, ribose biosynthesis, and the hexosamine biosynthesis pathway (HBP).

Glutamine Metabolism, Mitochondrial Metabolism, and Redox Homeostasis

The role of glutamine metabolism in cancer cells has been a subject of interest in the field of cancer metabolism. In a model of glioblastoma, ^{13}C nuclear magnetic resonance spectroscopy was used to examine metabolism in detail [37]. In this model, although features of aerobic glycolysis were observed, there was also evidence of an active tricarboxylic acid (TCA) cycle, particularly for substrate use in fatty acid synthesis. Interestingly, a high rate of glutamine metabolism supported these biosynthetic pathways, through both generation of reductive power in the form of nicotinamide adenine dinucleotide phosphate (NADPH) and support of anaplerosis through restoration of oxaloacetate. The importance of glutamine metabolism in cancer cells has also been demonstrated in a fibroblast inducible model of MYC [38], in which glutamine deprivation induced apoptosis in an MYC-dependent manner. Of interest, oxaloacetate was able to rescue apoptosis in the setting of glutamine deprivation, also suggesting a role of glutamine metabolism in supporting anaplerosis in cancer cells. In the setting of *KRAS*-transformed fibroblasts, glutamine deprivation has also been shown to limit cell proliferation [39].

As previously mentioned, the role of oncogenic KRAS in re-programming glutamine metabolism was investigated in a study that performed a systems level metabolomic and transcriptomic analysis in transformed mouse fibroblast cell lines and human cancer cell lines [33]. With regards to glutamine metabolism, in the transformed cells glutamine had a more substantial contribution to precursors of cellular biomass, such as aspartate, than in normal cells. A combination of transcriptomic analysis and non-targeted tracer fate detection using amino-labeled glutamine also demonstrated the role of glutamine in supporting anabolic processes in the transformed cells. In addition to examining normal and transformed cells, this study used a model of reversion of cellular transformation. This was achieved by enforced expression of a dominant-negative guanine exchange factor, which attenuates KRAS activation, thus providing a comparison with both normal and transformed cells. When the normal, transformed, and reverted cells were all treated with an inhibitor of aminotransferase activity (aminooxyacetate, AOA), only the transformed cells showed a decrease in cell proliferation that could be rescued by the addition of dimethyl aspartate. This study thus confirmed the importance of oncogenic KRAS in altering glutamine metabolism, with the particular role of supporting anabolic processes required for cancer cell growth.

Further support of the concept that glutamine metabolism has a role in KRAS-mutant cells comes from a study of HCT116 colon cancer cells [40]. The use of AOA as an inhibitor of aminotransferase (to block the conversion of glutamate into α-ketoglutarate and thus entry to the TCA cycle) resulted in inhibition of soft agar colony growth of the HCT116 cells, which could be rescued with the addition of cell-permeable dimethyl α-ketoglutarate. This finding was replicated in two other RAS-mutant models. This study was also able to demonstrate that oncogenic KRAS induced an increase in mitochondrial reactive oxygen species (ROS) with an effect on regulation of cellular proliferation via extracellular signal-regulated kinase 1/2 (ERK1/2) signaling. Specifically, mitochondrial-targeted nitroxides (but not control nitroxides) reduced colony growth. Cell lines treated with the mitochondrial-targeted antioxidants (but not control agents) had a reduced rate of proliferation and increased phosphorylation of ERK1/2. An MEK inhibitor was able to decrease this phosphorylation of ERK1/2 to a level similar to that in cells that were not treated with the mitochondrial-targeted nitroxides showing the dependence of MEK for the effects of ROS on ERK-activation. The role of ROS in cellular proliferation was further investigated using models deficient in mitochondrial DNA, which were unable to generate ROS and had impaired growth. In an in vivo model of *KRAS*-driven lung adenocarcinoma, reduction in mitochondrial transcription factor A (TFAM, which is required for mitochondrial DNA replication and transcription) resulted in the formation of smaller tumors, highlighting the essential role of mitochondrial metabolism in KRAS-driven cancers.

Although there is an increasing appreciation of the presence of increased levels of mitochondrial ROS in cancer cells, there is also an elucidation of the factors that play a role in the prevention of excessive ROS accumulation [41]. NADPH is one co-factor, which plays an important role in the maintenance of redox homeostasis [41]. Important sources of NADPH are the pentose phosphate pathway (PPP), generated by the oxidation of glucose-6 phosphate [41], and folate [42] and serine metabolism [43]. Interestingly, there is evidence to suggest that in the setting of a pancreatic cancer model, oncogenic KRAS re-programs glutamine metabolism to support NADPH production and maintain the cellular redox state [44]. In this study, glutamine was demonstrated to be a critical carbon source for the production of NADPH via a non-canonical pathway. That is, rather than using glutamine dehydrogenase (GLUD1) to convert glutamate into α-ketoglutarate to then enter the TCA cycle, KRAS-mutant pancreatic cancer relies on a pathway in which aspartate derived from glutamine is converted by aspartate transaminase (GOT1) into oxaloacetate. The oxaloacetate is subsequently converted by malate dehydrogenase 1 (MDH1) to malate. Finally, the malic enzyme (ME1) uses malate to produce both NADPH and pyruvate. Of particular importance, knockdown of key enzymes in this pathway resulted in suppression of pancreatic cancer growth both in vitro and in vivo, as well as being associated with increasing ROS and reduced glutathione. However, this pathway seemed to be dispensable in normal pancreatic ductal cells and human fibroblasts, suggesting potential therapeutic implications for this novel pathway of glutamine metabolism. However, it is unclear whether this pathway is specific to KRAS-mutant pancreatic cancer or has a role more generally in other KRAS-mutant tumor types [1] (Fig. 11.1).

Finally, there is one other mechanism by which KRAS affects redox homeostasis—through the Nrf2 (or Nfe2l2) transcription factor [45,46]. In a study using murine cells expressing $KRAS^{G12D}$ [45], a reduction in ROS was associated with increased transcription of *NRF2*. Conversely, small interfering ribonucleic acid (siRNA) knockdown of KRAS in human pancreatic cancer cells resulted in a decrease in *NRF2* and an increase in ROS. Investigation in vivo demonstrated that genetic targeting of the Nrf2 pathway results in impaired proliferation and tumorigenesis in $KRAS^{G12D}$-driven cancer. Another study [46], which employed models of both knockdown and over-expression of Nrf2 in cancer cells, demonstrated a role of Nrf2 in re-programming both glucose and glutamine metabolism, particularly under sustained PI3K-AKT signaling. Although the role of Nrf2 in tumorigenesis is not fully understood, these studies demonstrate the importance of this transcription factor in the metabolism of RAS-mutant cells.

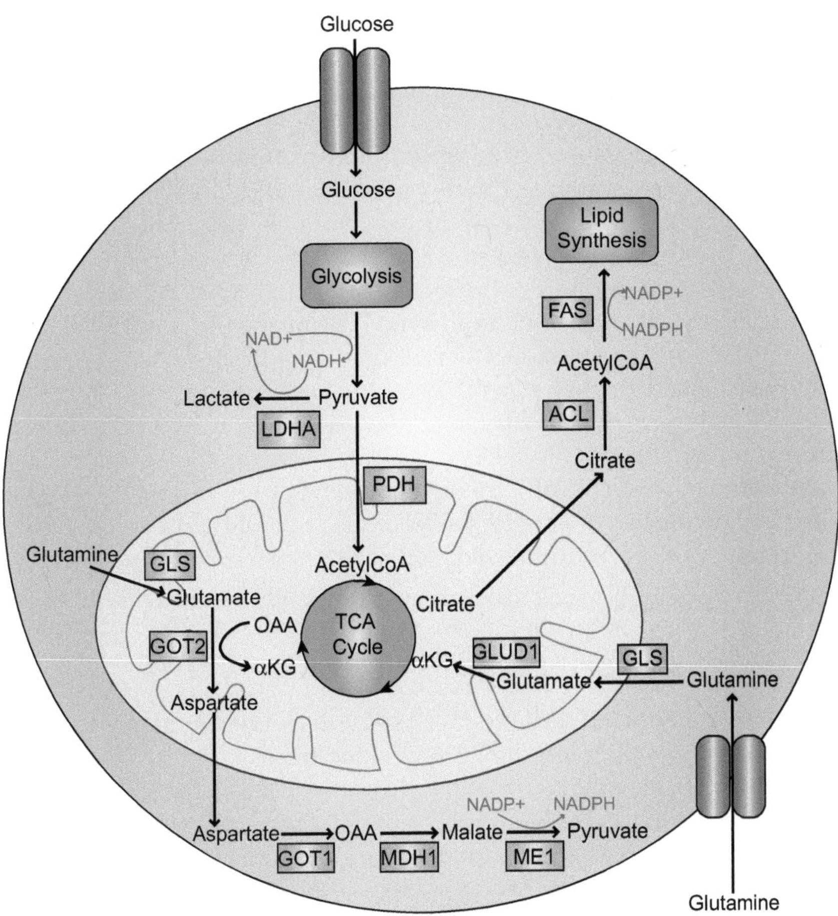

FIGURE 11.1 Key metabolic pathways in RAS-mutant cells.

In RAS-mutant cells, there is increased conversion of glucose to pyruvate by glycolysis. A majority of pyruvate is converted to lactate by lactate dehydrogenase A (LDHA). However, pyruvate dehydrogenase (PDH) converts some pyruvate into acetyl-CoA that enters the tricarboxylic acid (TCA) cycle. Note, not all steps of the TCA cycle are depicted here. Key intermediates, such as citrate, are used in macromolecular synthetic processes such as lipid synthesis. In KRAS-mutant pancreatic cancer, altered glutamine metabolism results from increased aspartate transaminase (GOT1) expression, with decrease of the usual conversion of glutamate into α-ketoglutarate by glutamine dehydrogenase (GLUD1). The re-programming supports NADPH production, by the eventual conversion of malate by malic enzyme (ME1) to generate NADPH and pyruvate. *Adapted from references Kimmelman AC. Metabolic dependencies in RAS-driven cancers. Clin Cancer Res April 15, 2015;21(8):1828–34; Ward PS, Thompson CB. Metabolic reprogramming: a cancer hallmark even Warburg did not anticipate. Cancer Cell March 20, 2012;21(3):297–308; and DeBerardinis RJ, Lum JJ, Hatzivassiliou G, Thompson CB. The biology of cancer: metabolic reprogramming fuels cell growth and proliferation. Cell Metab January 2008;7(1):11–20.*

In summary, research efforts have revealed a number of important alterations in the mitochondrial metabolism of RAS-mutant cells, particularly highlighting the role of glutamine metabolism in addition to that of glucose.

Ribose Biosynthesis and the Pentose Phosphate Pathway

In addition to the oxidative arm that produces NADPH, the PPP has a non-oxidative arm that has a role in producing ribose for RNA and DNA synthesis. In the previously described study of KRAS^{G12D}-mutant pancreatic cancer [36], metabolic flux studies demonstrated a decoupling of the oxidative and non-oxidative arms of the PPP in this model. In particular, KRAS^{G12D} extinction was associated with a reduced incorporation of labeled glucose into DNA and RNA, suggesting a key role for Ras in the non-oxidative arm of the PPP. KRAS^{G12D} extinction was also associated with a decreased expression of genes, namely *Rpia* and *Rpe*, that encode enzymes important in the carbon exchange of the non-oxidative arm of the PPP. Knockdown of Rpia and Rpe not only had effects on the flux of labeled glucose into RNA and DNA but also resulted in a decrease in the clonogenic potential of *KRAS*-mutant cells, particularly in low-glucose conditions. These findings were confirmed in vivo, with Rpia and Rpe knockdown resulting in decreased tumor formation in a xenograft model, confirming the importance of this arm of the PPP in Ras-mutant cancer.

Protein Glycosylation and the Hexosamine Biosynthesis Pathway

The HBP is responsible for the production of precursors, such as *N*-acetylglucosamine (GlcNAc), required for protein glycosylation. The importance of the HBP in the context of mutant RAS has been investigated in both in vitro and in vivo models of hypoxia in KRAS-mutant pancreatic ductal carcinoma [47]. In addition to demonstrating the importance of glycolysis and glutamine metabolism in these models, this study demonstrated an up-regulation of genes encoding enzymes activated after HBP activation, such as glutamine fructose-6-phosphate-aminotransferase (*Gfpt*) 1 and *Gfpt2*. When these Gfpt enzymes were inhibited with azaserine, a decrease in cell number was observed, suggesting the importance of HBP activation in the survival of hypoxic RAS-mutant pancreatic cancer cells. In another study, shRNA knockdown of Gfpt resulted in inhibition of tumor cell growth in clonogenic and soft agar assays, as well as in a tumor xenograft model [36], thus confirming the role of Kras in stimulating the HBP and subsequent protein glycosylation.

Altered Scavenging Pathways in RAS-Mutant Cancer

To meet the requirements of tumor growth, RAS-mutant cells employ a variety of cellular adaptations that allow the scavenging of required nutrients from intra- and extra-cellular sources. Such adaptations are integral to the altered metabolic state and these include alterations in the processes of autophagy and macropinocytosis.

Autophagy

Autophagy (also known as macroautophagy) is a catabolic cellular process by which intra-cellular components are captured into vesicles, called autophagosomes, and delivered to lysosomes where they are degraded, to be used as inputs to cellular metabolism [48]. Autophagy is increased in *RAS*-driven cancers and appears to play an important role in tumor growth [49–51]. In a study of murine cells transformed with $HRas^{V12}$ or $Kras^{V12}$, basal autophagy was increased [49]. Genetic defects in autophagy resulted in decreased cell survival and tumor growth in an in vivo model, suggesting a role for autophagy in Ras-mediated tumorigenesis. This study also examined a panel of human cancer cell lines, treating them with the autophagy inhibitor choloroquine (which inhibits lysosomal acidification), which suppressed or slowed growth in those cells with high basal autophagy (and RAS mutations). Furthermore, defects in autophagy resulted in the accumulation of abnormal mitochondria, reduced mitochondria, and TCA cycle metabolite and energy depletion. These findings suggest that in the context of RAS-mutant cells, autophagy plays a role in both tumorigenesis and mitochondrial metabolism.

Macropinocytosis

Macropinocytosis is an endocytic process, which involves the engulfment of extra-cellular content in vesicles known as macropinosomes. In a fibroblast model, expression of mutant RAS was shown to promote macropinocytosis [52]. In a study by Commisso et al. [53], KRAS-mutant human pancreatic cancer cells were shown to have an increase in macropinocytosis in vitro and in vivo (in a tumor xenograft model). KRAS knockdown was associated with a decrease in macropinocytosis. In Ras-transformed cells, macropinocytosis resulted in the internalization of albumin, which was shown to result in increased intra-cellular concentrations of glutamine and α-ketoglutarate. This study indicated that in Ras-mutant cells, the process of macropinocytosis results in the degradation of extra-cellular protein into amino acids, which can then have a role in anaplerosis. In a tumor xenograft model the use of 5-*N*-ethyl-*N*-isopropylamiloride (EIPA), which is known to inhibit macropinosome formation, was able to slow growth in the *KRAS*-mutant rather than *KRAS*-wild-type mice.

Taken together, these findings highlight the importance of scavenging pathways in the metabolism of RAS-mutant cells.

Are All RAS Mutations Equal?

An important and yet to be fully answered question is whether all *RAS* mutations have the same effects on cellular metabolism. As mentioned in Section "Glutamine Metabolism, Mitochondrial Metabolism and Redox Homeostasis" in the review of glutamine metabolism, it is unclear whether findings observed in a particular tissue type (in that case KRAS-mutant pancreatic cancer) will be consistent in other tissues of origin. The role of particular oncogenic mutations is also a question for further investigation. One study [54] that has attempted to shed light on this matter used a panel of isogenic non–small-cell lung cancer cell lines, over-expressing different forms of mutated KRAS at codon 12 (G12C, G12D, and G12V) alongside a matched wild type. The panel of cell lines was investigated using steady state metabolomics. Although there was a clear distinction between the wild-type and mutant lines, there were also notable differences between each of the mutant clones. In this model, glycolytic metabolism was not significantly altered, with the main alterations being associated with glycerolphospholipids and amino acids. Between the mutant clones, dependence on exogenous glutamine and the importance of the glutathione/ophthalmate redox buffering system varied. This study highlights the need for further comparative investigation between different *RAS* mutations, potentially also in different tumor or tissue types, when considering metabolic vulnerabilities as a therapeutic target in RAS-mutant cells.

POTENTIAL THERAPEUTIC APPROACHES TO TARGET METABOLISM IN *RAS*-DRIVEN CANCER

It is clear there are a number of potential metabolic vulnerabilities in *RAS*-mutant malignancies. It is tantalizing to consider that these metabolic alterations may prove to be an Achilles heel for cancer cells and provide opportunities for therapeutic intervention, particularly in combination strategies. However, it must be noted that, although a renewed interest in cancer metabolism has led to an increased interest in pursuing metabolism as a target for cancer therapeutics [55], the field has been limited by toxicity concerns relating to effect on metabolism in normal tissues and the potential lack of a safe therapeutic window [56]. In the following sections, the most promising strategies for each of the metabolic vulnerabilities observed in *RAS*-driven cancers are discussed.

Targeting Glucose Metabolism

Given the well-recognized alterations in glucose metabolism in cancer, many efforts have been directed toward targeting various key steps of glycolysis and mitochondrial metabolism. However, to date the clinical utility of these strategies has been limited by toxicity. An overview of such potential strategies that target glucose metabolism in cancer generally is provided in Table 11.1. In the setting of *RAS*-mutant cancers, the most promising strategies of targeting glucose metabolism have centered on the targeting of signaling pathways that control glycolysis and the use of inhibitors of OXPHOS.

The previously described study of $KRAS^{G12D}$-mutant pancreatic cancer [36] demonstrated the importance of the MAPK pathway in the regulation of

Table 11.1 Potential Targets of Glucose Metabolism in Cancer

Type of Metabolism	Target	Agents	Findings
Glycolysis	Glucose transport (GLUT 1 & 4)	Pre-clinical only, eg, phloretin	In vitro/pre-clinical findings only
	Hexokinase	2-Deoxyglucose, 3-bromopyruvate	Unacceptable toxicity in clinical trials
	PFK2	Tool compounds only	Inhibits growth of xenograft tumors
	PKM2	Pre-clinical	shRNA, ongoing pre-clinical work
	LDHA	Pre-clinical, eg, oxamate	Inhibits growth in xenograft tumors
	Lactate excretion (MCT4, MCT1)	Pre-clinical	Inhibitors block cell proliferation
Mitochondrial metabolism/ TCA cycle	PDK	Dichloroacetate	Limited solitary effect in clinical trials
	Mitochondrial complex 1	Metformin (& phenformin, pre-clinical)	Clinically available for diabetes, under investigation in cancer context

LDHA, *lactate dehydrogenase A;* PFK2, *phosphofructokinase 2;* TCA, *tricarboxylic acid.*
Adapted from references Zhao Y, Butler EB, Tan M. Targeting cellular metabolism to improve cancer therapeutics. Cell Death Dis 2013;4:e532; Vander Heiden MG. Targeting cancer metabolism: a therapeutic window opens. Nat Rev Drug Discov September 2011;10(9):671–84.

glycolysis in RAS-mutant cells. Use of the MEK inhibitor AZD8330 resulted in metabolic re-programming, as evidence by altered expression of key glycolytic genes, as well as HBP and non-oxidative PPP genes. Such findings have led to the suggestion that MEK inhibition may be a useful approach to suppress glucose metabolism in RAS-mutant cancers [1]. MEK inhibitors are currently being investigated in a variety of settings—a promising example is the MEK inhibitor trametinib, which has been approved for the treatment of patients with BRAF-mutant metastatic melanoma alone or in combination with a BRAF inhibitor [57]. However, to date, predominantly due to modest efficacy, single-agent MEK inhibitors have not yet been approved for the treatment of patients with RAS-mutant malignancies.

The approach of targeting signaling pathways that control glucose metabolism has also led to an appreciation of adaptive responses and resistance mechanisms and the need to consider potential combination approaches. One study [58], which also utilized an inducible mouse model of $KRAS^{G12D}$-mutant pancreatic cancer, highlighted these concepts. In this study, following oncogene ablation (achieved through either genetic means or the use of inhibitors targeting both MEK and PI3K/mTOR) the population of surviving cells was investigated using transcriptomic and metabolomic methods. The cells that survived oncogene ablation were no longer dependent on glycolysis for the metabolism, but seemed to have an increased reliance on OXPHOS. There was an observed increase in genes relating to mitochondrial function, as well as autophagy and lysosome activity. Most remarkably, the use of OXPHOS inhibitors, particularly the ATP synthetase inhibitor oligomycin, was able to reduce survival in these cells. In the murine model, improved survival was observed when oligomycin was combined with oncogene ablation. Furthermore, in a model of spheres formed from human patient-derived xenografts, a combination of oligomycin with the MEK inhibitor and PI3K/mTOR inhibitor significantly reduced sphere formation. These findings suggest a role for the combined targeting of OXPHOS and the signaling pathways that control glucose metabolism in RAS-mutant cells. Although this is a promising strategy, it must be noted that the OXPHOS inhibitors used in this study were pre-clinical or tool compounds. A significant challenge lies ahead in finding combinations that utilize clinically tolerable OXPHOS inhibitors given the importance of OXPHOS to a number of normal tissues including heart, kidney, and brain.

One such compound, which is already in clinical use, is the biguanide metformin. Metformin is an anti-diabetic drug, which is considered to be a non-specific OXPHOS inhibitor with effects on mitochondrial complex I and TCA cycle anaplerosis [59], as well as being considered an inhibitor of the mTOR pathway, through AMP-activated protein kinase (AMPK) activation [60]. In a pre-clinical study [61], the combination of the MEK inhibitor trametinib with metformin (with its proposed rationale being its use as inhibitor of mTOR, which is clinically well tolerated) resulted in decreased cell viability in a panel of NRAS-mutant cancer cells. The in vitro models included melanoma (including some lines that had acquired resistance to trametinib), lung cancer, and neuroblastoma cell lines. In an in vivo melanoma xenograft model, the combination of trametinib and metformin slowed tumor growth. Although this study did not detail the metabolic implications of combining these two inhibitors, their synergy and known mechanisms of action support the concept that a promising approach to targeting glucose metabolism in RAS-mutant cancers may be to combine inhibition of glycolysis (under the control of MAPK signaling) and other key metabolic pathways (in this instance OXPHOS).

Targeting Glutamine Metabolism

Given its critical role in *RAS*-driven cancers, glutamine metabolism is an attractive candidate for therapeutic intervention. In the case of pancreatic cancer, the novel pathway of glutamine metabolism [44] may provide rational therapeutic candidates, particularly as genetic targeting of key components resulted in reduced growth in pancreatic cancer cells but had only modest effects on normal cells. Unfortunately, there are currently no clinically relevant molecules available that selectively inhibit GOT1, MDH1, or ME1. However, inhibitors of glutaminase (the enzyme that converts glutamine to glutamate prior to eventual entry into the TCA cycle) exist. Compounds currently available for pre-clinical studies include bis-2-(5-phenylacetamido-1,2,4-thiadiazol-2-yl) ethyl sulfide (BPTES) [62] or compound 968 [63]. The most advanced compound in terms of clinical development is the glutaminase inhibitor CB-839, which has shown promise in pre-clinical models of triple-negative breast cancer [64] and acute myeloid leukemia [65]. This selective, orally bioavailable inhibitor is currently being studied in phase 1 clinical trials in patients with solid and hematologic malignancies (NCT02071862, NCT02071888, ClinicalTrials.gov).

A notable strategy, when considering the targeting of *RAS*-driven cancers, is the combination of inhibitors of glutamine metabolism with conventional therapies that increase intra-cellular ROS (such as radiotherapy or chemotherapy). Pre-clinical evidence, in the setting of KRAS-mutant pancreatic adenocarcinoma, demonstrated synergy between the glutaminase inhibitors BPTES and 968 with the ROS hydrogen peroxide (suggesting that cells have an increased sensitivity to ROS when glutamine metabolism is impaired) [44].

Thus glutamine metabolism represents a potential therapeutic target in *RAS*-driven cancer, with glutaminase inhibitors being the most advanced clinically relevant option. Future efforts to explore inhibitors of downstream targets specifically observed in *RAS*-mutant cancers (such as GOT1, MDH1, or ME1) and investigation into rational combinations (with other metabolic or conventional therapies) may shed further light on the potential of glutamine metabolism as a therapeutic target.

Targeting Scavenging Pathways

The requirement for RAS-mutant cancer cells to scavenge nutrients, through processes such as autophagy and macropinocytosis, represents another vulnerability that might be therapeutically targeted to target metabolism in cancer cells. The lysosome inhibitor hydroxychloroquine, an agent currently used in the treatment of both malaria and arthritis, has been shown to affect cancer growth in a number of pre-clinical models, including a study in patient-derived xenograft models of pancreatic cancer [66]. Given the known safety of this drug (and a similar agent chloroquine), it is currently being investigated in a number of clinical trials in a variety of cancers, including in combination with conventional chemotherapy, radiotherapy, and newer targeted therapies (ClinicalTrials.gov). There is some concern in the field regarding the efficacy of hydroxychloroquine in inhibiting the process of autophagy in patients [1], a question that ongoing clinical trials will begin to answer. At present there are no specific inhibitors of macropinocytosis being used clinically, although in pre-clinical models EIPA has been used effectively [53,67]. However, lysosome inhibitors, such as hydroxychloroquine, may also have a role in inhibiting macropinocytosis. Thus ongoing investigation into this agent may hold promise as a strategy that effectively limits the ability of mutant-RAS to alter scavenging pathways (Fig. 11.2).

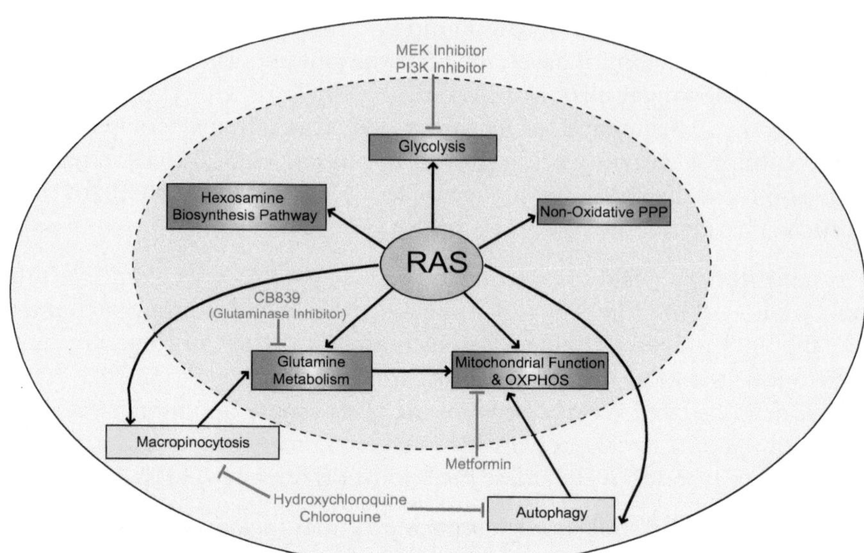

FIGURE 11.2 Potential targetable metabolic vulnerabilities in RAS-mutant cells.
RAS-mutant cells harbor metabolic alterations (indicated in purple) such as glucose metabolism; glutamine metabolism; redox homeostasis; ribose biosynthesis and protein glycosylation; and alterations in nutrient scavenging (indicated in yellow). These alterations represent vulnerabilities that have the potential to be targeted therapeutically; indicated in red are the inhibitors that are most advanced in clinical development and investigation. *Adapted from reference Kimmelman AC. Metabolic dependencies in RAS-driven cancers. Clin Cancer Res April 15, 2015;21(8):1828–34.*

CONCLUSION

Recent discoveries in the field of cancer metabolism research have defined an important role for oncogenes in the metabolic re-programming of cancer cells. These alterations confer an advantage beyond simple energy provision, playing a critical role in meeting the demand for macromolecules and intermediates required for rapid cell proliferation. In the case of *RAS*-mutant cancer cells, a number of metabolic adaptations have been observed that affect processes such as glucose metabolism, glutamine metabolism, redox homeostasis, ribose biosynthesis, protein glycosylation, and nutrient scavenging. *RAS*-mutant cancer cells seem to have a strong dependence on a number of these altered pathways, revealing a series of metabolic vulnerabilities that have the potential to be targeted therapeutically.

Like many aspects of *RAS* biology, there are still a number of unanswered questions in this field and a clinically successful therapeutic combination is still elusive. A key issue to be resolved in future research endeavors is the role of individual *RAS* mutations and the importance of tissue of origin; it is possible that prior observations may be attributed only to a specific mutation or more importantly be relevant only in certain types of cancer. Another concern of the work discussed in this chapter is that many of the models employed are in vitro and the most sophisticated analyses are often done in this context. With ongoing advancement in metabolomics and related technologies, development of techniques to assess metabolic changes in vivo (particularly in patients) need to be prioritized. Ultimately, the greatest challenge that lies ahead is finding agents for which a clinically safe therapeutic window exists. Thus a focus on metabolic defects that are truly cancer specific is essential. Strategies that employ combined approaches to target metabolic vulnerabilities will also need to be explored in greater detail.

In summary, in spite of a number of unresolved issues, recent advances in the field of cancer metabolism open up an exciting opportunity to inhibit this hallmark of cancer—through the targeting of metabolic vulnerabilities in *RAS*-mutant cancers. It is hoped that future research will realize this potential, to eventually develop a novel therapeutic strategy for patients with *RAS*-driven malignancies.

Glossary

Anabolic Description of a metabolic process that involves the construction of larger molecules from smaller molecules.

Anaplerosis Replenishing of intermediates for metabolic reactions (such as the TCA cycle in the setting of intermediates being extracted for biosynthetic reactions).

Autophagy A catabolic cellular process by which intra-cellular components are captured into vesicles, called autophagosomes and delivered to lysosomes where they are degraded, to be used as inputs to cellular metabolism (also known as macroautophagy).

Catabolic Description of a metabolic process that breaks down larger molecules into smaller molecules for conversion into energy or use in subsequent anabolic reactions.

Glycolysis An oxygen-independent metabolic process that converts glucose into pyruvate via a series of intermediate steps, resulting in net energy production.

Macropinocytosis An endocytic process, which involves the engulfment of extra-cellular content in vesicles known as macropinosomes.

Oxidative phosphorylation A cellular metabolic process in which mitochondrial respiratory enzymes synthesize ATP from phosphate and ADP. This process is coupled with electron transport.

Reactive oxygen species Molecules generated by the reduction of oxygen, such as by one electron (superoxide) or two electrons (hydrogen peroxide).

Warburg effect Glycolysis, occurring in the presence of oxygen, in rapidly proliferating cells (also known as "aerobic glycolysis").

List of Acronyms and Abbreviations

ACL ATP citrate lyase
AOA Aminooxyacetate
ATP Adenosine triphosphate

BPTES Bis-2-(5-phenylacetamido-1,2,4-thiadiazol-2-yl) ethyl sulfide
EIPA 5 *N*-ethyl-*N*-isopropylamiloride
FAS Fatty acid synthetase
Gfpt Glutamine fructose-6-phosphate-aminotransferase
GlcNAc *N*-acetylglucosamine
GLS Glutaminase
GLUD1 Glutamine dehydrogenase
GLUT Glucose transporter
GOT1 Cytosolic aspartate transaminase
GOT2 Mitochondrial aspartate transaminase
HBP Hexosamine biosynthesis pathway
HIF Hypoxia-inducible factor
HK Hexokinase
LDH Lactate dehydrogenase
MAPK Mitogen-activated protein kinase
MDH1 Malate dehydrogenase 1
ME1 Malic enzyme
mTORC Mammalian target of rapamycin
NADPH Nicotinamide adenine dinucleotide phosphate
OAA Oxaloacetate
OXPHOS Oxidative phosphorylation
PDH Pyruvate dehydrogenase
PDK Pyruvate dehydrogenase kinase
PFK2 Phosphofructokinase 2
PI3K Phosphoinositide 3-kinase
PPP Pentose phosphate pathway
ROS Reactive oxygen species
TCA Tricarboxylic acid
TFAM Mitochondrial transcription factor A
TXNIP Thioredoxin-interacting protein
αKG Alpha ketoglutarate

References

[1] Kimmelman AC. Metabolic dependencies in RAS-driven cancers. Clin Cancer Res April 15, 2015;21(8):1828–34.

[2] Qiu B, Simon MC. Oncogenes strike a balance between cellular growth and homeostasis. Semin Cell Dev Biol August 13, 2015;43:3–10.

[3] Hanahan D, Weinberg RA. Hallmarks of cancer: the next generation. Cell March 04, 2011;144(5):646–74.

[4] Vander Heiden MG, Cantley LC, Thompson CB. Understanding the Warburg effect: the metabolic requirements of cell proliferation. Science (New York, NY) May 22, 2009;324(5930):1029–33.

[5] Ward PS, Thompson CB. Metabolic reprogramming: a cancer hallmark even Warburg did not anticipate. Cancer Cell March 20, 2012;21(3):297–308.

[6] Warburg O. On respiratory impairment in cancer cells. Science (New York, NY) August 10, 1956;124(3215):269–70.

[7] Moreno-Sanchez R, Rodriguez-Enriquez S, Marin-Hernandez A, Saavedra E. Energy metabolism in tumor cells. FEBS J March 2007;274(6):1393–418.

[8] DeBerardinis RJ, Lum JJ, Hatzivassiliou G, Thompson CB. The biology of cancer: metabolic reprogramming fuels cell growth and proliferation. Cell Metab January 2008;7(1):11–20.

[9] Buzzai M, Bauer DE, Jones RG, Deberardinis RJ, Hatzivassiliou G, Elstrom RL, et al. The glucose dependence of Akt-transformed cells can be reversed by pharmacologic activation of fatty acid beta-oxidation. Oncogene June 16, 2005;24(26):4165–73. PMID: 15806154.

[10] Elstrom RL, Bauer DE, Buzzai M, Karnauskas R, Harris MH, Plas DR, et al. Akt stimulates aerobic glycolysis in cancer cells. Cancer Res June 01, 2004;64(11):3892–9.

[11] Robey RB, Hay N. Is Akt the "Warburg kinase"?-Akt-energy metabolism interactions and oncogenesis. Semin Cancer Biol February 2009;19(1):25–31.

[12] Khatri S, Yepiskoposyan H, Gallo CA, Tandon P, Plas DR. FOXO3a regulates glycolysis via transcriptional control of tumor suppressor TSC1. J Biol Chem May 21, 2010;285(21):15960–5.

[13] Guertin DA, Sabatini DM. Defining the role of mTOR in cancer. Cancer Cell July 2007;12(1):9–22.

[14] Duvel K, Yecies JL, Menon S, Raman P, Lipovsky AI, Souza AL, et al. Activation of a metabolic gene regulatory network downstream of mTOR complex 1. Mol Cell July 30, 2010;39(2):171–83.

[15] Inoki K, Corradetti MN, Guan KL. Dysregulation of the TSC-mTOR pathway in human disease. Nat Genet January 2005;37(1):19–24.

[16] Kapitsinou PP, Haase VH. The VHL tumor suppressor and HIF: insights from genetic studies in mice. Cell Death Differ April 2008;15(4):650–9.

[17] Selak MA, Armour SM, MacKenzie ED, Boulahbel H, Watson DG, Mansfield KD, et al. Succinate links TCA cycle dysfunction to oncogenesis by inhibiting HIF-alpha prolyl hydroxylase. Cancer Cell January 2005;7(1):77–85.

[18] O'Rourke JF, Pugh CW, Bartlett SM, Ratcliffe PJ. Identification of hypoxically inducible mRNAs in HeLa cells using differential-display PCR. Role of hypoxia-inducible factor-1. Eur J Biochem/FEBS October 15, 1996;241(2):403–10.

[19] Semenza GL, Roth PH, Fang HM, Wang GL. Transcriptional regulation of genes encoding glycolytic enzymes by hypoxia-inducible factor 1. J Biol Chem September 23, 1994;269(38):23757–63.

[20] Denko NC. Hypoxia, HIF1 and glucose metabolism in the solid tumour. Nat Rev Cancer September 2008;8(9):705–13.

[21] Kim JW, Tchernyshyov I, Semenza GL, Dang CV. HIF-1-mediated expression of pyruvate dehydrogenase kinase: a metabolic switch required for cellular adaptation to hypoxia. Cell Metab March 2006;3(3):177–85.

[22] Papandreou I, Cairns RA, Fontana L, Lim AL, Denko NC. HIF-1 mediates adaptation to hypoxia by actively downregulating mitochondrial oxygen consumption. Cell Metab March 2006;3(3):187–97.

[23] Dang CV, Kim JW, Gao P, Yustein J. The interplay between MYC and HIF in cancer. Nat Rev Cancer January 2008;8(1):51–6.

[24] Osthus RC, Shim H, Kim S, Li Q, Reddy R, Mukherjee M, et al. Deregulation of glucose transporter 1 and glycolytic gene expression by c-Myc. J Biol Chem July 21, 2000;275(29):21797–800.

[25] Shim H, Dolde C, Lewis BC, Wu CS, Dang G, Jungmann RA, et al. c-Myc transactivation of LDH-A: implications for tumor metabolism and growth. Proc Natl Acad Sci USA June 24, 1997;94(13):6658–63.

[26] Gao P, Tchernyshyov I, Chang TC, Lee YS, Kita K, Ochi T, et al. c-Myc suppression of miR-23a/b enhances mitochondrial glutaminase expression and glutamine metabolism. Nature April 09, 2009;458(7239):762–5.

[27] Parmenter TJ, Kleinschmidt M, Kinross KM, Bond ST, Li J, Kaadige MR, et al. Response of BRAF mutant melanoma to BRAF inhibition is mediated by a network of transcriptional regulators of glycolysis. Cancer Discov January 27, 2014;4(4):423–33.

[28] Haq R, Shoag J, Andreu-Perez P, Yokoyama S, Edelman H, Rowe GC, et al. Oncogenic BRAF regulates oxidative metabolism via PGC1alpha and MITF. Cancer Cell March 18, 2013;23(3):302–15.

[29] Land H, Parada LF, Weinberg RA. Tumorigenic conversion of primary embryo fibroblasts requires at least two cooperating oncogenes. Nature August 18–24, 1983;304(5927):596–602.

[30] Castellano E, Downward J. RAS interaction with PI3K: more than just another effector pathway. Genes Cancer March 2011;2(3):261–74.

[31] Yun J, Rago C, Cheong I, Pagliarini R, Angenendt P, Rajagopalan H, et al. Glucose deprivation contributes to the development of KRAS pathway mutations in tumor cells. Science (New York, NY) September 18, 2009;325(5947):1555–9.

[32] Chiaradonna F, Sacco E, Manzoni R, Giorgio M, Vanoni M, Alberghina L. RAS-dependent carbon metabolism and transformation in mouse fibroblasts. Oncogene August 31, 2006;25(39):5391–404.

[33] Gaglio D, Metallo CM, Gameiro PA, Hiller K, Danna LS, Balestrieri C, et al. Oncogenic K-Ras decouples glucose and glutamine metabolism to support cancer cell growth. Mol Syst Biol 2011;7:523.

[34] Chen C, Pore N, Behrooz A, Ismail-Beigi F, Maity A. Regulation of glut1 mRNA by hypoxia-inducible factor-1. Interaction between H-ras and hypoxia. J Biol Chem March 23, 2001;276(12):9519–25.

[35] Telang S, Yalcin A, Clem AL, Bucala R, Lane AN, Eaton JW, et al. Ras transformation requires metabolic control by 6-phosphofructo-2-kinase. Oncogene November 23, 2006;25(55): 7225–34. PMID: 16715124.

[36] Ying H, Kimmelman AC, Lyssiotis CA, Hua S, Chu GC, Fletcher-Sananikone E, et al. Oncogenic Kras maintains pancreatic tumors through regulation of anabolic glucose metabolism. Cell April 27, 2012;149(3):656–70.

[37] DeBerardinis RJ, Mancuso A, Daikhin E, Nissim I, Yudkoff M, Wehrli S, et al. Beyond aerobic glycolysis: transformed cells can engage in glutamine metabolism that exceeds the requirement for protein and nucleotide synthesis. Proc Natl Acad Sci USA December 04, 2007;104(49):19345–50.

[38] Yuneva M, Zamboni N, Oefner P, Sachidanandam R, Lazebnik Y. Deficiency in glutamine but not glucose induces MYC-dependent apoptosis in human cells. J Cell Biol July 02, 2007;178(1):93–105.

[39] Gaglio D, Soldati C, Vanoni M, Alberghina L, Chiaradonna F. Glutamine deprivation induces abortive s-phase rescued by deoxyribonucleotides in k-ras transformed fibroblasts. PLoS One 2009;4(3):e4715.

[40] Weinberg F, Hamanaka R, Wheaton WW, Weinberg S, Joseph J, Lopez M, et al. Mitochondrial metabolism and ROS generation are essential for Kras-mediated tumorigenicity. Proc Natl Acad Sci USA May 11, 2010;107(19):8788–93.

[41] Sabharwal SS, Schumacker PT. Mitochondrial ROS in cancer: initiators, amplifiers or an Achilles' heel? Nat Rev Cancer November 2014;14(11):709–21.

[42] Fan J, Ye J, Kamphorst JJ, Shlomi T, Thompson CB, Rabinowitz JD. Quantitative flux analysis reveals folate-dependent NADPH production. Nature June 12, 2014;510(7504):298–302.

[43] Lewis CA, Parker SJ, Fiske BP, McCloskey D, Gui DY, Green CR, et al. Tracing compartmentalized NADPH metabolism in the cytosol and mitochondria of mammalian cells. Mol Cell July 17, 2014;55(2):253–63.

[44] Son J, Lyssiotis CA, Ying H, Wang X, Hua S, Ligorio M, et al. Glutamine supports pancreatic cancer growth through a KRAS-regulated metabolic pathway. Nature April 04, 2013;496(7443):101–5.

[45] DeNicola GM, Karreth FA, Humpton TJ, Gopinathan A, Wei C, Frese K, et al. Oncogene-induced Nrf2 transcription promotes ROS detoxification and tumorigenesis. Nature July 07, 2011;475(7354):106–9.

[46] Mitsuishi Y, Taguchi K, Kawatani Y, Shibata T, Nukiwa T, Aburatani H, et al. Nrf2 redirects glucose and glutamine into anabolic pathways in metabolic reprogramming. Cancer Cell July 10, 2012;22(1):66–79.

[47] Guillaumond F, Leca J, Olivares O, Lavaut MN, Vidal N, Berthezene P, et al. Strengthened glycolysis under hypoxia supports tumor symbiosis and hexosamine biosynthesis in pancreatic adenocarcinoma. Proc Natl Acad Sci USA March 05, 2013;110(10):3919–24.

[48] Rabinowitz JD, White E. Autophagy and metabolism. Science (New York, NY) December 03, 2010;330(6009):1344–8.

[49] Guo JY, Chen HY, Mathew R, Fan J, Strohecker AM, Karsli-Uzunbas G, et al. Activated Ras requires autophagy to maintain oxidative metabolism and tumorigenesis. Genes Dev March 01, 2011;25(5):460–70.

[50] Kim MJ, Woo SJ, Yoon CH, Lee JS, An S, Choi YH, et al. Involvement of autophagy in oncogenic K-Ras-induced malignant cell transformation. J Biol Chem April 15, 2011;286(15):12924–32.

[51] Lock R, Roy S, Kenific CM, Su JS, Salas E, Ronen SM, et al. Autophagy facilitates glycolysis during Ras-mediated oncogenic transformation. Mol Biol Cell January 15, 2011;22(2): 165–78.

[52] Bar-Sagi D, Feramisco JR. Induction of membrane ruffling and fluid-phase pinocytosis in quiescent fibroblasts by ras proteins. Science (New York, NY) September 05, 1986;233(4768):1061–8.

[53] Commisso C, Davidson SM, Soydaner-Azeloglu RG, Parker SJ, Kamphorst JJ, Hackett S, et al. Macropinocytosis of protein is an amino acid supply route in Ras-transformed cells. Nature May 30, 2013;497(7451):633–7.

[54] Brunelli L, Caiola E, Marabese M, Broggini M, Pastorelli R. Capturing the metabolomic diversity of KRAS mutants in non-small-cell lung cancer cells. Oncotarget July 15, 2014;5(13):4722–31.

[55] Zhao Y, Butler EB, Tan M. Targeting cellular metabolism to improve cancer therapeutics. Cell Death Dis 2013;4:e532.

[56] Vander Heiden MG. Targeting cancer metabolism: a therapeutic window opens. Nat Rev Drug Discov September 2011;10(9):671–84.

[57] Robert C, Karaszewska B, Schachter J, Rutkowski P, Mackiewicz A, Stroiakovski D, et al. Improved overall survival in melanoma with combined dabrafenib and trametinib. N Engl J Med 2015;372(1):30–9.

[58] Viale A, Pettazzoni P, Lyssiotis CA, Ying H, Sanchez N, Marchesini M, et al. Oncogene ablation-resistant pancreatic cancer cells depend on mitochondrial function. Nature October 30, 2014;514(7524):628–32.

[59] Andrzejewski S, Gravel SP, Pollak M, St-Pierre J. Metformin directly acts on mitochondria to alter cellular bioenergetics. Cancer Metab 2014;2:12.

[60] Zhou G, Myers R, Li Y, Chen Y, Shen X, Fenyk-Melody J, et al. Role of AMP-activated protein kinase in mechanism of metformin action. J Clin Invest October 2001;108(8):1167–74.

[61] Vujic I, Sanlorenzo M, Posch C, Esteve-Puig R, Yen AJ, Kwong A, et al. Metformin and trametinib have synergistic effects on cell viability and tumor growth in NRAS mutant cancer. Oncotarget January 20, 2015;6(2):969–78.

[62] Robinson MM, McBryant SJ, Tsukamoto T, Rojas C, Ferraris DV, Hamilton SK, et al. Novel mechanism of inhibition of rat kidney-type glutaminase by bis-2-(5-phenylacetamido-1,2,4-thiadiazol-2-yl)ethyl sulfide (BPTES). Biochem J September 15, 2007;406(3): 407–14.

[63] Stalnecker CA, Ulrich SM, Li Y, Ramachandran S, McBrayer MK, DeBerardinis RJ, et al. Mechanism by which a recently discovered allosteric inhibitor blocks glutamine metabolism in transformed cells. Proc Natl Acad Sci USA January 13, 2015;112(2):394–9.

[64] Gross MI, Demo SD, Dennison JB, Chen L, Chernov-Rogan T, Goyal B, et al. Antitumor activity of the glutaminase inhibitor CB-839 in triple-negative breast cancer. Mol Cancer Ther April 2014;13(4):890–901.

[65] Jacque N, Ronchetti AM, Larrue C, Meunier G, Birsen R, Willems L, et al. Targeting glutaminolysis has anti-leukemic activity in acute myeloid leukemia and synergizes with BCL-2 inhibition. Blood July 17, 2015;126(11):1346–56.

[66] Yang A, Rajeshkumar NV, Wang X, Yabuuchi S, Alexander BM, Chu GC, et al. Autophagy is critical for pancreatic tumor growth and progression in tumors with p53 alterations. Cancer Discov August 2014;4(8):905–13.

[67] Koivusalo M, Welch C, Hayashi H, Scott CC, Kim M, Alexander T, et al. Amiloride inhibits macropinocytosis by lowering submembranous pH and preventing Rac1 and Cdc42 signaling. J Cell Biol February 22, 2010;188(4):547–63.

CHAPTER 12

Blocking SIAH Proteolysis, an Important K-RAS Vulnerability, to Control and Eradicate K-RAS-Driven Metastatic Cancer

R.E. Van Sciver[1,a], M.M. Njogu[1,a], A.J. Isbell[1], J.J. Odanga[1], M. Bian[1], E. Svyatova[1], L.L. Siewertsz van Reesema[1], V. Zheleva[1], J.L. Eisner[1], J.K. Bruflat[2], R.L. Schmidt[3], A.M. Tang-Tan[4], A.H. Tang[1,b]

[1]Eastern Virginia Medical School, Norfolk, VA, United States; [2]Cellular and Molecular Immunology Laboratory, Rochester, MN, United States; [3]Upper Iowa University, Fayette, IA, United States; [4]Princess Anne High School, Virginia Beach, VA, United States

INTRODUCTION

Attacking the "Achilles' Heel" of Oncogenic K-RAS Signaling Pathway in Cancer

The dismal prognosis of patients diagnosed with invasive and metastatic cancer points to our limited arsenal of effective anti-cancer therapies. Metastatic cancer is responsible for >90% of all cancer-related deaths in the United States [1]. Over the past decade, cancer drug discovery has undergone a paradigm shift from targeting "organ-specific tumors" to targeting "specific oncogenes/tumor suppressors." These precision-driven and personalized approaches are possible because of the discovery of oncogenes and tumor suppressor genes; however, the complex biological system alteration, rapid signaling network adaptation, and dynamic change in tumor heterogeneity, immune cell infiltration, and tumor microenvironment have thwarted many molecular targeted therapies [2–5]. A better molecular understanding of the pivotal oncogenic signaling events and concerted cancer signaling network cross talk that promote tumor initiation, progression, and metastasis should give rise to novel therapeutic interventions [6]. One of the major breakthroughs that led to this paradigm shift was the discovery of "RAS proto-oncogene and RAS oncogenes" in cancer biology [7–14].

CONTENTS

---

[a] These two authors share co-first authorship with equal contribution.

[b] Financial Support: A.H.T and this work are supported by National Institutes of Health (R01-CA140550), Pancreatic Cancer Action Network-AACR Innovative Grant (#169458). M.M.N. is supported by UNCF/Merck Graduate Science Research Dissertation Fellowship.

Conquering RAS. http://dx.doi.org/10.1016/B978-0-12-803505-4.00012-6

Rat sarcoma (RAS) proteins (K-RAS, H-RAS, or N-RAS) are evolutionarily conserved small GTPases that act as molecular switches (GTP-bound active to a GDP-bound inactive state) to transmit signals from receptor tyrosine kinases (RTKs), such as epidermal growth factor receptor (EGFR)/human epidermal growth factor receptor 2 (HER2), which control cell proliferation, differentiation, motility, longevity, and survival in all multi-cellular organisms during development [15–23]. Constitutively active forms of RAS are oncogenic and are among the most common genetic alterations detected in human cancers [24–26]. Oncogenic RAS signaling is detected in ~90% of pancreatic adenocarcinomas, ~50% of colorectal and thyroid cancers, ~80% malignant neoplasias, and ~30% of all human cancers [24,27,28]. For the past three decades, inhibition of oncogenic RAS proteins has remained the subject of intense study and a coveted prize in science and medicine. Despite the central importance of K-RAS signaling in human cancer, a direct inhibition strategy of the seemingly invincible oncogenic K-RAS has been difficult to achieve with clinical efficacy after 30 years of intense research [29–32]. Direct inhibitors targeting the RAS small GTPase itself remain elusive due to RAS enzymatic kinetics, the lack of binding pockets for small molecule inhibitors, and the high binding affinity for GDP and GTP with slow removal rates [32]. As an alternative anti-K-RAS approach, many research efforts have focused on targeting the downstream RAS effector pathways, such as rapidly accelerated fibrosarcoma (RAF)-MAPK/ERK kinase (MEK)-mitogen-activated protein kinase (MAPK), phosphoinositide 3-kinase (PI3K)-protein kinase B (AKT)-mammalian target of rapamycin (mTOR), ras related protein (Ral) guanine nucleotide dissociation stimulator (GDS), Rac/Rho/Cdc42 small GTPase pathways in the hope of controlling aberrant K-RAS signaling [33–35]. However, the approach of targeting these kinases downstream of K-RAS has generally been disappointing. Such efforts tend to result in compensatory activation and feedback mechanisms that appear to further activate RAS signaling in a context-dependent manner, and increased toxicity in combination therapies [36]. Moreover, there have been intense efforts to identify and target additional oncogenic K-RAS signaling "bottlenecks" or pathway vulnerabilities using synthetic lethal screenings, which have shown limited efficacy [37–39]. Since K-RAS-driven tumors often exhibit drug resistance to conventional therapies and carry a poor prognosis [40–43], it is imperative to find effective therapies to control highly aggressive oncogenic K-RAS-driven tumors in the absence of p53 tumor suppressor function.

Here, we provide an overview and the scientific rationale of how we developed an innovative anti-K-RAS and anti-cancer strategy by attacking the "Achilles' heel" of the K-RAS signaling pathway. Similar to its counterpart in the *Drosophila* RAS signaling pathway, we found that Seven In Absentia Homolog (SIAH) E3 ligase is a logical and effective drug target whose enzymatic function is a new and critically important K-RAS "vulnerability" that we can target to treat and control high-grade, relapsed, and resistant human cancer. By attacking this highly evolutionarily conserved signaling "gatekeeper" mechanism, we are able to achieve impressive anti-tumor efficacy against oncogenic K-RAS-driven cancers in pre-clinical studies.

Historic Overview: RAS Signaling During *Drosophila* Eye Development

Genetic studies in *Drosophila melanogaster* and *Caenorhabditis elegans* have identified many key components of the RAS pathway and characterized their functions in mediating RAS signals [44,45]. The developing *Drosophila* retina is a well-established and genetically tractable system in which the RAS signal transduction cascade can be delineated [46–48] (Fig. 12.1). The *Drosophila* compound eye is composed of 800 unit eyes or ommatidia, each comprising eight photoreceptor cells designated R1–R8, four lens-secreting cone cells and eight accessory cells [49]. The *Drosophila* eye is a dispensable and non-essential organ that provides a sensitive in vivo assay system to measure the strength of RAS activation by scoring the size, neuron number, and exterior morphology of the adult eye. This approach has worked well for deciphering the RAS signal transduction pathway because expression of RAS pathway components in the developing eye often results in dose-sensitive rough eye phenotypes, providing a starting point for genetic modifier screens [50–55]. During the third larval instar, proper neuronal differentiation of the developing eye imaginal discs (precursors to the adult eyes) depends on several well-orchestrated spatio-temporal signaling cascades that control proper cell fate determination [46,56]. Photoreceptor cells are recruited sequentially and acquire their distinctive cell fates through a series of local inductive events [57]. Although the development of all photoreceptors requires the RAS signaling pathway, the development of the R7 photoreceptor has shown exquisite sensitivity to RAS signaling, and thus it is the most extensively characterized cell model system to study RAS activation and inactivation in vivo [45,58]. The five precursor cells of the R7 equivalence group choose between two alternative fates: they will develop into an R7 photoreceptor if the RAS pathway is appropriately activated; otherwise these cells will adopt a non-neuronal cone cell fate without sufficient RAS activation signal [46,59].

R7 specification is initiated by Sevenless (SEV) Receptor Tyrosine Kinase (RTK) activation upon contact with its ligand, Bride of Sevenless (BOSS), which is expressed on the surface of the neighboring R8 photoreceptor [57,60,61]. SEV-RTK in turn induces RAS activation, which then initiates a cascade of successive phosphorylation events that result in MAPK activation [50]. MAPK accumulates in the nucleus, where it phosphorylates the E26 transformation-specific sequence (ETS) family of transcriptional factors, Jun proto-oncogene (JUN) and Pointed (Pnt), whose activities are positively required for photoreceptor development, and anterior open (YAN), a negative regulator of photoreceptor fate [62–64]. These three proteins are thought to regulate the transcription of another nuclear factor, Phyllopod (PHYL). Two other nuclear components have been identified that act downstream of PHYL, Seven-In-Absentia (SINA) and Tramtrack (TTK^{88}) [65–67] (Fig. 12.1).

Genetic studies of *Drosophila* photoreceptor development have identified the major key signaling components downstream of the RAS pathway [46,58]. From an elegant genetic screen, four key signaling components of the RAS pathway emerged—Sevenless (EGFR receptor), Ras1, son of sevenless (SOS), and Sina [50,57,65,68,69]. These important genetic screens demonstrated for the first time that Ras1 functions downstream of RTK, and provided an intellectual framework for constructing the RAS signal transduction cascade [50,70]. Interestingly, Sevenless receptor and SINA E3 ligase share identical mutant phenotypes (as their names imply), leading to disruption of the normal trapezoidal assembly of the ommatidium due to the lack of an R7 photoreceptor cell [45]. Both these mutant phenotypes result in similarly misaligned arrays of ommatidial structure, even though SEV (ie, the EGFR membrane receptor) is the most upstream signaling component, and SINA is the most downstream signaling component identified in the *Drosophila* RAS pathway [50,65]. The genetic epistasis established that SINA is the most downstream signaling component identified in the *Drosophila* RAS pathway [45,46]. In the absence of proper SINA function, activation of upstream RAS signal from Sevenless/Ras/Raf/Mek/Erk cannot be transmitted to support *Drosophila* R7 photoreceptor cell development, underscoring SINA's most downstream "gatekeeper" function in *Drosophila* RAS signal transmission [46,71]. Initially, SINA was "misclassified" as a subclass of zinc-finger transcriptional factors that were believed to bind DNA and regulate gene transcription. In 1997, we showed that SINA interacts with an E2, UbcD1, and targets a key transcriptional neuronal repressor TTK^{88} for ubiquitin-mediated proteolysis and degradation, demonstrating that SINA is not a zinc-finger transcriptional factor but a really interesting new gene (RING)-domain ubiquitin E3 ligase [53,72]. Collectively, the *Drosophila* studies demonstrated that the SINA-dependent proteolytic machinery is the most conserved signaling "gatekeeper" whose function is absolutely required for proper RAS signal transduction and R7 neuronal cell fate specification in *Drosophila* eye development (Fig. 12.1).

Regulated Proteolysis in RAS Signaling Pathway

Our previous work has shown that degradation of TTK^{88}, a neuronal repressor, is induced by PHYL and SINA [72]. We demonstrated that these three proteins physically interact with each other, and that SINA genetically and physically interacts with UbcD1, a ubiquitin E2 conjugating enzyme [72]. In addition, EBI physically interacts with SINA, PHYL, and TTK^{88} [73]. These results support a model in which RAS activation induces the transcription of a pro-proteolytic factor PHYL, which then recruits SINA (a ubiquitin E3 ligase)

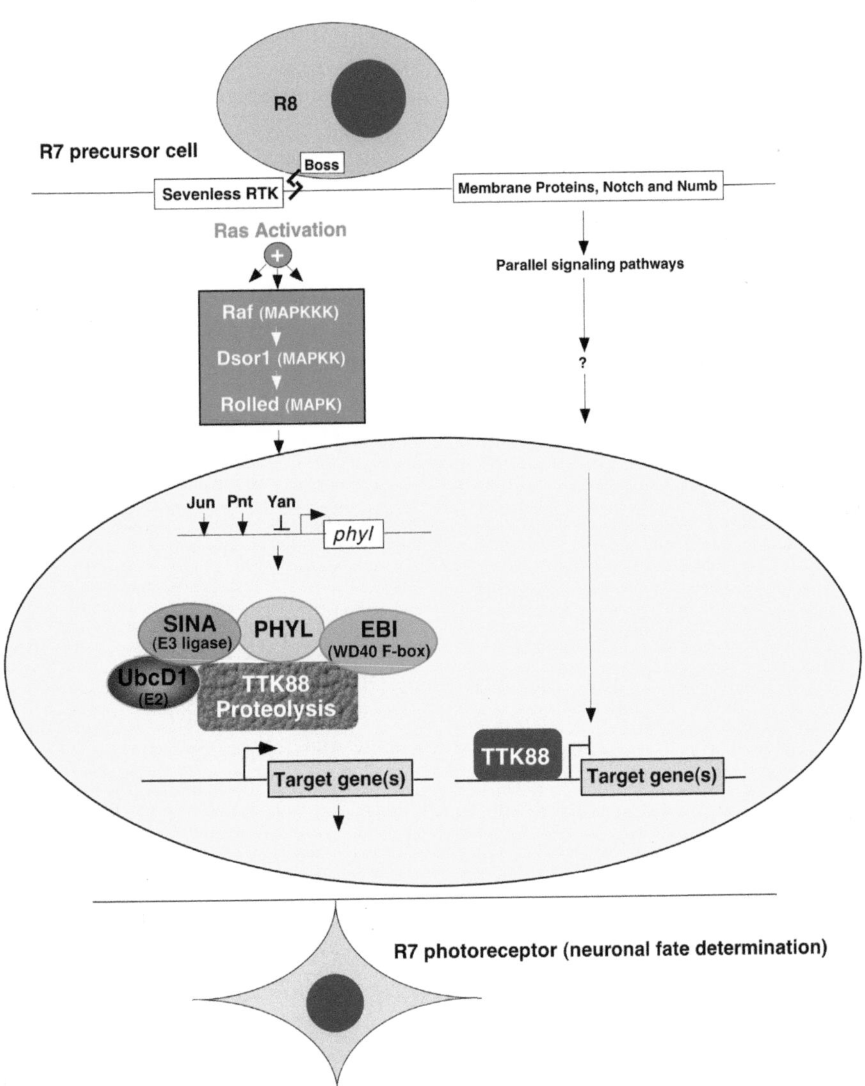

FIGURE 12.1 Schematic representation of RAS-mediated proteolysis in R7 cell fate determination.
The *Drosophila* RAS signaling pathway. The *phyl* gene encodes a nuclear protein that is required for R1, R6, and R7 cell fate determination, as well as for proper development of the embryonic peripheral nervous system and external sensory organ development [66,67,105]. The *phyl* gene is one of the earliest transcriptional targets to be up-regulated by RAS signaling during photoreceptor determination [66,67]. PHYL acts as an adaptor protein to link SINA to its substrate TTK [72,74]. The *sina* gene encodes a conserved RING domain E3 ligase that is essential for R7 photoreceptor cell fate specification [65,106]. SINA is one of the most downstream components in the RAS pathway. Loss-of-function mutations in *sina* block the ability of activated RAS1 [71], activated MAPK [107], ectopically expressed PHYL [66,67], or loss of the repressor YAN [108] to induce R7 cell formation. Thus SINA acts genetically downstream of these RAS pathway components. The *ebi* gene encodes an evolutionarily conserved WD40 and F-box protein [109]. EBI promotes the EGFR-dependent degradation of TTK^{88} and the Su(H)/SMRTER complex [109,110]. It interacts with SINA and PHYL to promote TTK^{88} degradation [73]. The *ttk* gene encodes two alternatively spliced zinc-finger transcription repressors (TTK^{88} and TTK^{69}) [111–114]. TTK^{69} and Pointed act as an EGFR-dependent transcriptional switch to regulate mitosis [115]. Ectopic expression of the TTK^{69} isoform blocks both central nervous system cell proliferation and neuronal development [116], whereas ectopic expression of the TTK^{88} isoform specifically represses the R7 cell fate [72].

and EBI (a WD40/F-Box protein) proteins to target the transcriptional repressor TTK[88] for ubiquitin-dependent proteolysis, thereby promoting neuronal cell fate specification due to a loss of TTK[88] (Fig. 12.1) [72]. First, RAS activation induces *phyl* gene transcription. Then PHYL, in conjunction with SINA, EBI, and UbcD1, targets a specific neuronal repressor, TTK[88], for degradation, and therefore sets off the neuronal cell fate determination program in the compound eye in response to RAS activation [72,74]. Owing to the extensive evolutionary conservation of the RAS-SINA signaling pathway, we hypothesize that this scheme of regulated proteolysis required for proper RAS signal transduction uncovered in *Drosophila* is likely to be conserved and applicable to mammalian systems including K-RAS-driven human cancer (Fig. 12.2).

SINA belongs to an evolutionarily conserved family of RING-domain E3 ubiquitin ligases; the human genome has two SIAHs, SIAH-1 and SIAH-2, which share 76% and 68% amino acid identity to *Drosophila* SINA, respectively [75]. As an E3 ligase, SINA/SIAH confers specificity to proteosome degradation of substrates and is required for the ubiquitin-dependent regulated proteolysis in the RAS signaling pathway [72]. The evolutionarily conserved nature of the K-RAS signaling pathway makes SIAH an excellent therapeutic drug target; the evolutionary constraint suggests that SIAH must have a critically important function in relaying RAS activation signal in human proliferating cells. We hypothesize that by targeting the most conserved and most downstream signaling module in the K-RAS pathway, SIAH E3 ligase, we may be able to circumvent the elaborate K-RAS signaling cross talk and rapid systematic adaptations to develop novel and potent therapeutics to shutdown oncogenic K-RAS signaling. By deploying anti-SIAH blockade downstream of RAS activation, metastatic cancer cells may not be able to escape or overcome such a strong chokehold at SIAH signaling "gatekeeper" (Figs. 12.2 and 12.3).

Breakthrough: Developing Anti-SIAH-Based Anti-K-RAS Cancer Therapy

As a major growth-promoting pathway, the central importance of the EGFR/HER2/K-RAS signal transduction cascade has been well established in human cancers [26,76–79]. Hyper-activated RAS is known to drive neoplastic transformation, tumorigenesis, and metastasis in 30% of all human cancers and 90% of pancreatic cancers. Thus, how best to contravene activated EGFR family (ERBB)/RAS signaling has been an intense area of investigation in the field of human cancer biology for three decades [29]. To date, covalent inhibitors that directly bind to hyper-activated K-RAS have shown conceptual promise, and hopefully clinical successes will be forthcoming [32,80]. Currently, many conventional and targeted anti-cancer therapies do not work in human tumors with both oncogenic K-RAS activation and p53 loss of function. Discovering alternative

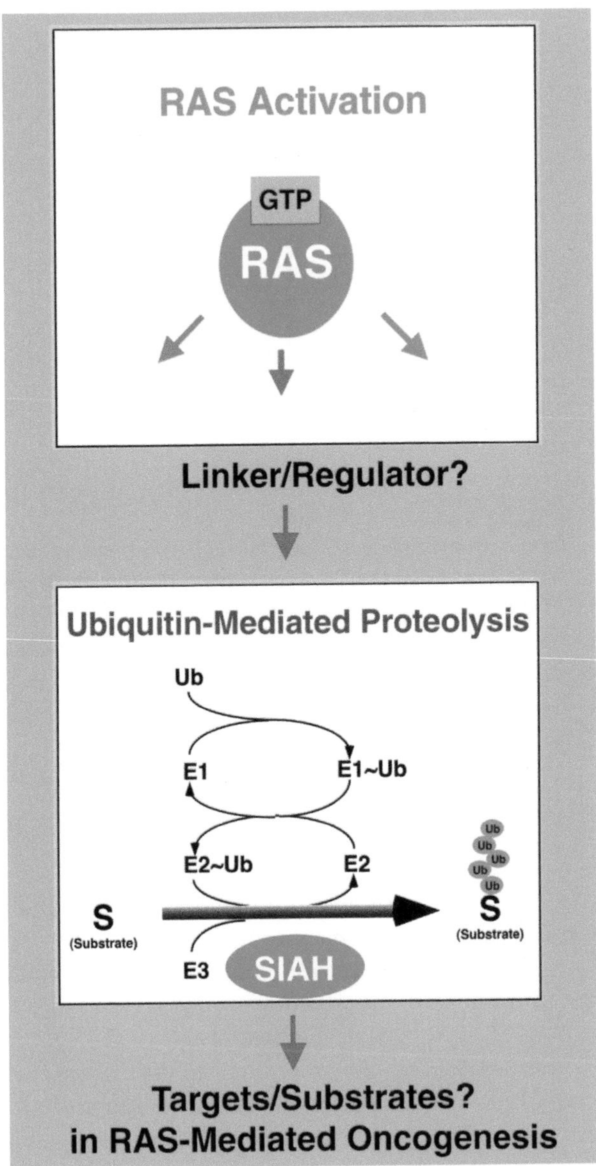

FIGURE 12.2 Our working model of SIAH-dependent proteolysis in K-RAS signal transduction. We hypothesize that RAS activation plays an active role in regulating the activity of SIAH-dependent proteolytic machinery in development and cancer biology. Our data support the following model for the function of SIAH-dependent proteolysis downstream of the RAS signaling pathway. Using *Drosophila* eye development, we would like to delineate the regulatory mechanisms by which the RAS pathway and SINA pathway interact, and translate this knowledge to RAS-driven tumorigenesis and metastasis in human cancer.

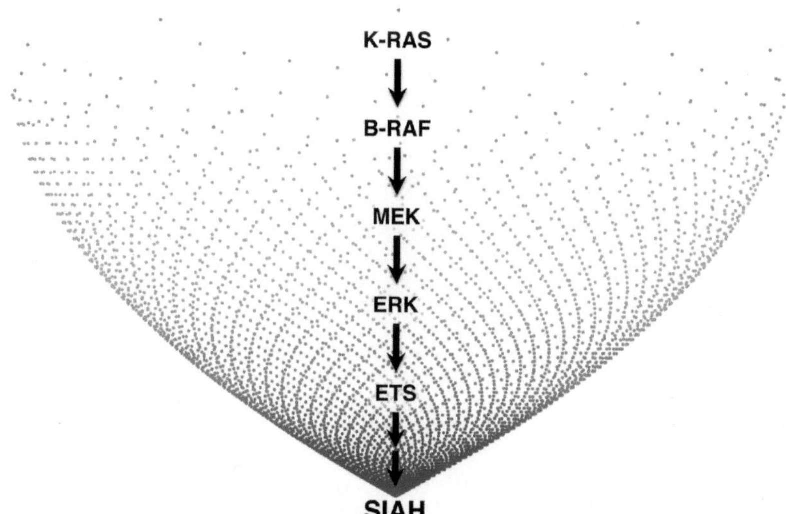

FIGURE 12.3 SIAH is a downstream "gatekeeper" in the K-RAS signaling pathway.
Schematic illustration to show the "gatekeeper" function of SIAH E3 ligase in the oncogenic K-RAS signaling pathway. The *dots* illustrate the complex and dynamic interactions between RAS and other signaling pathways. The critical role of SIAH proteolysis and the funneling constraint to control the K-RAS signal transduction is shown.

approaches, identifying additional K-RAS targets, and developing more durable and effective therapies are urgently needed to control "undruggable" oncogenic K-RAS signal in these high-grade malignant cancers through innovative, targeted, and combinatory therapies.

Proper SIAH function is an integral part of cell proliferation and normal developmental program, ie, the growth-promoting RAS pathway. Cancer cells hijack the K-RAS pathway to enhance its growth potential and proliferation rate and to increase cell motility and survival. Given the central importance and evolutionary conservation of EGFR/RAS/SINA pathway, we sought to determine whether blocking the conserved RAS signaling module with the highest evolutionary constraint impedes K-RAS-mediated cellular transformation and tumorigenesis in human cancer with both K-RAS activation and loss of p53 function. The human homolog of SINA, SIAH, is the most conserved and most downstream signaling "gatekeeper" in the mammalian RAS signal transduction pathway [45,65,72,75].

Guided by the molecular insights and principles obtained from *Drosophila* studies, we examined the biochemical roles and mechanisms of SIAH function in the context of K-RAS activation in human cancer. Instead of targeting an upstream signaling component such as EGFR or HER2 receptors and/or the midstream RAS/RAF/MEK/PI3K effector kinase cascades, we proposed to target the RAS signaling "gatekeeper," the SIAH-dependent proteolytic machinery, to block EGFR/

HER2/K-RAS activation at its downstream signaling bottleneck (Fig. 12.3). We discovered that by inhibiting SIAH E3 ligase function, we could completely abolish tumor growth and cancer metastasis in several of the most aggressive and invasive human cancer cells in soft agar assays as well as in nude mice [81,82]. Thus, we were the first group to demonstrate that SIAH is a key K-RAS vulnerability in human cancer. SIAH is expressed in both proliferating normal and cancer cells, but absent in non-dividing cells independent of tumorigenic potential and cellular stress [81,82]. Through shutting down SIAH proteolysis and constraining the K-RAS signaling pathway at its most downstream signaling "relay center," we can successfully eradicate K-RAS-driven pancreatic and lung tumors in xenograft mouse models [81,82]. Human SIAH2 is essential in regulating the cellular response of K-RAS pathway activation and controlling cellular behaviors such as cell proliferation, motility, invasion, and metastasis [81,82]. Without proper SIAH function, oncogenic K-RAS cannot transmit its signal in human cancer cells. Thus, acting as a critical "relay center" and/or key "bottleneck" of the RAS signaling network "funnel" (Fig. 12.3), proper SIAH2 function is crucial to relay and transduce the hyper-activated K-RAS signal that fuels aggressive tumor growth and cancer metastasis in K-RAS-driven malignant tumors.

Mammalian SIAH Function

Mammalian SIAHs have been implicated in tumorigenesis by interacting with and modulating the stability of potent signaling molecules in oncogenesis including β-catenin, prolyl-4-hydroxylases that control the stability of the hypoxia inducible factor-1α (HIF-1α), tumor necrosis factor receptor 2–associated factor (TRAF2), a NOTCH-interacting membrane protein and cell fate regulator (NUMB), a cyclin-dependent kinase activator called Rapid Inducer of G2/M progression in Oocytes (RINGO), a serine–threonine kinase homeodomain-interacting protein kinase 2 (HIPK2) that functions as a key regulator of DNA damage-induced cell death, and a negative regulator of RTK signaling, Sprouty [83–94]. Decreased expression of Sprouty accelerates tumor malignancy in non–small-cell lung cancer, confirming that RAS pathway activation promotes lung tumor progression [92]. Nonetheless, except for Sprouty [91–93,95], none of the diverse SIAH-interacting proteins identified thus far has been demonstrated to encode a bona fide signaling component in the RAS signaling pathway. Thus, the biological function, regulation, and substrate specificity of SIAHs in the context of K-RAS activation remains to be defined in mammalian systems. Importantly, how SIAH selectively interacts with its binding partners and degrades its substrates in response to normal and oncogenic EGFR/HER2/RAS activation, and how SIAH-dependent proteolysis promotes and facilitates EGFR/HER2/RAS-mediated cell transformation and tumorigenesis remain to be elucidated.

K-RAS-driven tumor cells often develop oncogene addiction and activate cellular adaptation mechanisms to protect themselves from oncogenic K-RAS-induced cellular stress, endoplasmic reticulum (ER) and mitochondrial stress, reactive oxygen and nitrogen species production, and hypoxia. Increased SIAH2 transcription is induced by severe ER stress and SIAH1/2 substrates TRAF2, Sprouty2, and PHD3 have been implicated in the unfolded protein response as part of the ER stress response [96]. SIAH2 has been shown to play an important role in oxidative stress by targeting NRF2, a key regulator of oxidative stress response, in hypoxia [97]. SIAH2 was shown to play a pro-survival role in oral squamous cell carcinoma where its knockdown led to growth suppression and induction of apoptosis in a p53-dependent manner [98]. This cellular adaptation mechanism may be exploited for drug target therapy against SIAH2 E3 ligase as SIAH2 plays a role in oxidative and hypoxia stress-activated signaling pathways [88,99–102]. Understanding SIAH biology and SIAH-dependent stress response in the context of oncogenic K-RAS activation, oncogene addiction, oncogene-induced stress responses, HIF1α signaling, and hypoxia may allow researchers to identify additional "druggable" targets to shutdown K-RAS signaling addiction in K-RAS/B-RAF-driven human tumors.

Impact and Promises of Anti-SIAH-Based Anti-K-RAS Strategy Guided by *Drosophila* RAS Studies

In spite of the amazing successes of the *Drosophila* RAS pathway studies, SIAH did not appear on the radar screen in human cancer biology for 17 years [45,65,72]. In the past 7–8 years, our understanding of the essential roles of SIAH in the genesis of human cancer has advanced significantly. Our group in particular has made significant contributions in this area. To the best of our knowledge, our reports on SIAH's "gatekeeper" function in human pancreatic and lung cancer cells were the first series to demonstrate the essential function of SIAH-dependent proteolysis for K-RAS-driven transformation, tumorigenesis, and metastasis in cancer biology. Ours and other studies have established unequivocally that SIAH is a great biomarker and potent anti-K-RAS therapeutic target in human cancer. Currently, there are no effective ways to treat metastatic cancers that have oncogenic K-RAS activation combined with tumor suppressor losses that confer drug resistance, aggressive tumor growth, systemic metastasis, and poor clinical outcome. Using anti-SIAH molecules, we have successfully abolished both tumorigenesis and metastasis in several of the most lethal forms of human cancer known, such as pancreatic cancer, lung cancer, and invasive and metastatic breast cancer [81,82,103,104].

As a new K-RAS vulnerability, SIAH has every potential to become a highly effective therapeutic anti-K-RAS and anti-cancer drug target in human cancer. Using anti-SIAH molecules to block RAS signaling in human cancer is an excellent example of science going "from the bench (basic science in fruit flies) to the bedside (preclinical studies and future clinical application)." Arising from the extensive genetic studies of the RAS signal transduction pathway in *Drosophila* eye development [72], SIAH is a highly evolutionarily conserved protein (*Drosophila* SINA and human SIAH are almost identical in the SIAH substrate-binding domains) [81]. As an essential downstream "gatekeeper" signaling module required for proper RAS signaling, SIAH is ideally and logically positioned to become the next anti-RAS and anti-cancer target. Our work has laid a solid foundation that clearly establishes the critical role of SIAH, as a new drug target, in RAS signal transduction, tumor growth, and metastasis in human cancer.

We hypothesize that SIAH E3 ligase is a new K-RAS vulnerability, a key bottleneck of K-RAS signaling network, and a logical and potent anti-K-RAS and logical anti-cancer drug target against metastatic cancer (Fig. 12.3). As a necessary step forward, it is critically important that we understand the biochemical function of SIAH E3 ligases; identify SIAH partners, substrates, and regulators; and elucidate the fundamental molecular mechanisms by which SIAH modulates RAS signaling, facilitates cell growth, and promotes cell invasion and migration—properties that are pertinent to ERBB/RAS-mediated tumorigenesis and metastasis in high-grade and malignant cancer cells. SINA and SIAH proteins constitute an evolutionarily conserved family of E3 ligases whose substrates, interacting partners, and regulators/regulatory mechanisms are not well understood. Because many instances of tumorigenesis and malignancy are associated with aberrant K-RAS hyper-activation, it is important to understand how the regulated proteolysis of SINA substrates promotes cell growth and increases cell motility and survival during animal development, neoplastic transformation, and oncogenesis. The mechanism of RAS-directed and SINA-mediated proteolysis is just the beginning to be explored to increase our understanding of K-RAS/SIAH biology to develop anti-SIAH-based therapies to treat metastatic cancer.

Importantly, as an enzyme and ubiquitin E3 ligase, SIAH E3 ligase is a "druggable" target in the development of novel therapeutics for the eradication of oncogenic K-RAS-driven human cancers through the development of anti-SIAH molecules as anti-cancer drugs (Fig. 12.4). Because the substrate specificity of ubiquitin-mediated proteolysis is primarily determined by the E3 ligases, results from our laboratory and many other research groups will provide useful insights into the altered proteolysis associated with human cancer, and may lead to new avenues for cancer diagnosis and SIAH-based anti-K-RAS and anti-cancer therapy. If we can combine anti-EGFR/HER2/RAS therapy with novel anti-SIAH therapy, we may be able to control the dysregulated K-RAS signaling pathway, the major engine that drives cell proliferation and tumor growth in high-grade and malignant cancer. The successful investigation and additional validations will set the stage for a large-scale small molecule screening against SIAH E3 ligases that will hopefully identify effective small molecule inhibitors with high specificity and low toxicity for pre-clinical and clinical studies and ultimately in human clinical trials. We are optimistic that developing a new anti-SIAH-based strategy is likely to lead to rational and efficacious

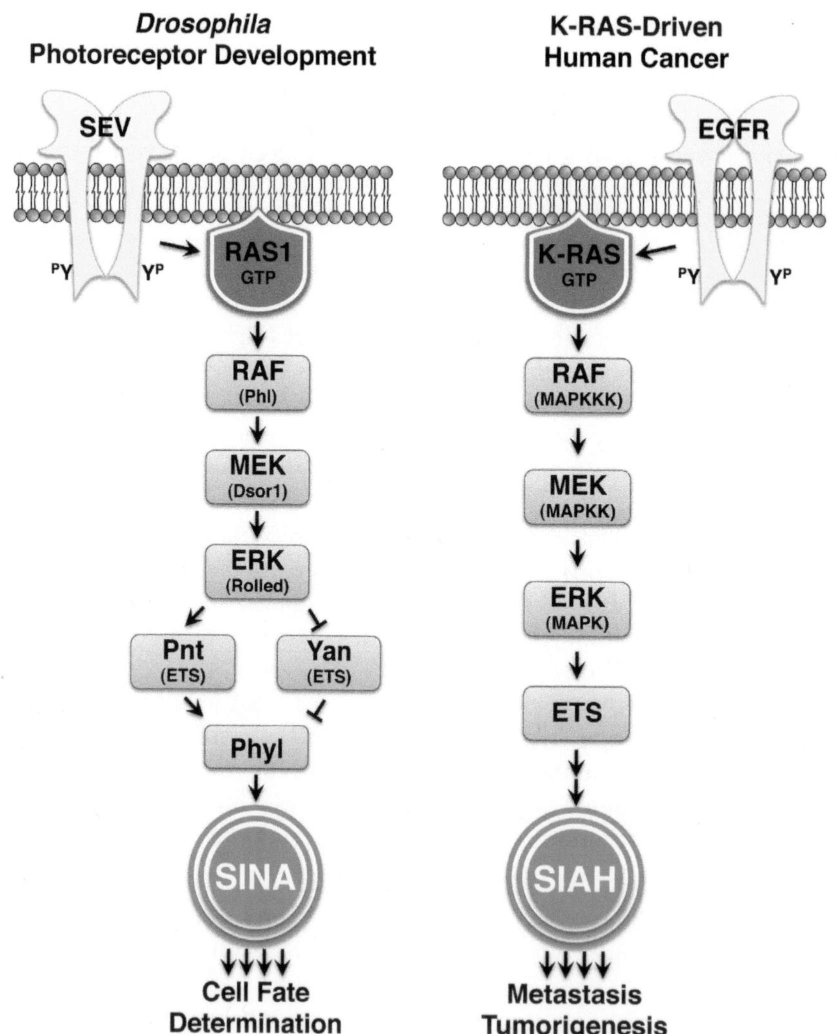

FIGURE 12.4 The evolutionary conservation of *Drosophila* and human RAS signaling pathway.
The human RAS signal transduction pathway is highly conserved with the *Drosophila* RAS pathway, and contains the same downstream signaling modules such as RAS, RAF, MEK, MAPK, and ETS family of transcription factors. (Left panel) Schematic illustration of the RAS signal transduction pathway in *Drosophila*. SINA is the most downstream component identified in the RAS pathway and it is absolutely required for transmission of activated RAS/RAF/MAPK signal in the R7 neuron cell fate determination in the fly. (Right panel) The human RAS pathway has the same signaling modules as the *Drosophila* RAS pathway. The extensive evolutionary conservation suggests that the "gatekeeper" function of the SINA/SIAH is likely to be conserved in humans as well. Thus, blocking SIAH function may impede RAS-mediated tumorigenesis and metastasis in cancer. Our results show that SIAH-dependent proteolysis is critical for proper K-RAS signaling in human cancer. By blocking SIAH function, we can inhibit RAS-dependent tumorigenesis and metastasis in human cancer [81,82]. Targeting SIAH may represent a logical and effective means to contravene oncogenic RAS signaling, providing a novel method of therapeutic intervention in the treatment of K-RAS-driven malignant cancer in the future.

treatment options to control K-RAS-driven malignant tumors in the pre-clinical and clinical settings. Furthermore, by combining anti-SIAH therapy with the existing strategy of targeting EGFR/RAS/MEK and PI3K/AKT/mTOR signaling pathway for inhibition, we may be in a position to develop effective therapeutic regimens to shutdown oncogenic K-RAS signaling, reduce tumor burden, and improve patient survival in the future.

Looking Into the Future: Conquering K-RAS-Driven Metastatic Cancer

SIAH is a new and logical drug target whose enzymatic function is an important RAS vulnerability. As an enzyme and ubiquitin E3 ligase, SIAH presents a prudent and pragmatic "druggable" target for the development of anti-K-RAS therapies against metastatic cancer. These evolutionarily conserved RING domain E3 ligases have begun to emerge as an important RAS signaling "gatekeeper" to control oncogenic K-RAS activation and to halt the most aggressive human cancers in many pre-clinical studies. Our ability to utilize multiple and complementary systems to study molecular details of the RAS signaling transduction from *Drosophila* eye development to human cancer biology is a great strength in our synergistic approaches. Hyper-activated K-RAS proteins are responsible for promoting un-checked cell proliferation, un-controlled tumor growth, and rapid cancer cell dissemination in human cancers. As the "Achilles' heel" and the fundamentally important K-RAS vulnerability identified in the K-RAS signaling network, SIAH is ideally positioned to become a promising and impactful anti-K-RAS drug target to control K-RAS-driven tumors.

We will continue to investigate the roles and regulation of SIAH-dependent proteolysis in the context of K-RAS activation in cancer. We hope that the knowledge and insight will assist to expedite and facilitate bench-to-bedside translation and attract biotech and pharmaceutical companies to invest and develop new anti-SIAH-based drugs to control metastatic human cancer in their drug development pipelines. This will provide a robust research platform from which we can move on to determine the efficacy of anti-SIAH-based therapies against the oncogenic K-RAS activation, malignant tumor growth, and cancer metastasis in the multiple model systems to study RAS-SIAH biology. This project exemplifies the translational research "from the bench to the bedside," fitting perfectly into the new National Cancer Institute (NCI) RAS initiative and National Cancer Moonshot Initiative, as a high-risk and high-reward project. Most importantly, if successful, the impact of this SIAH-centered study will be unprecedented: anti-SIAH molecules have the potential to become effective anti-K-RAS and anti-cancer agents that may benefit many patients with cancer, specifically ones with metastatic cancer carrying both K-RAS and p53 mutations, independent of their cancer subtypes.

List of Acronyms and Abbreviations

AKT Protein kinase B
BOSS Bride of sevenless
EGFR Epidermal growth factor receptor
ERBB Epidermal growth factor receptor (EGFR) family
ETS E26 transformation-specific sequence
HER2 Human epidermal growth factor receptor 2
HIF-1α Hypoxia inducible factor-1α
HIPK2 Homeodomain-interacting protein kinase 2
JUN Jun proto-oncogene
MAPK Mitogen-activated protein kinase
MEK MAPK/ERK kinase
mTOR Mammalian target of rapamycin
NCI National Cancer Institute
NUMB Notch-interacting membrane protein and cell fate regulator
PHYL Phyllopod
PI3K Phosphoinositide 3-kinase
Pnt Pointed
RAF Rapidly accelerated fibrosarcoma
RalGDS Ral (ras related protein) guanine nucleotide dissociation stimulator
RAS Rat sarcoma
RING Really interesting new gene
RINGO Rapid inducer of G2/M progression in oocytes
ROS Reactive oxygen species
RTK Receptor tyrosine kinase
SEV Sevenless
SIAH Seven-in-absentia homolog
SINA Seven-in-absentia
SOS Son of sevenless
TRAF2 Tumor necrosis factor receptor 2-associated factor
TTK Tramtrack
YAN Anterior open

References

[1] Siegel RL, Miller KD, Jemal A. Cancer statistics, 2015. CA Cancer J Clin 2015;65(1):5–29.

[2] Parker JS, Perou CM. Tumor heterogeneity: focus on the leaves, the trees, or the forest? Cancer Cell 2015;28(2):149–50.

[3] Alizadeh AA, Aranda V, Bardelli A, Blanpain C, Bock C, Borowski C, et al. Toward understanding and exploiting tumor heterogeneity. Nat Med 2015;21(8):846–53.

[4] Kitamura T, Qian BZ, Pollard JW. Immune cell promotion of metastasis. Nat Rev Immunol 2015;15(2):73–86.

[5] Almendro V, Marusyk A, Polyak K. Cellular heterogeneity and molecular evolution in cancer. Annu Rev Pathol 2013;8:277–302.

[6] Hanahan D, Weinberg RA. Hallmarks of cancer: the next generation. Cell 2011;144(5):646–74.

[7] Shih C, Weinberg RA. Isolation of a transforming sequence from a human bladder carcinoma cell line. Cell 1982;29(1):161–9.

[8] Goldfarb M, Shimizu K, Perucho M, Wigler M. Isolation and preliminary characterization of a human transforming gene from T24 bladder carcinoma cells. Nature 1982;296(5856):404–9.

[9] Pulciani S, Santos E, Lauver AV, Long LK, Aaronson SA, Barbacid M. Oncogenes in solid human tumours. Nature 1982;300(5892):539–42.

[10] Der CJ, Cooper GM. Altered gene products are associated with activation of cellular rasK genes in human lung and colon carcinomas. Cell 1983;32(1):201–8.

[11] Der CJ, Krontiris TG, Cooper GM. Transforming genes of human bladder and lung carcinoma cell lines are homologous to the ras genes of Harvey and Kirsten sarcoma viruses. Proc Natl Acad Sci USA 1982;79(11):3637–40.

[12] Tabin CJ, Bradley SM, Bargmann CI, Weinberg RA, Papageorge AG, Scolnick EM, et al. Mechanism of activation of a human oncogene. Nature 1982;300(5888):143–9.

[13] Reddy EP, Reynolds RK, Santos E, Barbacid M. A point mutation is responsible for the acquisition of transforming properties by the T24 human bladder carcinoma oncogene. Nature 1982;300(5888):149–52.

[14] Santos E, Martin-Zanca D, Reddy EP, Pierotti MA, Della Porta G, Barbacid M. Malignant activation of a K-ras oncogene in lung carcinoma but not in normal tissue of the same patient. Science 1984;223(4637):661–4.

[15] Barbacid M. Ras genes. Annu Rev Biochem 1987;56:779–827.

[16] Lowy DR, Willumsen BM. Function and regulation of ras. Annu Rev Biochem 1993;62: 851–91.

[17] Bollag G, McCormick F. Regulators and effectors of ras proteins. Annu Rev Cell Biol 1991;7:601–32.

[18] Campbell SL, Khosravi-Far R, Rossman KL, Clark GJ, Der CJ. Increasing complexity of Ras signaling. Oncogene 1998;17(11 Reviews):1395–413.

[19] Hynes NE, Lane HA. ERBB receptors and cancer: the complexity of targeted inhibitors. Nat Rev Cancer 2005;5(5):341–54.

[20] Gschwind A, Fischer OM, Ullrich A. The discovery of receptor tyrosine kinases: targets for cancer therapy. Nat Rev Cancer 2004;4(5):361–70.

[21] Schubbert S, Shannon K, Bollag G. Hyperactive Ras in developmental disorders and cancer. Nat Rev Cancer 2007;7:295–308.

[22] Slack C, Alic N, Foley A, Cabecinha M, Hoddinott MP, Partridge L. The ras-Erk-ETS-signaling pathway is a drug target for longevity. Cell 2015;162(1):72–83.

[23] Ory S, Morrison DK. Signal transduction: implications for Ras-dependent ERK signaling. Curr Biol 2004;14(7):R277–8.

[24] Bos JL. Ras oncogenes in human cancer: a review. Cancer Res 1989;49(17):4682–9.

[25] Santarpia L, Lippman SM, El-Naggar AK. Targeting the MAPK-RAS-RAF signaling pathway in cancer therapy. Expert Opin Ther Targets 2012;16(1):103–19.

[26] Downward J. Targeting RAS signalling pathways in cancer therapy. Nat Rev Cancer 2003;3(1):11–22.

[27] Bardeesy N, DePinho RA. Pancreatic cancer biology and genetics. Nat Rev Cancer 2002;2(12):897–909.

[28] Jiang Y, Kimchi ET, Staveley-O'Carroll KF, Cheng H, Ajani JA. Assessment of K-ras mutation: a step toward personalized medicine for patients with colorectal cancer. Cancer 2009;115(16):3609–17.

[29] Malumbres M, Barbacid M. RAS oncogenes: the first 30 years. Nat Rev Cancer 2003;3(6): 459–65.

[30] Downward J. Cancer biology: signatures guide drug choice. Nature 2006;439(7074):274–5.

[31] Baker NM, Der CJ. Cancer: drug for an 'undruggable' protein. Nature 2013;497(7451):577–8.

[32] McCormick F. Kras as a therapeutic target. Clin Cancer Res 2015;21(8):1797–801.

[33] Tolcher AW, Khan K, Ong M, Banerji U, Papadimitrakopoulou V, Gandara DR, et al. Antitumor activity in RAS-driven tumors by blocking AKT and MEK. Clin Cancer Res 2015;21(4):739–48.

[34] Lito P, Rosen N, Solit DB. Tumor adaptation and resistance to RAF inhibitors. Nat Med 2013;19(11):1401–9.

[35] Chapman PB, Solit DB, Rosen N. Combination of RAF and MEK inhibition for the treatment of BRAF-mutated melanoma: feedback is not encouraged. Cancer Cell 2014;26(5):603–4.

[36] Pylayeva-Gupta Y, Grabocka E, Bar-Sagi D. RAS oncogenes: weaving a tumorigenic web. Nat Rev Cancer 2011;11(11):761–74.

[37] Downward JRAS. Synthetic lethal screens revisited: still seeking the elusive prize? Clin Cancer Res 2015;21(8):1802–9.

[38] Luo J, Emanuele MJ, Li D, Creighton CJ, Schlabach MR, Westbrook TF, et al. A genome-wide RNAi screen identifies multiple synthetic lethal interactions with the Ras oncogene. Cell 2009;137(5):835–48.

[39] Astsaturov I, Ratushny V, Sukhanova A, Einarson MB, Bagnyukova T, Zhou Y, et al. Synthetic lethal screen of an EGFR-centered network to improve targeted therapies. Sci Signal 2010;3(140):ra67.

[40] Wolfgang CL, Herman JM, Laheru DA, Klein AP, Erdek MA, Fishman EK, et al. Recent progress in pancreatic cancer. CA Cancer J Clin 2013;63(5):318–48.

[41] Maitra A, Hruban RH. Pancreatic cancer. Annu Rev Pathol 2008;3:157–88.

[42] De Raedt T, Walton Z, Yecies JL, Li D, Chen Y, Malone CF, et al. Exploiting cancer cell vulnerabilities to develop a combination therapy for ras-driven tumors. Cancer Cell 2011;20(3):400–13.

[43] Adjei AA. Blocking oncogenic Ras signaling for cancer therapy. J Natl Cancer Inst 2001;93(14):1062–74.

[44] Greenwald I, Rubin GM. Making a difference: the role of cell-cell interactions in establishing separate identities for equivalent cells. Cell 1992;68(2):271–81.

[45] Zipursky SL, Rubin GM. Determination of neuronal cell fate: lessons from the R7 neuron of *Drosophila*. Annu Rev Neurosci 1994;17:373–97.

[46] Rubin GM, Chang HC, Karim F, Laverty T, Michaud NR, Morrison DK, et al. Signal transduction downstream from Ras in *Drosophila*. Cold Spring Harb Symp Quant Biol 1997;62:347–52.

[47] Thomas BJ, Wassarman DA. A fly's eye view of biology. Trends Genet 1999;15(5):184–90.

[48] Kumar JP. My what big eyes you have: how the *Drosophila* retina grows. Dev Neurobiol 2011;71(12):1133–52.

[49] Wolff T, Ready DF. Pattern formation in the *Drosophila* retina. In: Bate M, Martins-Arias A, editors. Developmental biology of *Drosphila*. Cold Spring Harbor Press; 1993. p. 1277–325.

[50] Simon MA, Bowtell DD, Dodson GS, Laverty TR, Rubin GM. Ras1 and a putative guanine nucleotide exchange factor perform crucial steps in signaling by the sevenless protein tyrosine kinase. Cell 1991;67(4):701–16.

[51] Dickson B, Hafen E. Genetic dissection of eye development in *Drosophila*. In: Bate M, Martins-Arias A, editors. Developmental biology of *Drosophila*. Cold Spring Harbor Press; 1993. p. 1327–62.

[52] Karim FD, Chang HC, Therrien M, Wassarman DA, Laverty T, Rubin GM. A screen for genes that function downstream of Ras1 during *Drosophila* eye development. Genetics 1996;143(1):315–29.

[53] Neufeld TP, Tang AH, Rubin GM. A genetic screen to identify components of the sina signaling pathway in *Drosophila* eye development. Genetics 1998;148(1):277–86.

[54] Rebay I, Chen F, Hsiao F, Kolodziej PA, Kuang BH, Laverty T, et al. A genetic screen for novel components of the Ras/Mitogen-activated protein kinase signaling pathway that interact with the yan gene of *Drosophila* identifies split ends, a new RNA recognition motif-containing protein. Genetics 2000;154(2):695–712.

[55] Therrien M, Morrison DK, Wong AM, Rubin GM. A genetic screen for modifiers of a kinase suppressor of Ras-dependent rough eye phenotype in *Drosophila*. Genetics 2000;156(3):1231–42.

[56] Morante J, Desplan C, Celik A. Generating patterned arrays of photoreceptors. Curr Opin Genet Dev 2007;17(4):314–9.

[57] Tomlinson A, Ready DF. Sevenless: a cell-specific homeotic mutation of the *Drosophila* eye. Science 1986;231(4736):400–2.

[58] Wassarman DA, Therrien M, Rubin GM. The Ras signaling pathway in *Drosophila*. Curr Opin Genet Dev 1995;5(1):44–50.

[59] Dickson BJ. Photoreceptor development: breaking down the barriers. Curr Biol 1998;8(3):R90–2.

[60] Hafen E, Dickson B, Brunner D, Raabe T. Genetic dissection of signal transduction mediated by the sevenless receptor tyrosine kinase in *Drosophila*. Prog Neurobiol 1994;42(2):287–92.

[61] Reinke R, Zipursky SL. Cell-cell interaction in the *Drosophila* retina: the bride of sevenless gene is required in photoreceptor cell R8 for R7 cell development. Cell 1988;55(2):321–30.

[62] Bohmann D, Ellis MC, Staszewski LM, Mlodzik M. *Drosophila* Jun mediates Ras-dependent photoreceptor determination. Cell 1994;78:973–86.

[63] O'Neill EM, Rebay I, Tjian R, Rubin GM. The activities of two Ets-related transcription factors required for *Drosophila* eye development are modulated by the Ras/MAPK pathway. Cell 1994;78(1):137–47.

[64] Rebay I, Rubin GM. Yan functions as a general inhibitor of differentiation and is negatively regulated by activation of the Ras1/MAPK pathway. Cell 1995;81(6):857–66.

[65] Carthew RW, Rubin GM. Seven in absentia, a gene required for specification of R7 cell fate in the *Drosophila* eye. Cell 1990;63(3):561–77.

[66] Chang HC, Solomon NM, Wassarman DA, Karim FD, Therrien M, Rubin GM, et al. Phyllopod functions in the fate determination of a subset of photoreceptors in *Drosophila*. Cell 1995;80(3):463–72.

[67] Dickson BJ, Dominguez M, van der Straten A, Hafen E. Control of *Drosophila* photoreceptor cell fates by phyllopod, a novel nuclear protein acting downstream of the Raf kinase. Cell 1995;80(3):453–62.

[68] Hafen E, Basler K, Edstroem JE, Rubin GM. Sevenless, a cell-specific homeotic gene of *Drosophila*, encodes a putative transmembrane receptor with a tyrosine kinase domain. Science 1987;236(4797):55–63.

[69] Simon MA, Dodson GS, Rubin GM. An SH3-SH2-SH3 protein is required for p21Ras1 activation and binds to sevenless and Sos proteins in vitro. Cell 1993;73(1):169–77.

[70] Simon MA, Carthew RW, Fortini ME, Gaul U, Mardon G, Rubin GM. Signal transduction pathway initiated by activation of the sevenless tyrosine kinase receptor. Cold Spring Harb Symp Quant Biol 1992;57:375–80.

[71] Fortini ME, Simon MA, Rubin GM. Signaling by the sevenless protein tyrosine kinase is mimicked by Ras1 activation. Nature 1992;355(6360):559–61.

[72] Tang AH, Neufeld TP, Kwan E, Rubin GM. PHYL acts to down-regulate TTK88, a transcriptional repressor of neuronal cell fates, by a SINA-dependent mechanism. Cell 1997;90(3):459–67.

[73] Boulton SJ, Brook A, Staehling-Hampton K, Heitzler P, Dyson N. A role for Ebi in neuronal cell cycle control. Embo J 2000;19(20):5376–86.

[74] Li S, Li Y, Carthew RW, Lai ZC. Photoreceptor cell differentiation requires regulated proteolysis of the transcriptional repressor Tramtrack. Cell 1997;90(3):469–78.

[75] Hu G, Chung YL, Glover T, Valentine V, Look AT, Fearon ER. Characterization of human homologs of the *Drosophila* seven in absentia (sina) gene. Genomics 1997;46(1): 103–11.

[76] Hanahan D, Weinberg RA. The hallmarks of cancer. Cell 2000;100(1):57–70.

[77] Sebolt-Leopold JS, Herrera R. Targeting the mitogen-activated protein kinase cascade to treat cancer. Nat Rev Cancer 2004;4(12):937–47.

[78] Jones S, Zhang X, Parsons DW, Lin JC, Leary RJ, Angenendt P, et al. Core signaling pathways in human pancreatic cancers revealed by global genomic analyses. Science 2008;321(5897):1801–6.

[79] Comprehensive genomic characterization defines human glioblastoma genes and core pathways. Nature 2008;455(7216):1061–8.

[80] Ostrem JM, Peters U, Sos ML, Wells JA, Shokat KM. K-Ras(G12C) inhibitors allosterically control GTP affinity and effector interactions. Nature 2013;503(7477):548–51.

[81] Schmidt RL, Park CH, Ahmed AU, Gundelach JH, Reed NR, Cheng S, et al. Inhibition of RAS-mediated transformation and tumorigenesis by targeting the downstream E3 ubiquitin ligase seven in absentia homologue. Cancer Res 2007;67(24):11798–810.

[82] Ahmed AU, Schmidt RL, Park CH, Reed NR, Hesse SE, Thomas CF, et al. Effect of disrupting seven-in-absentia homolog 2 function on lung cancer cell growth. J Natl Cancer Inst 2008;100(22):1606–29.

[83] Liu J, Stevens J, Rote CA, Yost HJ, Hu Y, Neufeld KL, et al. Siah-1 mediates a novel beta-catenin degradation pathway linking p53 to the adenomatous polyposis coli protein. Mol Cell 2001;7(5):927–36.

[84] Matsuzawa SI, Reed JC. Siah-1, SIP, and Ebi collaborate in a novel pathway for beta-catenin degradation linked to p53 responses. Mol Cell 2001;7(5):915–26.

[85] Susini L, Passer BJ, Amzallag-Elbaz N, Juven-Gershon T, Prieur S, Privat N, et al. Siah-1 binds and regulates the function of Numb. Proc Natl Acad Sci USA 2001;98(26):15067–72.

[86] Habelhah H, Frew IJ, Laine A, Janes PW, Relaix F, Sassoon D, et al. Stress-induced decrease in TRAF2 stability is mediated by Siah2. Embo J 2002;21(21):5756–65.

[87] Polekhina G, House CM, Traficante N, Mackay JP, Relaix F, Sassoon DA, et al. Siah ubiquitin ligase is structurally related to TRAF and modulates TNF-alpha signaling. Nat Struct Biol 2002;9(1):68–75.

[88] Nakayama K, Frew IJ, Hagensen M, Skals M, Habelhah H, Bhoumik A, et al. Siah2 regulates stability of prolyl-hydroxylases, controls HIF1alpha abundance, and modulates physiological responses to hypoxia. Cell 2004;117(7):941–52.

[89] House CM, Hancock NC, Moller A, Cromer BA, Fedorov V, Bowtell DD, et al. Elucidation of the substrate binding site of Siah ubiquitin ligase. Structure 2006;14(4):695–701.

[90] Gutierrez GJ, Vogtlin A, Castro A, Ferby I, Salvagiotto G, Ronai Z, et al. Meiotic regulation of the CDK activator RINGO/Speedy by ubiquitin-proteasome-mediated processing and degradation. Nat Cell Biol 2006;8(10):1084–94.

[91] Nadeau RJ, Toher JL, Yang X, Kovalenko D, Friesel R. Regulation of Sprouty2 stability by mammalian Seven-in-Absentia homolog 2. J Cell Biochem 2007;100(1):151–60.

[92] Sutterluty H, Mayer CE, Setinek U, Attems J, Ovtcharov S, Mikula M, et al. Down-regulation of Sprouty2 in non-small cell lung cancer contributes to tumor malignancy via extracellular signal-regulated kinase pathway-dependent and -independent mechanisms. Mol Cancer Res 2007;5(5):509–20.

[93] Kim HJ, Taylor LJ, Bar-Sagi D. Spatial regulation of EGFR signaling by Sprouty2. Curr Biol 2007;17(5):455–61.

[94] Winter M, Sombroek D, Dauth I, Moehlenbrink J, Scheuermann K, Crone J, et al. Control of HIPK2 stability by ubiquitin ligase Siah-1 and checkpoint kinases ATM and ATR. Nat Cell Biol 2008;10(7):812–24.

[95] Shaw AT, Meissner A, Dowdle JA, Crowley D, Magendantz M, Ouyang C, et al. Sprouty-2 regulates oncogenic K-ras in lung development and tumorigenesis. Genes Dev 2007;21(6): 694–707.

[96] Scortegagna M, Kim H, Li JL, Yao H, Brill LM, Han J, et al. Fine tuning of the UPR by the ubiquitin ligases Siah1/2. PLoS Genet 2014;10(5):e1004348.

[97] Baba K, Morimoto H, Imaoka S. Seven in absentia homolog 2 (Siah2) protein is a regulator of NF-E2-related factor 2 (Nrf2). J Biol Chem 2013;288(25):18393–405.

[98] Hsieh SC, Kuo SN, Zheng YH, Tsai MH, Lin YS, Lin JH. The E3 ubiquitin ligase SIAH2 is a prosurvival factor overexpressed in oral cancer. Anticancer Res 2013;33(11):4965–73.

[99] Khurana A, Nakayama K, Williams S, Davis RJ, Mustelin T, Ronai Z. Regulation of the ring finger E3 ligase Siah2 by p38 MAPK. J Biol Chem 2006;281(46):35316–26.

[100] Qi J, Nakayama K, Gaitonde S, Goydos JS, Krajewski S, Eroshkin A, et al. The ubiquitin ligase Siah2 regulates tumorigenesis and metastasis by HIF-dependent and -independent pathways. Proc Natl Acad Sci USA 2008;105(43):16713–8.

[101] Nakayama K, Qi J, Ronai Z. The ubiquitin ligase Siah2 and the hypoxia response. Mol Cancer Res 2009;7(4):443–51.

[102] Qi J, Kim H, Scortegagna M, Ronai ZA. Regulators and effectors of Siah ubiquitin ligases. Cell Biochem Biophys 2013;67(1):15–24.

[103] Qin R, Smyrk TC, Reed NR, Schmidt RL, Schnelldorfer T, Chari ST, et al. Combining clinicopathological predictors and molecular biomarkers in the oncogenic K-RAS/Ki67/HIF-1alpha pathway to predict survival in resectable pancreatic cancer. Br J Cancer 2015;112(3):514–22.

[104] Behling KC, Tang A, Freydin B, Chervoneva I, Kadakia S, Schwartz GF, et al. Increased SIAH expression predicts ductal carcinoma in situ (DCIS) progression to invasive carcinoma. Breast Cancer Res Treat 2011;129(3):717–24.

[105] Pi H, Wu HJ, Chien CT. A dual function of phyllopod in *Drosophila* external sensory organ development: cell fate specification of sensory organ precursor and its progeny. Development 2001;128(14):2699–710.

[106] Lorick KL, Jensen JP, Fang S, Ong AM, Hatakeyama S, Weissman AM. RING fingers mediate ubiquitin-conjugating enzyme (E2)-dependent ubiquitination. Proc Natl Acad Sci USA 1999;96(20):11364–9.

[107] Brunner D, Oellers N, Szabad J, Biggs 3rd WH, Zipursky SL, Hafen E. A gain-of-function mutation in *Drosophila* MAP kinase activates multiple receptor tyrosine kinase signaling pathways. Cell 1994;76(5):875–88.

[108] Lai ZC, Rubin GM. Negative control of photoreceptor development in *Drosophila* by the product of the yan gene, an ETS domain protein. Cell 1992;70(4):609–20.

[109] Dong X, Tsuda L, Zavitz KH, Lin M, Li S, Carthew RW, et al. Ebi regulates epidermal growth factor receptor signaling pathways in *Drosophila*. Genes Dev 1999;13(8):954–65.

[110] Tsuda L, Nagaraj R, Zipursky SL, Banerjee U. An EGFR/Ebi/Sno pathway promotes delta expression by inactivating Su(H)/SMRTER repression during inductive notch signaling. Cell 2002;110(5):625–37.

[111] Harrison SD, Travers AA. The tramtrack gene encodes a *Drosophila* finger protein that interacts with the ftz transcriptional regulatory region and shows a novel embryonic expression pattern. Embo J 1990;9(1):207–16.

[112] Brown JL, Sonoda S, Ueda H, Scott MP, Wu C. Repression of the *Drosophila* fushi tarazu (ftz) segmentation gene. Embo J 1991;10(3):665–74.

[113] Read D, Levine M, Manley JL. Ectopic expression of the *Drosophila* tramtrack gene results in multiple embryonic defects, including repression of even-skipped and fushi tarazu. Mech Dev 1992;38(3):183–95.

[114] Brown JL, Wu C. Repression of *Drosophila* pair-rule segmentation genes by ectopic expression of tramtrack. Development 1993;117(1):45–58.

[115] Baonza A, Murawsky CM, Travers AA, Freeman M. Pointed and Tramtrack69 establish an EGFR-dependent transcriptional switch to regulate mitosis. Nat Cell Biol 2002;4(12): 976–80.

[116] Badenhorst P. Tramtrack controls glial number and identity in the *Drosophila* embryonic CNS. Development 2001;128(20):4093–101.

CHAPTER 13

Extracellular Signal-Regulated Kinase (ERK1 and ERK2) Inhibitors

A.A. Samatar

TheraMet Biosciences LLC, Princeton Junction, NJ, United States

INTRODUCTION

The RAS-ERK pathway (also termed MAPK pathway) is one of the major signal transduction pathways that govern fundamental cellular processes that include proliferation, cell survival, and cell differentiation. The RAS-ERK pathway is activated in response to growth factor binding and regulates cellular growth, differentiation, and survival in a variety of cell types [1–3]. Activation of this pathway occurs via a cascade of protein phosphorylation events which culminates in the phosphorylation and activation of ERK1, 2, which is the most downstream module (Fig. 13.1). The pathway consists of RAS and a three-tiered kinase module made of RAF, MEK, and ERK. RAS (H-, K-, and N-RAS isoforms) is a family of small GTPases that act as a switch to initiate the signal transduction pathway. Activation of RAS by mutations or by receptor tyrosine kinases induces conformational change in RAS from a GDP-bound to a GTP-bound form. The conformational change in RAS leads to binding to RAF (A-, B-, and RAF-1 isoforms) kinase, which is recruited to the membrane and gets activated [4,5]. Activated RAF leads to phosphorylation of MEK (MEK1 and MEK2 isoforms), a dual serine/threonine kinase that phosphorylates and activates ERK. Once ERK (ERK1 and ERK2 isoforms) is activated, it phosphorylates its cytoplasmic and nuclear substrates leading to alterations in gene expression profiles, increase in proliferation, differentiation, and cell survival [6–8]. Even though the RAS-ERK cascade is thought of as a linear pathway, its signaling network has multiple inputs and outputs that include positive and negative feedback mechanisms. The feedback mechanisms which are dynamic control the activity of the pathway [9].

Constitutive activation of the RAS-ERK pathway is frequently observed in human cancers and is associated with high rates of cancer cell proliferation [10]. Pathway activation occurs as a consequence of gain-of-function mutations in one of the RAS isoforms or in BRAF, a member of the RAF kinase family.

CONTENTS

Conquering RAS. http://dx.doi.org/10.1016/B978-0-12-803505-4.00013-8

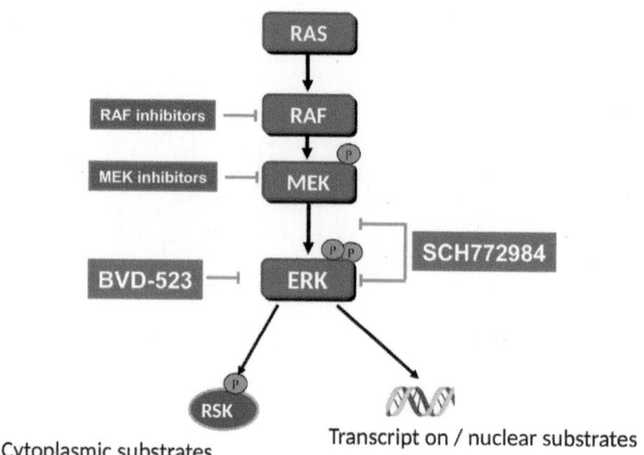

FIGURE 13.1

Current approaches to inhibit RAS-ERK pathway. Strategies to inhibit RAS-ERK signaling with small molecules include targeting RAF with vemurafenib or dabrafenib, inhibition of MEK with trametenib and inhibition of ERK with SCH772984 or BVD-523. SCH772984 has a unique dual mechanism of inhibiting phosphorylation of ERK and its kinase activity.

The high frequency of BRAF (V600E) mutations in melanoma (70%), thyroid (50%), and colorectal (10%) cancers and K- or N-RAS mutations in melanoma (20%), pancreatic (90%), colorectal (50%), and non–small-cell lung (30%) cancers make targeting this pathway an attractive strategy for cancer therapy [11–15]. There is a large unmet medical need for improved therapies in these diseases especially in the advanced, refractory setting.

RAS INHIBITORS

In the past decades the pharmaceutical industry has put intensive efforts toward the identification and development of selective inhibitors of the RAS-ERK pathway as anti-cancer drugs. The observation that activated RAS is a key driver of several human cancers has resulted in efforts to develop RAS inhibitors. The first concept was developing drugs that blocked RAS farnesylation, which led to the clinical development of farnesyltransferase inhibitors [16,17]. The result of clinical trials of the farnesyltransferase inhibitors was disappointing, showing very minimal anti-cancer activity. The reason for the ineffectiveness of the farnesyltransferase inhibitors was ascribed to the fact that in the presence of farnesyltransferase inhibitors, *KRAS* and *NRAS* isoforms (most commonly mutated in tumors) were still active because of the lipid modification by geranylgeranyltransferase [18,19]. In addition to these efforts of preventing RAS medications by farnesyltransferase inhibitors, several other approaches of targeting RAS were pursued. These include efforts to block the interaction between RAS-GTP and RAF, preventing the interaction of RAS with son of sevenless and inhibition of phosphodiesterase-δ and KRAS interaction [20]. Efforts to directly target RAS with small molecules that covalently bind to KRAS G12C mutant were described [21]. However, all these efforts are still in early pre-clinical proof-of-concept studies and it is unknown whether this approach will lead to effective cancer therapies. Therefore, direct inhibitors of RAS with clinical utility have yet to be identified.

RAF INHIBITORS

Since direct inhibitors of RAS have been challenging to develop, an alternative approach is to target the downstream effector signaling responsible for RAS-mediated activation. RAF is a family of protein kinases, which consists of three isoforms (ARAF, BRAF, and RAF1). RAF is directly downstream of RAS and is activated by recruitment to the plasma membrane and forms a RAS–RAF complex [19]. Formation of this complex induces a conformational change that activates RAF. Activation of RAF leads to phosphorylation and activation of MEK.

The discovery of BRAF mutations in a number of human cancers sparked interest in targeting BRAF with small molecule inhibitors. Mutations in the BRAF isoform occur with high frequency than in the ARAF or RAF1 isoforms. This is due to the differences in their regulation and it appears that BRAF is more prone to become activated through a single point mutation than the other RAF isoforms. The most common mutation found in BRAF is the substitution of valine to glutamic acid at residue 600 (V600E). This mutation allows the kinase to adopt into an active conformation and becomes constitutively active kinase. Mutations in BRAF have been identified in a significant percentage of cancers with melanoma having one of the highest prevalence. There are several small molecule RAF inhibitors that have been developed including sorafenib, Raf265, and XL281. Even though sorafenib was developed to target RAF, it gained US Food and Drug Administration (FDA) approval for renal cell carcinoma, which lacks mutations in RAS and BRAF. This is because sorafenib is a more potent inhibitor of VEGFR in vivo than RAF and is effective in tumors that rely on VEGFR such as renal cancers.

Vemurafenib and dabrafenib, which are selective BRAF inhibitors, have demonstrated remarkable activity in patients with melanoma with the BRAF V600E mutation. The response rates observed and the overall survival benefit (vemurafenib) of these inhibitors changed the landscape of melanoma. The high response rate was associated with a near-complete inhibition of ERK phosphorylation in tumors treated with the BRAF inhibitors. This observation suggests that potent inhibition of ERK phosphorylation may be required for robust clinical activity. The side effects of the BRAF inhibitors are tolerable and they seem to have a high window of therapeutic index. An intriguing observation with vemurafenib and dabrafenib is the paradoxical activation of

ERK in non-BRAFV600-mutant tumors while inhibiting ERK in tumors cells harboring the BRAF V600E mutation. Patients with melanoma with BRAF V600E mutation who were treated with the BRAF inhibitors (vemurafenib or dabrafenib) developed keratoacanthomas and squamous cell carcinomas of the skin, which was associated with activation of the RAS-ERK pathway. The mechanisms of the paradoxical activation of the RAS-ERK pathway were later found to involve BRAF inhibitor–induced dimerization of RAF in the presence of RAS. These observations suggest that only patients known to harbor the BRAF V600E mutation would derive benefit from these BRAF inhibitors. It also raises the concern that these BRAF inhibitors might accelerate progression in RAS-mutant tumors. This paradoxical activation phenomenon limits the clinical utility of the BRAF inhibitors.

Despite the impressive initial clinical responses to these BRAF inhibitors, most patients develop resistance while on drug within 6–7 months [22,23]. Research from multiple laboratories has shown that the resistance mechanisms include the development of activating NRAS mutations, BRAF amplification, generation of a BRAF splice variant, and mutations in MEK [24–28]. All these resistance mechanisms lead to the re-activation of ERK. One strategy to overcome the acquired BRAF inhibitor resistance is to combine BRAF inhibitor with MEK inhibitor. One study demonstrated that combining BRAF inhibitor with MEK inhibitor prolonged the suppression of phosphorylated ERK and inhibited proliferation in BRAF-resistant cells [29]. In support of this approach, a randomized clinical trial comparing the combination of BRAF and MEK to BRAF inhibitor monotherapy showed a significant prolongation of progression-free survival with the combination (9.4 months as compared with 5.8 months in the monotherapy group) in patients with BRAF-mutant melanoma [30]. It is anticipated that the combination of BRAF and MEK inhibitors will become a new standard treatment for patients with advanced BRAF-mutant melanoma. However, based on published pre-clinical studies, there is an indication that acquired resistance to the combination therapy can also be developed [29].

MEK INHIBITORS

MEK1 and MEK2 are dual-specificity protein kinases that are very closely related sharing 80% overall amino acid identity. They lie downstream of RAF and mediate the transmission of growth factor signaling from activated RAF to ERK. MEK activation occurs as a result of phosphorylation by RAF kinase and activated MEK in turn phosphorylates the threonine and tyrosine residues within the activation loop of ERK kinase (Thr202 and Try204 in ERK1) [31]. Activating mutations in MEK are very rare and happen at a very low frequency in human cancers [32,33]. This is perhaps because activation of MEK is due to the activation of upstream effectors such as mutant RAS or mutant BRAF.

Activation of MEK modulates several cell functions through phosphorylation and subsequent activation of ERK. Intensive efforts have been directed toward the identification and pre-clinical and clinical development of MEK inhibitors, in addition to RAF inhibitors. Underneath are selected examples of some of the MEK inhibitors that have been evaluated.

The first MEK inhibitor, PD098059, an allosteric inhibitor, was discovered nearly two decades ago [34]. Since then, over 15 allosteric MEK inhibitors have been identified and several of them have progressed to clinical trials [34–38]. PD098059 was not progressed to the clinic because of its weak inhibitory effect in vitro and poor pharmaceutical properties. Optimization of PD098059 led to CI-1040, which demonstrated potent inhibition of tumor growth in human tumor xenografts mice models. CI-1040 was progressed into the clinic becoming the first small molecule MEK inhibitor to be evaluated in clinical trials. However, it failed to demonstrate sufficient anti-tumor activity in the clinical trials because of its low exposure and rapid metabolism. CI-1040 was further optimized to improve its pharmaceutical properties and this effort resulted in PD0325901, a structural analogue of CI-1040. PD0325901 with its increased potency and improved pharmaceutical properties was evaluated in a phase I study. The most common adverse events associated with PD0325901 included rash, diarrhea, fatigue, nausea, and neurotoxicity. A decrease in ERK phosphorylation level was demonstrated in some of the patients dosed with PD0325901. Even though PD0325901 demonstrated some clinical activity at its maximum tolerated dose, the clinical trials were terminated because of ocular and neurological toxicities. It is unclear whether these toxicities are mechanism based but many of the allosteric MEK inhibitors have shown a similar toxicity profile in clinical trial studies.

AZD6244 (selumetinib), an allosteric MEK inhibitor similar to PD0325901, has been evaluated in several clinical trials either as a monotherapy or in combination with other therapeutic agents. However, results from the clinical trials show that AZD6244 has modest clinical activity as a single agent but additional clinical trials are ongoing in combination with other cancer drugs. The clinical trials with PD0325901 and AZD6244 showed that treatment with these inhibitors at their maximum tolerated doses achieved minimal clinical activity and was associated with toxicities [37].

In 2013, the FDA approved the first MEK inhibitor, trametinib, for the treatment of metastatic melanoma with the *BRAF*(V600E/K) mutation. Trametenib is a potent inhibitor of MEK that preferentially binds to un-phosphorylated MEK and prevents RAF-dependent MEK phosphorylation. The phase I study that evaluated the safety and pharmacokinetics of trametinib showed that the most common adverse events included rash, fatigue, and diarrhea. As in the case of AZD6244 and PD0325901, the dose limiting toxicities for trametinib were rash, diarrhea, and ocular and neurological toxicities. The recommended phase II dose was 2 mg per day because of the poor tolerability of trametinib. The efficacy of trametinib was demonstrated in a phase III study of patients with *BRAF*-mutant melanoma with advanced disease who were not previously treated with BRAF inhibitors [39]. Trametinib had an overall response rate of 22%, which is lower than that of BRAF inhibitors (vemurafenib 48.4%, or dabrafenib 52%). It is unclear as to why trametinib has a lower response rate than the BRAF inhibitors. Even though trametinib showed activity in patients who were not previously treated with BRAF inhibitors, it failed to demonstrate meaningful activity in patients who were previously treated with BRAF inhibitors. The reason for this observation is unclear but it could be due to the decreased dependence of the tumor on the RAS-ERK pathway.

Cobimetinib (XL-518), another potent allosteric inhibitor of MEK, was approved by the FDA in the beginning of November 2015. The approval of cobimetinib is in combination with vemurafenib for patients with melanoma with the BRAF V600E/K mutation. Unlike trametinib, cobimetinib is not approved as a single agent. The approval of cobimetinib in combination with vemurafenib was based on a randomized clinical study of patients with the BRAF V600E mutation who were previously un-treated. Patients treated with cobimetinib in combination with vemurafenib demonstrated a median progression-free survival and lived longer than those taking vemurafenib only. The most common side effects of the combination treatment include diarrhea, nausea, photosensitivity reaction, and pyrexia. Some of the severe side effects of cobimetinib are cardiomyopathy, primary cutaneous malignancies, hepatotoxicity, and hemorrhage. The clinical trials with MEK inhibitors demonstrate some clinical activity, but their utility as therapeutics is mostly limited by toxicity [37]. In addition to the MEK inhibitors mentioned earlier, a number of other MEK inhibitors are undergoing early clinical evaluations. These include E6201, GDC-0623, TAK-733, and WX-554.

Targeting BRAF and MEK has become a powerful therapeutic intervention for patients with melanoma with the BRAF mutation. Nevertheless, the evidence that has accumulated to date suggests the inevitable development of acquired resistance to BRAF and MEK inhibitors. This development of resistance is mostly due to re-activation of ERK despite continued presence of the drugs. Inhibition of ERK, which is the primary downstream module of RAS-ERK, may provide a strategy for how to overcome BRAF and MEK inhibitor resistance. This approach may lead to a high percentage of durable responses in patients who develop resistance to BRAF and MEK inhibitors. However, there has been limited progress in the discovery and development of ERK inhibitors. An overview of the current ERK inhibitors that are being developed for clinical trials is provided later in the discussion. For most of these ERK inhibitors, there is no publically available literature data on their pre-clinical activities. The only small molecule kinase inhibitor of ERK that has been extensively characterized pre-clinically is SCH772984.

DIRECT SMALL MOLECULE INHIBITORS OF ERK

The mitogen-activated protein kinases ERK1 and ERK2 are ubiquitously expressed and share a high degree of homology with 85% overall sequence identity. Stimulation of the RAS-ERK pathway leads to the parallel activation of ERK1 and ERK2. These two proteins have identical substrates and it seems there are no differences between the two ERK isoforms in their role as signal transducers. However, knockout mouse studies have provided evidence for differential functions of the ERK1,2 isoforms in developmental biology. The knockout mouse studies showed that deletion of erk1 gene resulted in mice that were viable but had defects in thymocyte maturation. In contrast, deletion of the erk2 gene is embryonic lethal because of impaired placental development. Apart from the differential roles of the ERK isoforms in developmental studies, they seem to perform the same functions. ERK is activated by phosphorylation on tyrosine and threonine residues by MEK. In their un-phosphorylated and inactivated states, both ERK1 and ERK2 are localized in the cytoplasm. Activation of ERK promotes translocation to the nucleus and phosphorylates several proteins including transcription factors. Unlike RAF kinase and MEK, which have narrow substrates specificity, ERK has over 200 substrates. This broad substrate specificity feature in ERK might prevent the development of acquired resistance, which has been an issue with RAF and MEK inhibitors.

The activation of ERK through the RAS-RAF-MEK-ERK pathway leads to the regulation of several cellular processes including cell cycle progression, cell migration, and cell survival. Aberrant activation of ERK plays a critical role in the initiation and progression of several human cancers. In contrast to the efforts dedicated to develop RAF and MEK inhibitors in the past decades, there has been little interest in the development of selective ERK inhibitors. This is due to the assumption that the ERK pathway is linear and because ERK is downstream of RAF and MEK, there will be no added advantage in inhibiting ERK. This perception changed when the emergence of acquired resistance to RAF and MEK inhibitors became apparent. Moreover, many of the mechanisms leading to the acquisition of resistance to BRAF and MEK inhibitors involve re-activation of ERK. Because of these observations, efforts to identify and develop ERK inhibitors were initiated (Table 13.1). In addition, there has been an appreciation of the complexity and diversity of the biochemical effects of small molecule inhibitors targeting different components of the same pathway. Finally, the feedback loops that are promoted by small molecule inhibitors of RAF, MEK, and ERK may show differences depending on the components of the pathway that is targeted. For the past decades the pharmacological inhibition of ERK has always been inferred from inhibiting MEK with small molecules because there were no selective and potent ERK inhibitors. Thus, ERK inhibitors could have the potential to overcome the acquired BRAF and MEK inhibitors, avoid acquired drug resistance processes, and shed light on the

Table 13.1 ERK Inhibitors

Compound	Structure	Phase	Mechanism
SCH746514	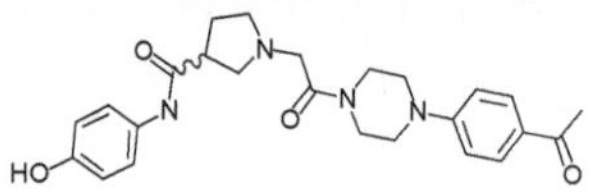	Pre-clinical	ATP-competitive, dual mechanism
SCH772984	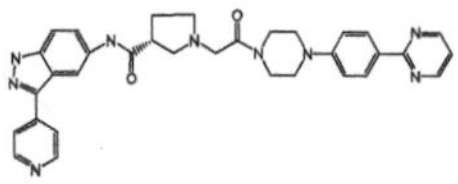	Pre-clinical	ATP-competitive, dual mechanism
MK8353	Structure not disclosed	Phase 1	ATP-competitive, dual mechanism
BVD-523	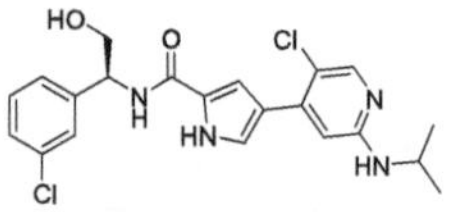	Phase 1	ATP-competitive, single mechanism
GDC-0994	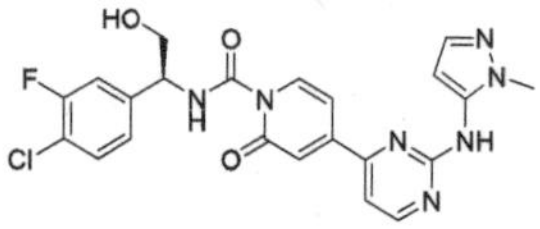	Phase 1	ATP-competitive, single mechanism
CC-90003	* 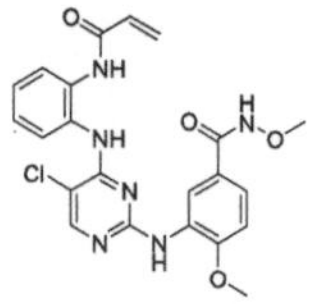	Phase 1	Irreversible inhibitor
KO-947	* 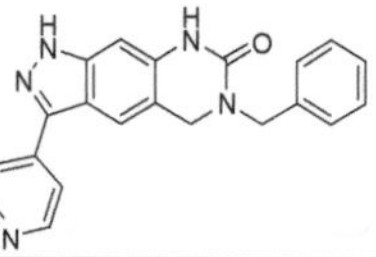	Pre-clinical	ATP-competitive, single mechanism

**Representative structures from the patent literature.*

effects of pharmacological inhibition of ERK in vitro and in vivo. Therefore, there is a need for novel ERK inhibitors for patients with BRAF and MEK inhibitor-resistant tumors and patients with tumors that are driven by activated ERK as a result of RAS or RAF mutations. Indeed, as described later, SCH772984, a selective and potent ERK inhibitor, demonstrates activity in tumors with RAS-ERK–activated pathway and BRAF and MEK inhibitor-resistant cell lines.

FR180204

FR180204, which is a pyrazolopyridazine derivative (Fig. 13.2C), was identified from a chemical library that was screened with active ERK2 [40]. FR180204, which is an ATP-competitive inhibitor, is potent against active ERK1 and ERK2 with IC_{50} of 0.51 and 0.33 nM, respectively. The crystal structure of FR180204 complexed to human ERK2 shows that it occupies the hinge region and

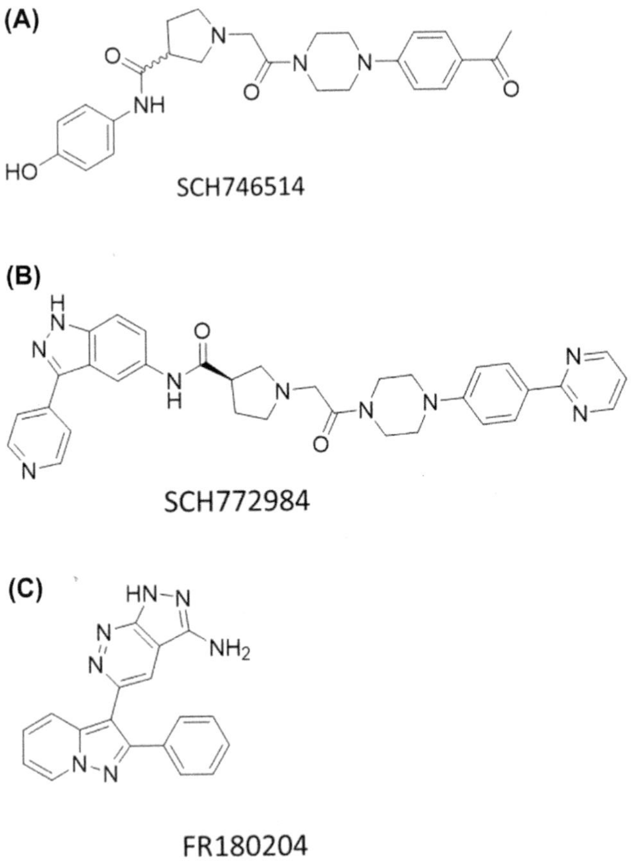

(A)

SCH746514

(B)

SCH772984

(C)

FR180204

FIGURE 13.2
Chemical structures of three ERK inhibitors. (A) SCH746514, the initial screening hit identified through screening a diverse chemical library of compounds. (B) Optimization of SCH746514 resulted in SCH772984, a potent and selective ERK inhibitor with a dual mechanism of inhibition. (C) Structure of FR180204, another small molecule ERK inhibitor that represents a single mechanism inhibitor.

hydrogen bonds with Gln105, Asp106, and Met108. FR148083, which is an irreversible inhibitor, inhibits the kinase activity of ERK2 with an IC_{50} of 80 nM [41]. X-ray structural studies of human ERK2 bound to FR148083 reveal that the inhibitor binds to the hinge region through a hydrogen bond with Met108. It also makes contact with the carbonyl group of Ser153 (through a hydrogen bond) and hydrophobic interactions with Val39, Ala52, Ile31, and Leu156. In addition, it forms a covalent bond with the thiol group of C166. This covalent bond formation differentiates FR148083 from FR180204, SCH772984, BVD-523, and GDC-0994. CC-90003, which is an irreversible inhibitor like FR148083, is the only irreversible ERK inhibitor that is in early clinical trials.

An issue with the ERK2 irreversible inhibitors is that they will also inhibit MEK1, MEK2, and MKK7 because of the cysteine residues that correspond to C166 that are found in these proteins.

SCH77298

The key screening strategy used to identify a selective ERK inhibitor involved the use of an affinity-based screen of a diverse chemical library. The chemical library was screened for binding to the un-phosphorylated form of the ERK2 protein. This approach identified SCH746514 (Fig. 13.2A), an ATP-competitive inhibitor that demonstrated good kinase selectivity but had lower potency. SCH746514 bound to un-phosphorylated ERK2 as well as to phosphorylated ERK2 but showed a higher affinity for the un-phosphorylated form of ERK2. To understand the structural basis of ERK2 inhibition by SCH746514, the X-ray crystal structure of the non-phosphorylated rat ERK2 protein bound to SCH746514 was determined. SCH746514 bound at the hinge region (ATP-binding pocket) of ERK2, with the hydroxyl of the phenol group forming two hydrogen bonds with the hinge region, a hydrogen bond donor to the Asp104 backbone amide NH, and a hydrogen bond acceptor to the backbone carbonyl of Asp104. Binding of SCH746514 to ERK2 induced a conformational change that opened up a new side pocket. These observations suggested that these binding modes of SCH746514 to ERK2 might confer selectivity to SCH746514 [42].

Because SCH746514 was not potent enough, the goal was to increase the potency while retaining the selectivity profile and the unique binding mode. An iterative optimization process through synthetic chemistry efforts led to significant improvements in enzymatic potency and selectivity, culminating in the synthesis of an ATP-competitive compound, SCH772984 (Fig. 13.2B). SCH772984 is a potent and selective compound that inhibits ERK1 and ERK2 proteins with an IC_{50} of 4 and 2 nM, respectively [29]. In addition, it displays a remarkable selectivity profile against a panel of over 300 kinases. The cellular activity of SCH772984 was evaluated in a BRAF (V600E)-mutant melanoma tumor cell line by measuring its potency for inhibiting ERK phosphorylation of p90 ribosomal S6 kinase (RSK). SCH772984 completely inhibits RSK phosphorylation in a dose-dependent manner (Fig. 13.3) indicating that SCH772984 inhibits the intrinsic kinase activity of ERK1/2.

Chaikuad et al. reported the crystal structure of SCH772984 bound to human ERK2 protein, which revealed similar binding modes to SCH746514 [43]. The X-ray crystallography data show that SCH772984 occupies the ATP-binding pocket, with its indazole N atoms forming two hydrogen bonds with the hinge, a hydrogen bond donor to the Asp106 backbone carbonyl oxygen, and a hydrogen bond acceptor to the backbone amide NH of Met108. The indazole 3-position substituent pyridine N atom makes a hydrogen bond with the

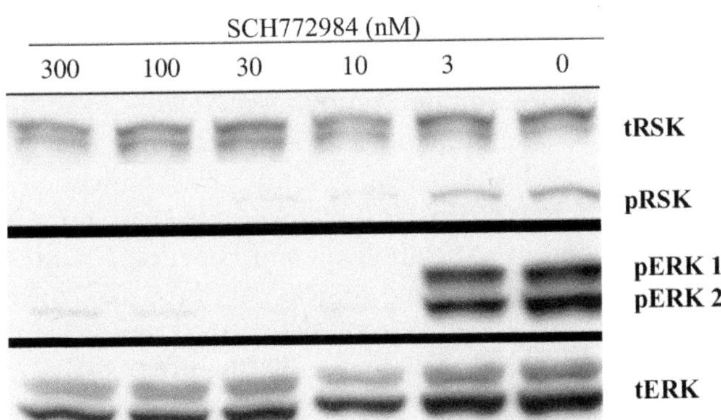

FIGURE 13.3
Effects of SCH772984 in cells. Exposure of SCH772984 to BRAF mutant cells results in the inhibition of phosphorylation of ERK and its downstream substrate RSK, displaying the dual mechanism of inhibition.

Lys114 side chain. The carbonyl oxygen of the 5-aminodoindazole is involved in an extended hydrogen bond network through a water molecule with the side chains of Gln105, the catalytic Lys54, and the backbone NH of Asp167, whereas the NH of the amide H-bonds with a water molecule. The pyrrolidine is in the neutral form and acts as a hydrogen bond acceptor to Lys54. In order for the extended long piperazine-phenyl-pyrimidine motif of SCH772984 to bind ERK2, a large conformational change at Tyr36 was observed in the Gly-rich loop. An intriguing property of SCH772984 is its inhibition of phosphorylation of ERK (Fig. 13.3). MEK phosphorylates ERK1 and ERK2 on the activation loop residues Thr202/Tyr204 (ERK1) and Thr185/Tyr187 (ERK2). Based on the results from a kinase counter-screen and binding assays, SCH772984 does not bind or inhibit the upstream kinases MEK or RAF [29]. So why does SCH772984 inhibit the phosphorylation of ERK2? Based on the X-ray structural studies, SCH772984 binds to the active site of ERK2 and inhibits the intrinsic kinase activity of ERK. At the same time a large conformational change at Tyr36 is induced in the Gly-rich loop. Unlike most known kinase structures, the highly conserved Tyr36 flips its side chain under the Gly-rich loop toward the adenine-binding site and opens up a new side pocket (the Tyr36 side pocket). In this new conformation, the pyrrolidine ring of SCH772984 is now positioned directly under Tyr36, making a favorable hydrophobic interaction with the flipped aromatic ring of Tyr36. These conformational changes induced on SCH772984 binding to ERK may prevent MEK access to the phosphorylation sites of ERK1,2 (Thr202/Tyr204 and Thr185/Tyr187) on the ERK activation loop. This unique ability of SCH772984 to simultaneously inhibit the intrinsic kinase activity of ERK and prevent the phosphorylation of ERK is termed dual mechanism of inhibition. Currently, SCH772984 and a clinical analogue (MK8353) are the only ERK inhibitors with dual mechanisms of inhibition.

The anti-tumor activity of SCH772984 in vivo was assessed in xenograft models established from human tumor cell lines [27]. SCH772984 induced tumor regressions in BRAF- or RAS-mutant xenograft models. SCH772984 also effectively suppressed ERK signaling and cell proliferation in the BRAF- or MEK-resistant cells. These observations support the development of SCH772984 and related compounds for clinical trials. Because of its poor pharmaceutical [absorption, distribution, metabolism, and excretion (ADME)] properties, SCH772984 was not progressed for clinical development. However, an analogue of SCH772984, MK8353/SCH900353, is in the early stages of clinical trials. There is no information available publically on MK8353.

BVD-523 (Ulixertinib)

BVD-523 is a small molecule inhibitor designed to inhibit ERK1/2 kinase. Data presented at the 2015 American Association for Cancer Research (AACR) annual meeting (http://cancerres.aacrjournals.org/content/75/15_Supplement/4693) showed that it is a potent and selective inhibitor of ERK1 and ERK2. It is a reversible ATP-competitive inhibitor that suppresses cell proliferation and cell survival in cell lines bearing mutations that activate RAS-ERK pathway signaling. It also inhibits tumor growth in vivo in BRAF-mutant or KRAS xenografts. Unlike SCH772984, BVD-523 lacks the dual mechanism of inhibition and inhibits only the intrinsic kinase activity of ERK (Fig. 13.4). Treatment of BRAF or KRAS-mutant cells with BVD-523 increases the baseline phosphorylation of ERK, whereas the phosphorylation of RSK is inhibited. The biological consequences of the increased ERK phosphorylation in the presence of BVD-523 are unknown. BVD-523 is in phase I clinical trials and a dose escalation in patients with advanced solid tumors was presented at the 2015 annual American Society of Clinical Oncology (ASCO) meeting (http://meetinglibrary.asco.org/content/145666-156). The dose escalation studies included 27 patients with advanced solid tumors. The study objectives included evaluation of maximum tolerated dose (MTD),

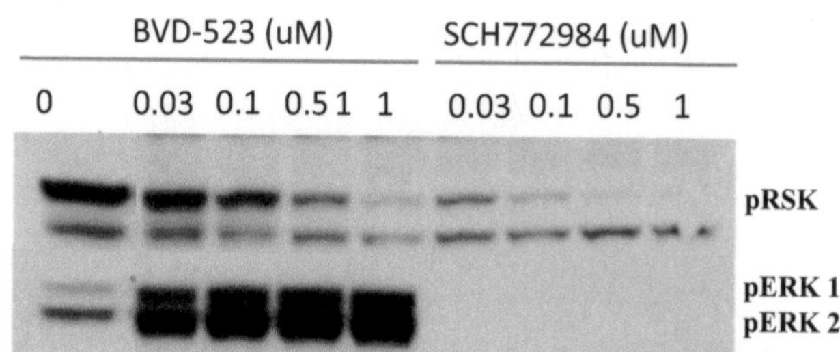

FIGURE 13.4

Comparison of ERK pathway inhibition by SCH772984 and BVD-523. Exposure of SCH772984 or BVD-523 to BRAF mutant cells inhibited phosphorylation of RSK in a dose dependent manner. Treatment of cells with SCH772984 inhibited phosphorylation of ERK but in the presence of BVD-523 phosphorylation of ERK increased.

dose-limiting toxicities, and determination of phase II recommended dose. The dose level ranged from 10 to 900 mg twice a day. Dose-limiting toxicities, which include grade 3 rash and grade 3 pruritus, were observed in five patients. The MTD was determined to be 600 mg twice a day and the most common adverse events included diarrhea, nausea, vomiting, and rash. One patient had a partial response and five patients showed metabolic response as assessed by fluorodeoxyglucose positron emission tomography. BVD-523 achieved pharmacologically relevant exposure in this trial and tolerability at 600 mg twice a day was manageable. A cohort expansion to evaluate safety and efficacy assessments are planned in additional phase I and phase II studies.

GDC-0994

GDC-0994 (RG7842) is another potent inhibitor of ERK with biochemical potency of IC_{50} values of 1.1 (ERK1) and 0.3 nM (ERK2). GDC-0994 demonstrated in vivo efficacy in several human xenograft tumors in mice including BRAF-mutant and KRAS-mutant models (http://cancerres.aacrjournals.org/content/74/19_Supplement/DDT02-03). Inhibition of phosphorylated RSK in tumor tissues derived from mice treated with GDC-0994 correlate with the in vivo anti-tumor activity. Like BVD-523, GDC-0994 lacks the dual mechanism of inhibition and inhibits only the kinase activity of ERK. However, unlike BVD-523, treatment of cells with GDC-0994 does not increase the basal level of phosphorylated ERK. These observations suggest that SCH772984, BVD-523, and GDC-0994 may affect the feedback loop differently. GDC-0994 is currently in phase I clinical development and no information is publically available on its clinical activity.

Other ERK inhibitors include KO–947, CC90003, AEZS-131/AEZS-134, and VTX11e. These ERK inhibitors are in early pre-clinical development and no public information is available on their pre-clinical activities or clinical development stages. In addition to the aforementioned ERK inhibitors, which are all kinase inhibitors, another novel ERK inhibitor, DEL-22379, was described [44]. Unlike the inhibitors described earlier, which inhibit the catalytic activity of ERK, DEL-22379 inhibits the dimerization of ERK. Characterization of DEL-22379 showed that it inhibited tumor growth in BRAF- or KRAS-mutant tumors. Interestingly, DEL-22379 is unaffected by resistance mechanisms that are observed with BRAF and MEK inhibitors. It is still too early to judge whether dimerization inhibitors of ERK will lead to effective anti-cancer drugs in patients.

CONCLUSIONS AND FUTURE DIRECTIONS

The importance of the RAS-ERK pathway in several cancers has intensified efforts directed toward the identification and clinical development of selective inhibitors of the pathway. Most of the discovery and clinical efforts for the past decades focused on RAF or MEK inhibitors. As a result, there are two RAF inhibitors, vemurafenib and dabrafenib, that have been approved by the FDA. Even though several MEK inhibitors have been developed, the only FDA-approved MEK inhibitors to date are trametinib (a single agent or in combination with dabrafenib) and cobimetinib (only in combination with vemurafenib). The MEK inhibitors have not demonstrated robust clinical activities as single agents like the BRAF inhibitors. However, although BRAF inhibitors have demonstrated remarkable clinical activities, they are only active in melanomas with BRAF V600E mutations, but not effective in RAS-mutant tumors. The importance of developing ERK inhibitors for use as anti-cancer therapies was just realized recently. As such there is interest in developing novel ERK inhibitors that can be clinically useful. The robust clinical efficacy observed with BRAF inhibitors or combination of BRAF and MEK inhibitors has validated the RAS-ERK pathway and generated enthusiasm for targeting different components of the pathway. Moreover, the development of ERK re-activation-resistant mechanisms to BRAF and MEK inhibitors has placed importance on the identification and development of ERK inhibitors. Some of the ERK inhibitors described earlier have been shown to be active in cancers with mutant RAS or mutant RAF and those with acquired resistance to BRAF and MEK inhibitors. ERK inhibition is a promising therapeutic strategy in many human cancers with constitutive activation of the ERK pathway and warrants further clinical development.

The clinical evaluation of the ERK inhibitors should shed light on their activities in patients with cancer. It would be important to understand in what context the ERK inhibitors demonstrate anti-tumor activity in the clinic. Will resistance be developed to ERK inhibitors? The fact that ERK has over 200 substrates suggests that it may be difficult to develop resistance to ERK inhibitors. The ERK inhibitors described earlier have subtle differences in their mechanisms of inhibition. Will a compound with dual mechanism of inhibition such as SCH772984 be superior to a compound with single mechanism of action such as BVD-523? It would be interesting to understand the effects of ERK inhibitors with different mechanisms on the translocation of ERK to the nucleus. Pre-clinical comparisons of ERK inhibitors with different mechanisms should shed light on the importance of dual versus single mechanism of inhibition. Based on the accumulated data, it is clear that different inhibitors of the same target demonstrate marked differences in effectiveness, depending on tumor type and mutational status. Further characterization of the biological and biochemical activities of these ERK inhibitors will help to improve therapeutic strategies. The design of optimal clinical trials including dosing schedule and combination studies of ERK inhibitors with other targeted therapies may provide lasting treatment to patients with RAS-ERK activated pathway. The discovery and development of the ERK inhibitors are in their infancy, but they do hold therapeutic potential either as a single agent or in combination in patients with cancer. It will be important to define the optimal point of intervention in this critical signaling pathway to aid the development of effective therapeutics for BRAF and RAS-mutant cancers. The ERK inhibitors described earlier provide potent and selective tools to help address this question. In addition to their potential as therapeutic agents, they may also be useful chemical tools in studying the biological functions of ERK.

References

[1] Montagut C, Settleman J. Targeting the RAF–MEK–ERK pathway in cancer therapy. Cancer Lett 2009;283(2):125–34.

[2] Roberts P, Der C. Targeting the Raf-MEK-ERK mitogen-activated protein kinase cascade for the treatment of cancer. Oncogene 2007;26(22):3291–310.

[3] Dhillon A, Hagan S, Rath O, Kolch W. MAP kinase signalling pathways in cancer. Oncogene 2007;26(22):3279–90.

[4] Chong H, Vikis H, Guan K. Mechanisms of regulating the Raf kinase family. Cell Signal 2003;15(5):463–9.

[5] Campbell S, Khosravi-Far R, Rossman K, Clark G, Der C. Increasing complexity of Ras signaling. Oncogene 1998;17(11):1395–413.

[6] Chen R, Sarnecki C, Blenis J. Nuclear localization and regulation of erk- and rsk-encoded protein kinases. Mol Cell Biol 1992;12(3):915–27.

[7] Ahn N. The MAP kinase cascade. Discovery of a new signal transduction pathway. Mol Cell Biochem 1993;127-128(1):201–9.

[8] Pearson G, Robinson F, Beers Gibson T, Xu B, Karandikar M, Berman K, et al. Mitogen-activated protein (MAP) kinase pathways: regulation and physiological functions 1. Endocr Rev 2001;22(2):153–83.

[9] Lito P, Pratilas C, Joseph E, Tadi M, Halilovic E, Zubrowski M, et al. Relief of profound feedback inhibition of mitogenic signaling by RAF inhibitors attenuates their activity in BRAFV600E melanomas. Cancer Cell 2012;22(5):668–82.

[10] Hingorani S, Wang L, Multani A, Combs C, Deramaudt T, Hruban R, et al. Trp53R172H and KrasG12D cooperate to promote chromosomal instability and widely metastatic pancreatic ductal adenocarcinoma in mice. Cancer Cell 2005;7(5):469–83.

[11] Arteaga C. Epidermal growth factor receptor dependence in human tumors: more than just expression? Oncologist 2002;7(90004):31–9.

[12] Samowitz W. Poor survival associated with the BRAF V600E mutation in microsatellite-stable colon cancers. Cancer Res 2005;65(14):6063–9.

[13] Prior I, Lewis P, Mattos C. A comprehensive survey of Ras mutations in cancer. Cancer Res 2012;72(10):2457–67.

[14] Rajasekharan S, Raman T. Ras and Ras mutations in cancer. Open Life Sci 2013;8(7).

[15] Vakiani E, Solit D. KRAS and BRAF: drug targets and predictive biomarkers. J Pathol 2010;223(2):220–30.

[16] Stephen A, Esposito D, Bagni R, McCormick F. Dragging Ras back in the ring. Cancer Cell 2014;25(3):272–81.

[17] Cox A, Fesik S, Kimmelman A, Luo J, Der C. Drugging the undruggable RAS: mission possible? Nat Rev Drug Discov 2014;13(11):828–51.

[18] Ahearn I, Haigis K, Bar-Sagi D, Philips M. Regulating the regulator: post-translational modification of RAS. Nat Rev Mol Cell Biol 2011;13(1):39–51.

[19] Berndt N, Hamilton A, Sebti S. Targeting protein prenylation for cancer therapy. Nat Rev Cancer 2011;11(11):775–91.

[20] Zimmermann G, Papke B, Ismail S, Vartak N, Chandra A, Hoffmann M, et al. Small molecule inhibition of the KRAS-PDEδ interaction impairs oncogenic KRAS signalling. Nature 2013;497(7451):638–42.

[21] Ostrem J, Peters U, Sos M, Wells J, Shokat K. K-Ras(G12C) inhibitors allosterically control GTP affinity and effector interactions. Nature 2013;503(7477):548–51.

[22] Bollag G, Hirth P, Tsai J, Zhang J, Ibrahim P, Cho H, et al. Clinical efficacy of a RAF inhibitor needs broad target blockade in BRAF-mutant melanoma. Nature 2010;467(7315):596–9.

[23] Abraham J. Vemurafenib in melanoma with the BRAF V600E mutation. Community Oncol 2012;9(3):85–6.

[24] Nazarian R, Shi H, Wang Q, Kong X, Koya R, Lee H, et al. Melanomas acquire resistance to B-RAF(V600E) inhibition by RTK or N-RAS upregulation. Nature 2010;468(7326):973–7.

[25] Shi H, Moriceau G, Kong X, Lee M, Lee H, Koya R, et al. Melanoma whole-exome sequencing identifies V600EB-RAF amplification-mediated acquired B-RAF inhibitor resistance. Nat Commun 2012;3:724.

[26] Villanueva J, Vultur A, Lee J, Somasundaram R, Fukunaga-Kalabis M, Cipolla A, et al. Acquired resistance to BRAF inhibitors mediated by a RAF kinase switch in melanoma can be overcome by cotargeting MEK and IGF-1R/PI3K. Cancer Cell 2010;18(6):683–95.

[27] Girotti M, Pedersen M, Sanchez-Laorden B, Viros A, Turajlic S, Niculescu-Duvaz D, et al. Inhibiting EGF receptor or SRC family kinase signaling overcomes BRAF inhibitor resistance in melanoma. Cancer Discov 2012;3(2):158–67.

[28] Johannessen C, Boehm J, Kim S, Thomas S, Wardwell L, Johnson L, et al. COT drives resistance to RAF inhibition through MAP kinase pathway reactivation. Nature 2010;468(7326):968–72.

[29] Morris E, Jha S, Restaino C, Dayananth P, Zhu H, Cooper A, et al. Discovery of a novel ERK inhibitor with activity in models of acquired resistance to BRAF and MEK inhibitors. Cancer Discov 2013;3(7):742–50.

[30] Long G, Stroyakovskiy D, Gogas H, Levchenko E, de Braud F, Larkin J, et al. Combined BRAF and MEK inhibition versus BRAF inhibition alone in melanoma. N Engl J Med 2014;371(20):1877–88.

[31] Chen J, Fujii K, Zhang L, Roberts T, Fu H. Raf-1 promotes cell survival by antagonizing apoptosis signal-regulating kinase 1 through a MEK-ERK independent mechanism. Proc Natl Acad Sci USA 2001;98(14):7783–8.

[32] Crews C, Alessandrini A, Erikson R. The primary structure of MEK, a protein kinase that phosphorylates the ERK gene product. Science 1992;258(5081):478–80.

[33] Bansal A, Ramirez R, Minna J. Mutation analysis of the coding sequences of MEK-1 and MEK-2 genes in human lung cancer cell lines. Oncogene 1997;14(10):1231–4.

[34] Favata M, Horiuchi K, Manos E, Daulerio A. Identification of a novel inhibitor of mitogen-activated protein kinase kinase. J Biol Chem 1998;273(29):18623–32.

[35] Yoshida T, Kakegawa J, Yamaguchi T, Hantani Y, Okajima N, Sakai T, et al. Identification and characterization of a novel chemotype MEK inhibitor able to alter the phosphorylation state of MEK1/2. Oncotarget 2013;3(12):1533–45.

[36] Yeh T, Marsh V, Bernat B, Ballard J, Colwell H, Evans R, et al. Biological characterization of ARRY-142886 (AZD6244), a potent, highly selective mitogen-activated protein kinase kinase 1/2 inhibitor. Clin Cancer Res 2007;13(5):1576–83.

[37] Kim K, Kong S, Fulciniti M, Li X, Song W, Nahar S, et al. Blockade of the MEK/ERK signalling cascade by AS703026, a novel selective MEK1/2 inhibitor, induces pleiotropic anti-myeloma activity in vitro and in vivo. Br J Haematol 2010;149(4):537–49.

[38] Domann F, Mitchen J, Clifton K. Restoration of thyroid function after total thyroidectomy and quantitative thyroid cell transplantation. Endocrinology 1990;127(6):2673–8.

[39] LoRusso P, Krishnamurthi S, Rinehart J, Nabell L, Malburg L, Chapman P, et al. Phase I pharmacokinetic and pharmacodynamic study of the oral MAPK/ERK kinase inhibitor PD-0325901 in patients with advanced cancers. Clin Cancer Res 2010;16(6):1924–37.

[40] Infante J, Papadopoulos K, Bendell J, Patnaik A, Burris H, Rasco D, et al. A phase 1b study of trametinib, an oral Mitogen-activated protein kinase kinase (MEK) inhibitor, in combination with gemcitabine in advanced solid tumours. Eur J Cancer 2013;49(9):2077–85.

[41] Ohori M, Kinoshita T, Okubo M, Sato K, Yamazaki A, Arakawa H, et al. Identification of a selective ERK inhibitor and structural determination of the inhibitor-ERK2 complex. Biochem Biophys Res Commun 2005;336(1):357–63.

[42] Ohori M, Kinoshita T, Yoshimura S, Warizaya M, Nakajima H, Miyake H. Role of a cysteine residue in the active site of ERK and the MAPKK family. Biochem Biophys Res Commun 2007;353(3):633–7.

[43] Deng Y, Shipps G, Cooper A, English J, Annis D, Carr D, et al. Discovery of novel, dual mechanism ERK inhibitors by affinity selection screening of an inactive kinase. J Med Chem 2014;57(21):8817–26.

[44] Chaikuad A, Tacconi EM, Zimmer J, Liang Y, Gray N, et al. A unique inhibitor binding site in ERK1/2 is associated with slow binding kinetics. Nat Chem Biol 2014;10(10):853–60.

CHAPTER 14

Targeting Rho, Rac, CDC42 GTPase Effector p21 Activated Kinases in Mutant K-Ras-Driven Cancer

A.S. Azmi, P.A. Philip
Wayne State University, Detroit, MI, United States

CONTENTS

INTRODUCTION

Ras proteins (K-ras, H-ras, and N-Ras) are ubiquitously expressed in almost all types of cells and have been considered as the holy grail of cancer due in part to their diverse roles in regulating various important cellular signaling pathways [1,2]. Ras mutations are perhaps the most prevalent among all somatic mutations in cancer (http://cancer.sanger.ac.uk/cosmic/) [3]. Belonging to the GTPases, ras proteins are activated through the binding of GTP by guanine exchange factors (GEFs) and GTPase-activating proteins (GAPs) resulting in transmitting signals in the cells [4]. The activated ras protein acts as a molecular switch that turns on various target proteins necessary for important cellular processes such as division and proliferation. In normal cells, a balanced cycling of the GTP to GDP through the inherent GTPase activity of ras keeps ras-mediated signaling in check. However, mutations in ras particularly K-ras, which is observed in the majority of cancers, abrogates this GTPase activity and disturbs the GTP–GDP cycling rendering the protein in a constant "on state"[5]. This results in hyper-activated ras-driven signaling in cells that is akin to a car without brakes. There are two hot spots for ras oncogenic mutations that are located at codons 12 and 61 of their highly conserved coding sequences. The glycine to valine mutation at residue 12 renders the GTPase domain of ras insensitive to inactivation by GAP. On the other hand, the residue 61 stabilizes the transition state for GTP hydrolysis. Mutation considerably suppresses the rate of intrinsic Ras GTP hydrolysis contributing to the permanent activation of Ras.

Given that ras signals directly impact cell growth and division, mutation-driven hyper-activated ras signaling can result in loss of growth control and cancer. It is not surprising to note that H-Ras, K-ras, and N-Ras are collectively considered the most common oncogenes in cancer. Activating mutations in K-ras are found in ~25% of all cancers and >95% in certain types of

Conquering RAS. http://dx.doi.org/10.1016/B978-0-12-803505-4.00014-X

difficult-to-treat cancers such as pancreatic cancer. These observations drove the initial interest in the development of ras-targeted therapeutics. However, ras protein structure lacks any putative drug-binding site and given its high affinity toward GTP (in the picomolar range), the design of any agent that can specifically attach to ras has been futile [6]. The search is still on for novel target sites within the ras structure or direct interacting partners downstream of this master oncogenic regulator (some discussed in the different chapters of this book).

RAS SUPERFAMILY

All of the proteins that are related to Ras are pooled into a superfamily of small GTPases [7]. Their number runs into hundreds of proteins and they are classified into five main families of GTPases that are based on structure, sequence, and function, namely Ras, Rho, Ran, Rab, and Arf GTPases, and each family member plays a distinct functional role [7]. For example, the Ras family is responsible for cell proliferation [8]; on the other hand the Rho family is responsible for structural integrity and cellular morphology [9]. Ran provides spatial regulation to proteins as it is the direct target of nuclear exporter protein XPO1/CRM1/exportin1 and import proteins (mainly importin α) and is responsible for nuclear transport of cargo proteins [10]. Emerging reports have attributed the Rab and Arf family for their role in vesicular (such as exosomal) transport [11]. All these sub-family members share a common G domain that is critical for the GTPase and nucleotide exchange activity. The surrounding sequence helps determine the functional specificity of the small GTPases.

RHO FAMILY GTPases AND CANCER

The Rho family of GTPases is a family of small signaling G proteins that have an approximate size of 21 KDa [12]. The members of the Rho GTPase family have been demonstrated to play critical roles in the biology of intra-cellular actin in most eukaryotes [13]. The three most studied members of the Rho family are Cdc42 (responsible for filopodia formation), Rac1 (known for its role in lamellipodia formation), and RhoA (best recognized for its role in stress fiber biology) [14]. All of these family members act as molecular switches, with common roles in cytoskeletal dynamics, organelle development, cell motility, and additional cellular functions.

Cells carry certain well-known regulatory mechanisms that govern the biology of Rho GTPases. Chief among them is the guanine nucleotide exchange factor (GEF) that catalyzes the exchange of GDP to GTP [15]. The GAPs regulate the precise hydrolysis of GTP to GDP by controlling the activity of GTPase, thereby guiding the changes in the conformation from the active to inactive form [16]. The guanine nucleotide dissociation inhibitors (GDIs) are responsible for the formation of a complex with the Rho protein, thereby preventing diffusion within the membrane and into the cytosol and thus acting as an anchor and allowing tight spatial control of Rho activation [17]. Aside from these three, other mechanisms have also been attributed to the regulation of RhoGTPases such as through microRNAs [18], phosphorylation, palmitoylation, nuclear targeting, transglutamination, and AMPylation as well as ubiquitination (reviewed in Ref. [19]).

Rho GTPase has been shown to play a prominent role in almost all of the steps of cancer initiation, progression, self-renewal (stemness), pro-survival, evasion from cell death/apoptosis, epithelial-to-mesenchymal transition (EMT) invasion, migration, and metastasis. Unlike the activating mutations in the three Ras isoforms, K-Ras, N-Ras, and H-Ras, that are observed in >40% of all human tumors, the Rho proteins are only rarely mutated in tumors. Nevertheless, their activity or expression status is found to be quite frequently altered. For example, the up-regulation of several Rho GTPases such as RhoA, RhoC, Rac (1,2 and 3), and Cdc42 has been consistently reported in various tumors [20]. Some Rho GTPase family members have been shown to act as pro-oncogenic through the stimulation of cell cycle progression and regulation of gene transcription [21]. Other Rho GTPases are considered as the regulators of neo-vascularization by secreting pro-angiogenic factors [22]. They are recognized to play a crucial role in cell polarity signaling in the plasticity of cancer cell invasiveness [23]. In addition to these mechanisms, the Rho GTPases regulate cancer signaling through many more diverse downstream effectors. These effectors have additional myriad targets resulting in an exponential number of mechanisms within the influence of these Rho GTPases and some are listed in the following sections.

RHO GTPase EFFECTORS

Research over the past several years has shed light onto the many downstream effectors of the Rho proteins and these effectors have been found to play a critical role in important cellular processes. There are more than 60 targets of the three common Rho GTPases defined so far (see Fig. 14.1). The most extensively studied Rho GTPase effectors are the p21 activated serine/threonine kinases (PAKs) [24]. PAKs are highly conserved across various species ranging from yeast to humans [25]. This chapter focuses on the GTPase p21 activated kinases (PAKs).

P21 ACTIVATED KINASES AS EFFECTORS OF GTPases

The PAKs belonging to the serine/threonine protein kinases are well recognized as downstream effectors of the Rho family of GTPases [26]. PAKs are

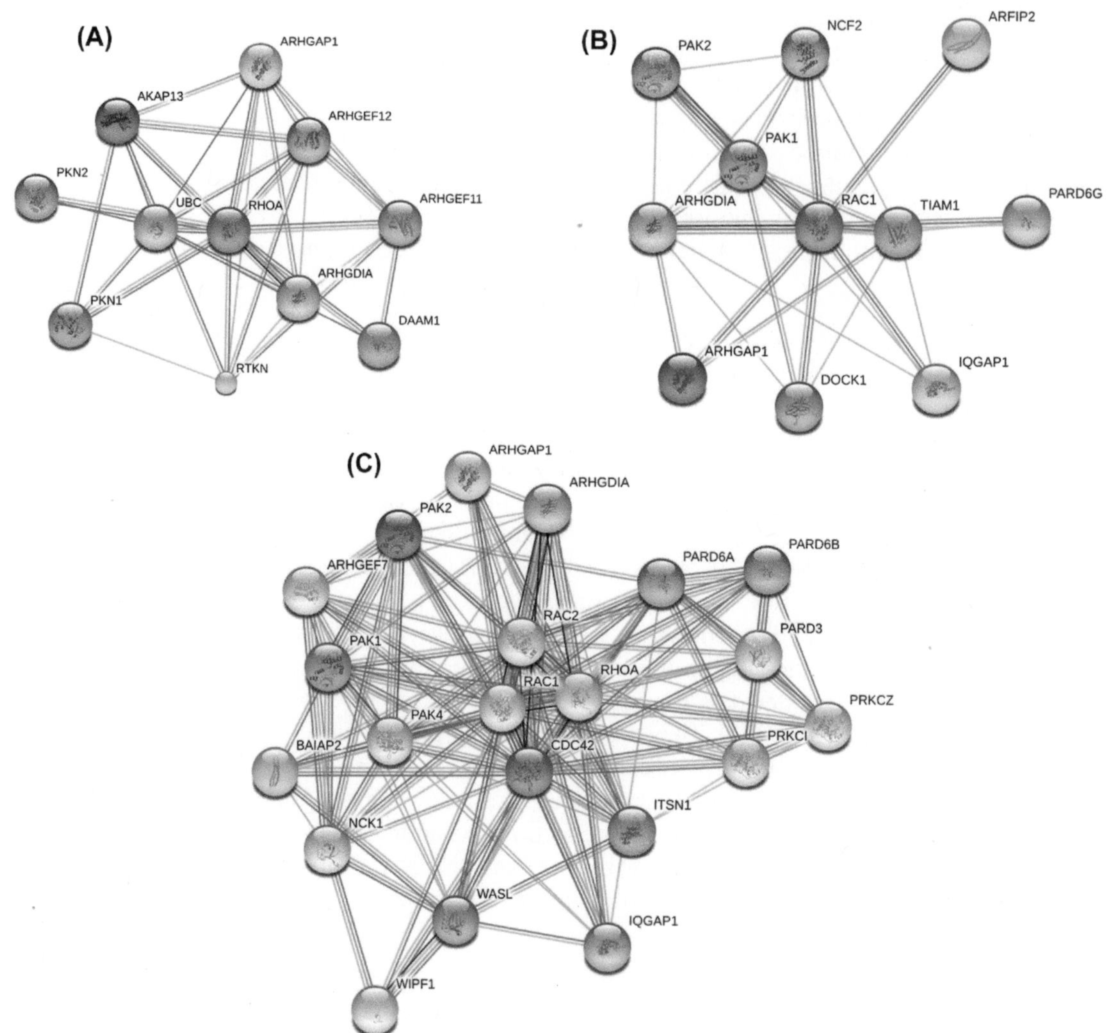

FIGURE 14.1 The multi-faceted interacting partners of Rho GTPases.
Rho A, Rac-1, CDC42 known interaction networks depicted through String protein–protein interaction program version 10 (http://string-db.org/).

frequently up-regulated in human diseases [27], including various cancers, and their over-expression correlates with disease progression [28]. A number of publications have validated the important roles of PAKs in cell proliferation, survival, gene transcription, transformation, and cytoskeletal re-modeling (reviewed in Ref. [29]). PAKs are recognized to act as a converging node for many signaling pathways that regulate these cellular processes and therefore have emerged as attractive targets for the treatment of diseases.

PAKs have been classified broadly into two groups: group I (PAK 1, 2, and 3) and group II (PAK 4, 5, and 6) [30]. Both group I and II PAKs contain a highly conserved C-terminal kinase domain (carrying one unique phosphorylation site). However, the differences lie in the N-terminal regulatory domain that is quite unique between group I and II PAKs (see domain structures in Fig. 14.2; see also Ref. [30]). Group I PAKs modulate actin dynamics through a proline-rich region that contains Nck adapter protein-binding motifs. Group I PAKs also possess an auto-inhibitory domain (AID) that is found overlapping with the p21-binding domain (PBD) [28]. Group I PAKs are recognized to form a transinhibited conformation that depends on dimerization with AID in a manner that one PAK attaches to and blocks the catalytic domain of the other PAK [25]. The activation of group I PAKs occurs when it is bound to the GTPases such as Cdc42 or Rac that lead to the disruption of PBD and lack of dimerization [25]. Aside from PAK4 the remaining group II PAKs lack AID and are considered to possess enhanced kinase activities than group I PAKs [25]. Attachment of GTP bound Rac or Cdc42 to group II PAKs has little effect on their kinase activities. However, such binding is proposed to cause localization-mediated regulation of group II PAKs. In the absence of a signal from GTPase, the autoinhibitory pseudo-substrate domain within the N-terminal regions of group II PAKs blocks their kinase activity. In addition, the N-terminal domain, that binds to hypoxia-regulated [31] PAK interacting exchange factors (PIX) [32], is lacking in group II PAKs that in group I family members has been well studied as an important downstream effector. Despite sharing numerous common substrates, PAKs belonging to each group (I and II) have distinct specificities toward their biological targets and each group member is regulated in a distinct manner.

Genetic studies have clearly defined the critical roles that PAKs play in the biology of organisms and deletion of the PAK gene in mice has varying effects on viability and phenotype depending on the isoform that has been depleted. PAK1, PAK5, and PAK6 have minimal impact as mice with either partial or complete knockout remain viable. On the other hand, PAK2 and PAK4 knockout mice are embryonically lethal, pointing to their necessary roles during development [33]. In cases where viability is not lost, certain key functions are impacted. For example, PAK3 knockout mice, despite being viable, demonstrate defects in synaptic plasticity, indicating its critical role in neural differentiation [34].

PAKs IN CANCER

The earliest studies on the role of PAKs in transformation came from the work of Satoh and colleagues in the 1990s who demonstrated that p21.GTP levels increased in cells treated with fetal bovine serum or platelet-derived growth factor to initiate DNA synthesis [35]. They further went on to show that epidermal

GROUP I PAKs

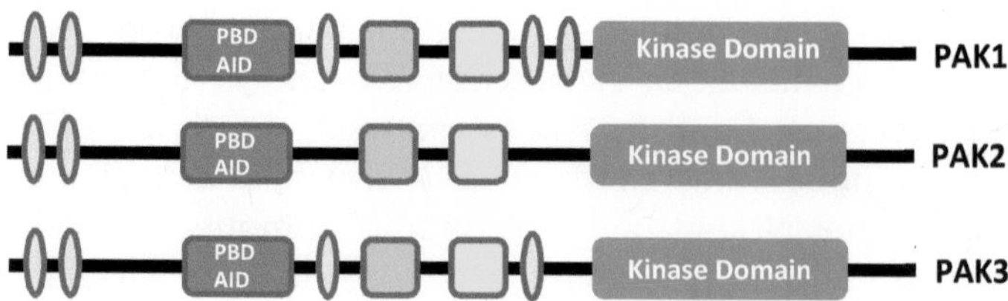

GROUP II PAKs

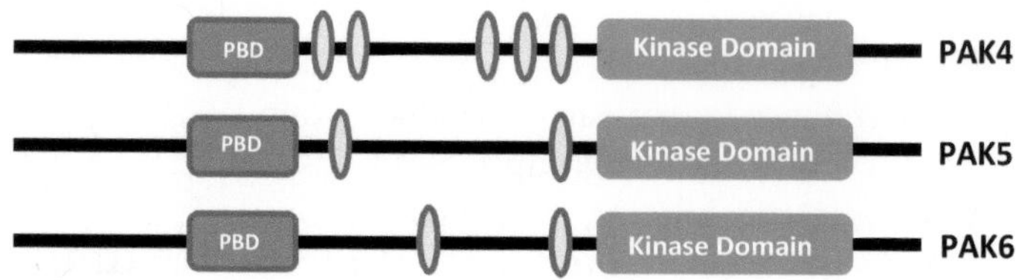

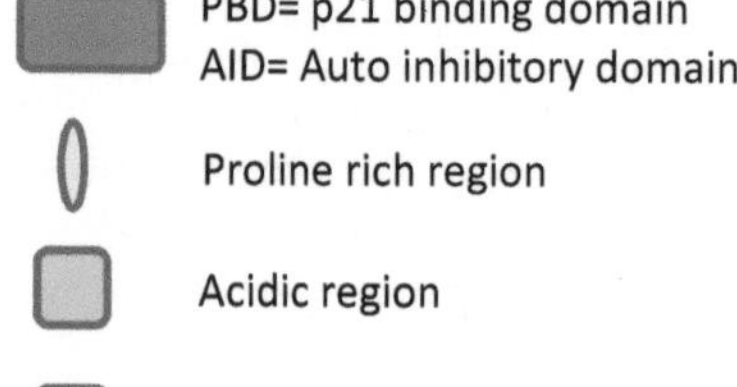

PIX biding region

FIGURE 14.2 Domain structure of PAKs.

Group I and group II PAKs share the p-21-binding domain (PBD) region for CDC42/rac interaction/binding domain for interacting with Rho family GTPases; auto-inhibitory domain that is overlapping in both group I and group II PAKs, a catalytic domain; phospholipid-binding domain consisting of a cluster of basic residues domain, and a proline-rich domain. The group I PAKs have an additional PAK interacting exchange factor (PIX)-binding domain that is absent in group II PAKs. *Figure adopted from the review article by Van Den Brueke C, et al. Trends Cell Biol 2010 March;20(3):160–9.*

growth factor can also increase the amounts of p21.GTP in the cells. Their results strongly suggested an important and previously un-explored role of p21 in transduction of signals for both normal proliferation and malignant transformation through growth factor receptors. Around the same time, Leevers and Marshall showed that p21ras kinases are activated following scrape loading in a quiescent cell model system. They observed a rapid activation of 42 and 46 kDa protein kinases that were mitogen and extra-cellular signal regulated kinase ERK2 (MAP2 kinase) [36]. These original papers laid the foundation for the expansion of PAK-related studies in many different cancers. Owing to their diverse roles in regulating important cellular process, it is not surprising to note that the majority of PAKs (belonging to both group and group II) are found to be de-regulated in cancer. In the following sections the role of PAKs in different cancers is highlighted.

GROUP I PAKs IN CANCER

Much of the original work relating to group I PAKs in oncogenesis was obtained in breast cancer models. In the late 1990s, Bekri and colleagues demonstrated that amplification of loci present on band q13 of human chromosome 11 is a feature of a subset of estrogen receptor–positive breast carcinomas prone to metastasis [37]. In this study, PAK was among the four new genes placed on the regional map (namely, CBP2, CLNS1A, UVRAG, and PAK1) as critical players in breast cancer progression. At the same time, Tang and colleagues from Chernoff's group showed that PAK1 kinase activity is critical for transformation of Rat-1 fibroblast by ras [38]. Later on, the same group demonstrated using site-directed mutagenesis that regulation of microfilament reorganization and invasiveness of breast cancer cells is governed by PAK1 [39]. In another study, PAK1 was shown to promote anchorage-independent growth and abnormal organization of mitotic spindles in human epithelial breast cancer cells [40]. Bagheri-Yarmand R et al. showed that Etk/Bmx tyrosine kinase activates Pak1 and regulates tumorigenic potential of breast cancer cells [41]. Li and colleagues showed that PAK1 can phosphorylate histone H3 and may thus influence the PAK1–histone H3 pathway, which in turn may influence mitotic events in breast cancer cells [42]. Adding on to these studies, Wang and colleagues demonstrated that hyper-activation of PAK1 is sufficient for mammary gland tumor formation [43]. Highlighting their role in therapy resistance, Yoon et al. have shown that the small GTPase Rac1 is involved in the maintenance of stemness in glioma stem-like cells [44]. Given that PAKs are directly under the influence of Rac1, it is logical to assume their important role in the maintenance of stemness characteristics, which will be detailed in the forthcoming passages.

Like PAK1, the other group I PAK (PAK2) is also involved in cancer invasion and metastasis [45]. In addition, the role of PAK2 in cell promoting cellular motility has been established [46]. Sato and colleagues demonstrated the role of PAK2 may be a critical mediator of transforming growth factor β–mediated cell migration in a hepatoma cell model [47]. Its role in stemness has also been investigated, although indirectly. For example, it was earlier shown that knockdown of Rac1 in adult mouse epidermis stimulated stem cells and commits them to divide and undergo terminal differentiation by negatively regulating c-Myc through PAK2 phosphorylation [48]. Studies have linked PAK2 to drug resistance as well. Li et al. have demonstrated that phosphorylation of caspase-7 by PAK2 inhibits chemically induced apoptosis in breast cancer cells [49]. In another study, Yan and colleagues demonstrated that prostasin may contribute to chemoresistance through CASP/PAK2-p34/actin signaling in ovarian cancer [50]. Similarly in gastric cancer models, Cho et al. showed using a proteomic approach the molecular mechanisms of RhoGDI2-downstream effector PAK2 induced metastasis and drug resistance [51]. Gopal et al. showed that oncogenic epithelial cell–derived exosomes are rich in Rac1 and PAK2 induces angiogenesis in recipient endothelial cells [52].

The third member of the group I PAK, although less studied, has also been shown to play an important role in cancer. For example, Liu and colleagues have demonstrated PAK3 to promote the progression of adrenocorticotropic hormone–producing thymic carcinoid [53]. In another study, Holdernesss et al. showed that the activating protein 1 regulated PAK4 to induce actin organization and migration of transformed fibroblasts [54]. Exemplifying its role as a target, Baldwin and group showed synthetic lethal interactions between p53 and the protein kinases SGK2 and PAK3, thereby making it a target for p53-specific drug development [55]. However, the synthetic lethality of PAK3 was challenged when Zhou and colleagues demonstrated that the HPV+ cervical cancer cell death was not associated with RNAi-induced PAK3 and SGK2 knockdown but was likely through off-target effects [56]. This shows that more work needs to be done to delineate the exact role of PAK3 in cancer initiation, progression, and therapy resistance. Nevertheless, these and other studies comprehensively attest to the significant role of group I PAKs in the biology of cancer and make them attractive therapeutic targets across a broad spectrum of malignancies.

GROUP II PAKs IN CANCER

The group II PAKs (5, 6, and 7) have also been studied for their roles in various cancers [57]. PAK4 was identified at the turn of the century when Abo et al. demonstrated that it is among the novel p21 activated kinases that are essential for re-organization of the actin cytoskeleton and the formation of filopodia [58]. Later on, Zarnegar et al. identified a novel androgen receptor–interacting protein, which was provisionally termed PAK6 and shared a high degree of sequence similarity with other PAKs [59]. Panday and colleagues were among the first groups to clone and characterize the group II PAK5, which was considered to be closely linked to PAK4 and PAK6 both structurally and functionally [60]. Their work also highlighted that PAK5 transcript was predominantly expressed in brain and this expression pattern was distinct from that of PAK4 and PAK6, suggesting a functional division among PAK-II sub-family kinases based on differential tissue distribution [60].

After these initial discoveries, the focus shifted toward the identification of the exact roles of PAKs in cancer and other diseases. Group II PAKs are recognized to impact some of the major pathways driving cancer. Numerous groups have verified the requirement of PAK4 in supporting anchorage-independent cell growth [61,62], supporting cell migration and invasion [63–68], metastasis (reviewed in Ref. [69]), drug resistance [70,71], and poor prognosis [72,73]. Similarly, PAK5 has also been linked to cancer cell survival [74,75], migration and invasion [76–78], EMT [79,80], therapy resistance [81], and poor prognosis [82]. In regards to PAK6, much of the work has linked its role to cancer metastasis [83], although studies have also shown PAK6 to have tumor suppressive function [84].

P21 ACTIVATED KINASE 4 IN PANCREATIC CANCER STEMNESS AND DRUG RESISTANCE

Studies on ras signaling pathways are a major focus in pancreatic research given that K-ras mutations are observed in >90% of patients suffering from this deadly disease. PAK4 has been quite well studied in the context of pancreatic cancer. The earliest investigations on this topic were performed by Mahlamaki et al., who captured using high resolution genomic and expression profiling PAK4 as one of the 105 putative genes amplified in 13 pancreatic cancer cell lines using a 12,232-clone cDNA microarray [85]. Few years later, Chen and colleagues confirmed recurrent PAK4 amplification as the major copy number alteration in 72 pancreatic adenocarcinoma patient samples [86]. This study focused on chromosome 19q13, a region frequently found amplified in pancreatic cancer. Almost simultaneously, Kimmelman and colleagues, while validating resident genes in highly recurrent and focal amplifications in pancreatic ductal adenocarcinoma, identified Rio Kinase 3 (RIOK3) as an amplified gene that modulates cytoskeletal architecture as well as promotes pancreatic ductal cell migration and invasion [87]. In this study the group also deduced the link between RIOK3 promoted small G protein Rac activation that is upstream of PAK4. Their analysis showed consistent PAK4 amplification in pancreatic ductal adenocarcinoma tumors and cell lines that was absent in normal pancreas tissue. These results clearly established the Rho family GTP-binding proteins to play an integral role in invasive pancreatic ductal adenocarcinoma. Tyagi and colleagues showed that PAK4 promotes proliferation and survival of pancreatic cancer cells through AKT- and ERK-dependent activation of the nuclear factor κB pathway [88].

Although controversial, the idea of the presence of a sub-population of pancreatic tumor cells, known as cancer stem cells (CSCs), responsible for therapy resistance and metastasis is gaining traction [89]. In this direction, the role of PAK4 in pancreatic cancer gemcitabine resistance was established [90].

Working further on this topic, our group was the first to present that PAK4 promotes pancreatic cancer stemness characteristics that is directly linked to gemcitabine resistance in pancreatic ductal adenocarcinoma cell line–derived (flow sorted) cancer stem-like cells (CSLCs) that are triple positive for stem cell markers CD133; CD44; EPCAM (Abstract 4688 Cancer Research, 2015 75;4688 doi:10.1158/1538-7445.AM2015-4688). Our studies showed consistent activation of Rho, Rac, CDC42, as well as PAK4 (protein and mRNA) in these CSLCs. Simultaneously another group verified our findings demonstrating that PAK4 is responsible for driving pancreatic cancer stemness [91]. Their data demonstrated that triple-positive [CD24(+)/CD44(+)/EpCAM(+)] sub-population of pancreatic CSCs exhibits greater level of PAK4 than triple-negative [CD24(−)/CD44(−)/EpCAM(−)] cells. Moreover, PAK4 silencing in pancreatic cancer cells leads to diminished fraction of CD24-, CD44-, and EpCAM-positive population alongside decreased sphere-forming ability and increased chemosensitivity to gemcitabine. These authors further verified that PAK4 expression is also directly linked to stemness markers such as (Oct4/Nanog/Sox2 and KLF4) that were associated with STAT3 expression and localization.

That PAK4 chemical modulation can impact pancreatic cancer growth was proved by Yeo and colleagues who showed that a natural product glaucarubinone (an anti-malaria drug) from *Simarouba glauca* can suppress pancreatic cancer survival through suppression of PAK4 [92]. However, given that glaucarubinone is a natural product with multi-targeted activities, the authors also observed suppression of PAK1. In our hands, certain PAK4-targeted agents (detailed later) could inhibit spheroid-forming ability of these pancreatic cancer cells (when grown in three-dimensional culture) through down-regulation of EMT and stemness markers (un-published work). The studies from our group and those of others certainly point to the key role of PAK4 in sustaining cancer stemness and therefore make this group II PAK an ideal drug target candidate to overcome therapy resistance in difficult-to-treat cancers. Collectively, such investigations have clearly demonstrated the importance of group I and group II PAKs in promoting cancer sustaining signaling making them as attractive therapeutic targets.

SMALL MOLECULE INHIBITORS TARGETING PAKs

Given the role of PAKs in the biology of various diseases, the pharmaceutical field has for long been interested in developing potent PAK inhibitors. Numerous PAK-targeted compounds have been evaluated in the pre-clinical setting for their potential anti-cancer activities. The reader is referred to some outstanding patent reviews that provide a comprehensive list of the available inhibitors in the field. Most of the initial attempts were restricted to the development of type ATP competitive inhibitors against group I PAKs. Nevertheless, the ATP-binding pocket in PAKs has been extensively investigated and is found to be flexible and open. This makes the development of a very-high-affinity ATP competitive molecule futile. Lei et al. have shown that there is a large gap between the N-lobe and C-lobe alongside the high mobility of N-lobe giving flexibility to the ATP-binding pocket within the PAK structure (Fig. 14.3) [93,94]. These structural and stearic problems have been the major reasons for the lack of the development of any agent with strong binding affinity and robust PAK inhibitory activity [95]. Despite this, a number of ATP-competitive type I PAK inhibitors were developed and tested pre-clinically. The oxindole or maleimide-based inhibitors such as the indolocarbazole-based natural product staurosporine and analogs, and the hydroxy derivative ST2001, are prototypical ATP-competitive kinase inhibitors with a high affinity for a broad range of kinases and activity against most PAKs.

Table 14.1 lists some of the ATP-competitive PAK inhibitors [93]. Among these, only PF-3578309 could make it to phase I clinical evaluation (Clinical

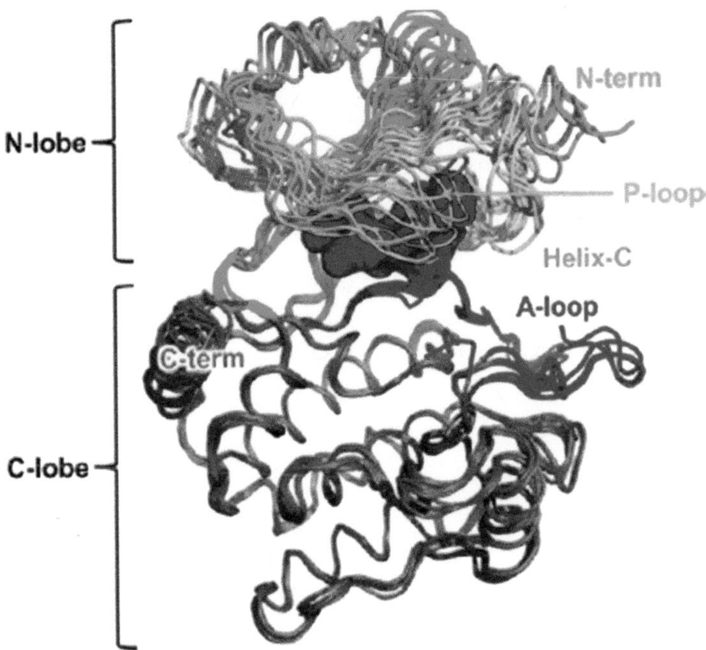

FIGURE 14.3 PAK4 kinase domain.
The available domain structures are superimposed at the C-terminal lobe where the structure is considered most conserved. The N-terminal lobe demonstrates structural heterogeneity. The structural variation includes a concerted N-lobe shift relative to the C-lobe and localized conformational shift more frequently seen in p loop, helix-C activation loop. *With permission from Rudolph J, Crawford JJ, Hoeflich KP, et al. Med. Chem 2015;58:111–29 Copyright (2014) American Chemical Society.*

Table 14.1 Currently Available PAK Inhibitors

Compound Name	Structure	PAK specificity (IC_{50})
ST2001	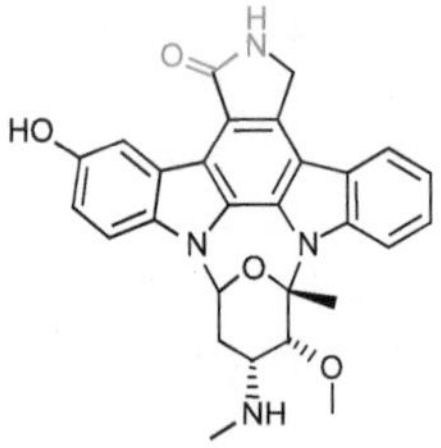	PAK1 (1 nM)
RDW12	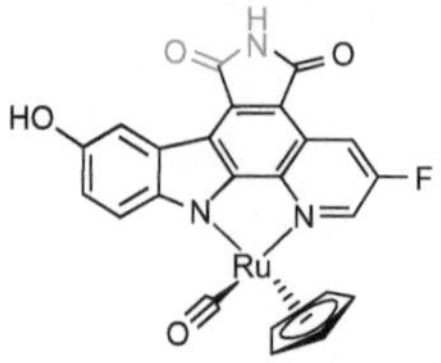	PAK1 (1090 nM)
A-FL172	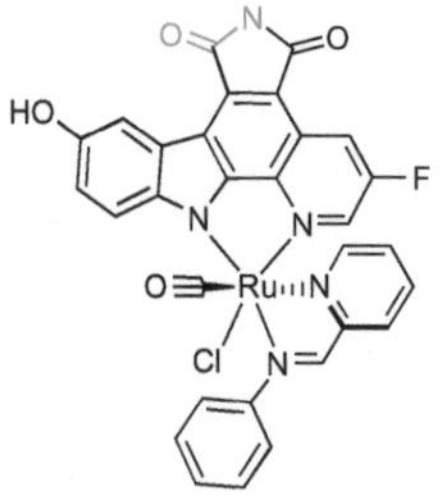	PAK1 (130 nM)
PF3578309	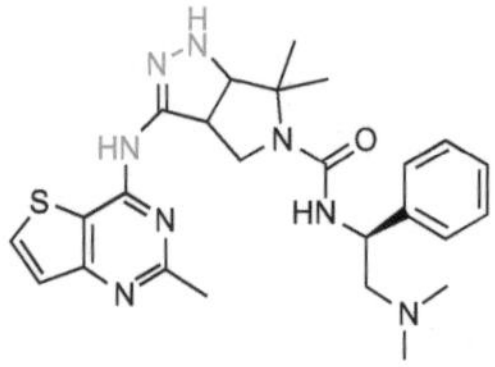	PAK1 (14 nM) PAK4 (19 nM)
FRAX597	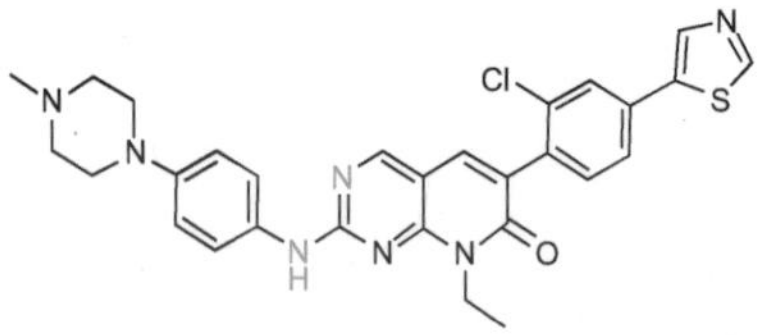	PAK1 (7.7 nM)
CGP74514	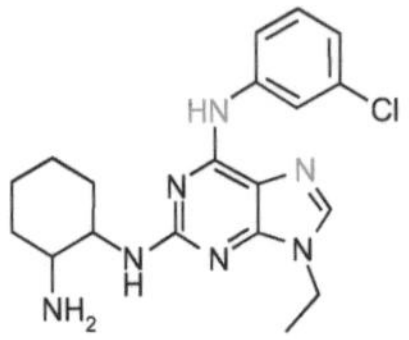	PAK4 (10 μM)
KY04031	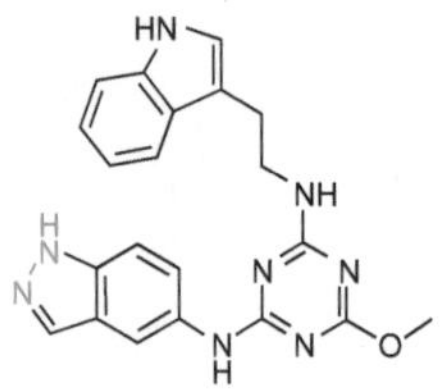	PAK4 (790 nM)

Adopted from Rudolph J, Crawford JJ, Hoeflich KP, et al. Med Chem 2015;58:111–29.

Trial Identifier NCT00932126). Nevertheless, lack of any objective response, as well as un-desirable Pk and Pd characteristics attributed to the compound being drug efflux protein pgp substrate, resulted in the abrupt withdrawal of the drug from this single clinical study. Since then there have not been any new ATP-competitive inhibitors against PAKs that could make it to the clinic. Another research group developed LCH-7749944 inhibitor, which was not only effective in suppressing PAK4 activity but could also interfere with cell plasticity through inhibition of filapodia formation cancer cell lines [96]. Nevertheless, till date this compound remains restricted to pre-clinical work. In view of these failures with type I inhibitors, the focus has shifted toward the development of type II or allosteric modulators of PAKs.

As mentioned in the previous section, glaucarubinone isolated from the seeds of the tree S. *glauca* was shown to inhibit PAK4 protein expression in mutant K-ras harboring pancreatic cancer cell lines and corresponding xenograft models [92]. Glaucarubinone could also synergize gemcitabine in a mechanism involving PAK4 inhibition. Nevertheless, PAK4 is not the primary target of glaucarubinone given that this agent was developed as an anti-malarial agent with pleiotropic/multi-targeted properties. Using high-throughput screening and structure-based drug design, a novel PAK4 inhibitor, KY-04031, was identified [97]. KY-04031 also belongs to the class of ATP-competitive inhibitor, and possesses weak binding affinity toward PAK4. The drug is required in high micromolar concentrations to effectively inhibit cell proliferation in a PAK4-dependent manner. Nevertheless, this agent has been projected to serve as a tool compound for the future development of newer more powerful ATP-competitive agents against PAKs.

ALLOSTERIC PAK MODULATORS

A number of different groups have attempted the development of allosteric modulators against both group I and group II PAKs (also known as type II allosteric modulators). The dibenzodiazepine PAK inhibitors and related compounds that bind to group I PAKs allosterically in proximity to the ATP-binding pocket are weakly active PAK1 binders [98]. These compounds are similar to the anti-psychotic drug clozapine as verified by high throughput screening and nuclear magnetic resonance fragment screening. Furthermore, to avoid kinase selectivity issues arising from the flexible nature of the PAK1 catalytic pocket, Deacon et al. developed non-ATP-like, un-competitive PAK1 inhibitor IPA-3 through high-throughput screening involving full-length PAK1 protein activated in vitro with recombinant Cdc42-GTPγS with ATP hydrolysis serving as a surrogate for PAK1 activity [98]. Another interesting set of PAK4 allosteric modulators (PAMs) has been developed at Karyopharm Therapeutics (http://karyopharm.com/drug-candidate/pak4-inhibitors/). PAMs were identified using small molecule library screening with follow-up surface plasmon resonance, isothermal calorimetry, and molecular analysis. These compounds (KPT-9274, KPT-7523, and KPT-7189) have specificity toward PAK4 as with minimal inhibitory activity against other group II PAKs (PAK5 and PAK6). PAMs have sub-micromolar IC_{50}s in most hematological and solid tumor cell lines. In pancreatic cancer models, the drugs demonstrate less than sub-micromolar IC_{50}s and do not inhibit the growth of normal pancreatic ductal epithelial (HPDE) cells (~10-fold differences in IC_{50}s) (J Clin Oncol 32, 2014 (Suppl. 3; abstr 233). PAMs also synergize with standard chemotherapeutics such as gemcitabine and oxaliplatin (doi:10.1158/1538-7445.AM2014-1771 Cancer Res October 1, 2014 74; 1771). Most significantly, the drugs show remarkable anti-tumor activity against pancreatic cancer xenograft and also show activity against therapy-resistant CSC-derived xenografts (doi:10.1158/1557-3125. RASONC14-A24 Mol Cancer Res December 2014 12; A24). The lead PAM KPT-9274 is currently being evaluated in a Phase I open-label study of the safety, tolerability and efficacy as a dual inhibitor of PAK4 and NAMPT, in patients with advanced solid malignancies or non-Hodgkin's lymphoma [ClinicalTrials.gov identifier NCT02702492 (https://clinicaltrials.gov/ct2/show/NCT02702492)].

CONCLUSIONS AND FUTURE DIRECTIONS

Despite being a topic of intense research focus, K-ras has remained an elusive target. Lack of an appropriate drug-binding pocket in the ras structure along with its extremely high affinity to bind to GTP has made the design of small molecule drugs against this master regulator futile. Therefore, novel targets within the ras structure or newer downstream targets need to be urgently identified. Unfortunately, the approaches targeting both upstream and downstream players within the ras network have also not showed clinical benefits. Redundancies and cross talks within ras signaling alongside the presence of numerous feedback loops keep these targets at bay and not eligible candidates for effective therapy. Rho GTPases belong to the ras superfamily of proteins that act as "molecular switches" and are recognized to play a role in organelle development, cytoskeletal dynamics, cell movement, and other common cellular functions. Unlike K-ras mutations that are quite common in almost all cancers, the Rho GTPases (Rho, Rac, and CDC42) are rarely found to be mutated. Their effectors especially the PAKs are uniquely placed at the nexus of various oncogenic signaling and act as the legitimate choke point downstream of ras. As presented in this chapter, the PAKs have important roles in various important cancer-sustaining pathways such as stemness as well as drug resistance. This makes them attractive therapeutic targets in oncogenic K-ras-driven tumors where other therapies have failed. Despite their significance, the pharmaceutical industry is yet to bring forward any suitable small molecule inhibitor that could show efficacy beyond pre-clinical stages. Despite these setbacks the efforts to develop PAK inhibitors continues and some novel approaches beyond the traditional ATP-competitive inhibitor–targeted approach are being evaluated. This renewed research interest is anticipated to lead to the discovery of potent PAK-targeted drugs with robust clinical utility.

List of Acronyms and Abbreviations

EGFR Epidermal Growth Factor Receptor
ITC Isothermal Calorimetry
KPT Karyopharm Therapeutics
MAPK Mitogen Activated Kinase
PAK P21 activated kinase
PAM P21 activated kinase allosteric modulators
PIX PAK interacting exchange factor
SPR Surface Plasmon Resonance

Acknowledgments

Work in the laboratory of ASA is supported by NIH Grant 1 R21 CA188818 01 A2. The authors gratefully acknowledge the generous support from SKY Foundation, Perri Foundation, and Qatar Foundation.

References

[1] McCormick F. Ras-related proteins in signal transduction and growth control. Mol Reprod Dev 1995;42:500–6.

[2] McCormick F. Signalling networks that cause cancer. Trends Cell Biol 1999;9:M53–6.

[3] Prior IA, Lewis PD, Mattos C. A comprehensive survey of Ras mutations in cancer. Cancer Res 2012;72:2457–67.

[4] McCormick F. GTP binding and growth control. Curr Opin Cell Biol 1990;2:181–4.

[5] Gibbs JB, Sigal IS, Poe M, Scolnick EM. Intrinsic GTPase activity distinguishes normal and oncogenic ras p21 molecules. Proc Natl Acad Sci USA 1984;81:5704–8.

[6] Vetter IR, Wittinghofer A. The guanine nucleotide-binding switch in three dimensions. Science 2001;294:1299–304.

[7] Colicelli J. Human RAS superfamily proteins and related GTPases. Sci STKE 2004;2004:RE13.

[8] Shapiro P. Ras-MAP kinase signaling pathways and control of cell proliferation: relevance to cancer therapy. Crit Rev Clin Lab Sci 2002;39:285–330.

[9] Schmitz AA, Govek EE, Bottner B, Van AL. Rho GTPases: signaling, migration, and invasion. Exp Cell Res 2000;261:1–12.

[10] Clarke PR, Zhang C. Spatial and temporal coordination of mitosis by Ran GTPase. Nat Rev Mol Cell Biol 2008;9:464–77.

[11] Nielsen E, Cheung AY, Ueda T. The regulatory RAB and ARF GTPases for vesicular trafficking. Plant Physiol 2008;147:1516–26.

[12] van AL, D'Souza-Schorey C. Rho GTPases and signaling networks. Genes Dev 1997;11:2295–322.

[13] Sit ST, Manser E. Rho GTPases and their role in organizing the actin cytoskeleton. J Cell Sci 2011;124:679–83.

[14] Heider D, Hauke S, Pyka M, Kessler D. Insights into the classification of small GTPases. Adv Appl Bioinforma Chem 2010;3:15–24.

[15] Cook DR, Rossman KL, Der CJ. Rho guanine nucleotide exchange factors: regulators of Rho GTPase activity in development and disease. Oncogene 2014;33:4021–35.

[16] Ten Klooster JP, Hordijk PL. Targeting and localized signalling by small GTPases. Biol. Cell 2007;99:1–12.

[17] Cherfils J, Zeghouf M. Regulation of small GTPases by GEFs, GAPs, and GDIs. Physiol Rev 2013;93:269–309.

[18] Liu M, Bi F, Zhou X, Zheng Y. Rho GTPase regulation by miRNAs and covalent modifications. Trends Cell Biol 2012;22:365–73.

[19] Visvikis O, Maddugoda MP, Lemichez E. Direct modifications of Rho proteins: deconstructing GTPase regulation. Biol Cell 2010;102:377–89.

[20] Gomez del PT, Benitah SA, Valeron PF, Espina C, Lacal JC. Rho GTPase expression in tumourigenesis: evidence for a significant link. Bioessays 2005;27:602–13.

[21] Benitah SA, Valeron PF, Van AL, Marshall CJ, Lacal JC. Rho GTPases in human cancer: an unresolved link to upstream and downstream transcriptional regulation. Biochim Biophys Acta 2004;1705:121–32.

[22] Merajver SD, Usmani SZ. Multifaceted role of Rho proteins in angiogenesis. J Mammary Gland Biol Neoplasia 2005;10:291–8.

[23] Aneta G, Tomas V, Daniel R, Jan B. Cell polarity signaling in the plasticity of cancer cell invasiveness. Oncotarget 2016, http://dx.doi.org/10.18632/oncotarget.7214.

[24] King H, Nicholas NS, Wells CM. Role of p-21-activated kinases in cancer progression. Int Rev Cell Mol Biol 2014;309:347–87.

[25] Bokoch GM. Biology of the p21-activated kinases. Annu Rev Biochem 2003;72:743–81.

[26] Bokoch GM. PAK'n it in: identification of a selective PAK inhibitor. Chem Biol 2008;15:305–6.

[27] Ma QL, Yang F, Calon F, Ubeda OJ, Hansen JE, Weisbart RH, et al. p21-activated kinase-aberrant activation and translocation in Alzheimer disease pathogenesis. J Biol Chem 2008;283:14132–43.

[28] Chan PM, Manser E. PAKs in human disease. Prog Mol Biol Transl Sci 2012;106:171–87.

[29] Radu M, Semenova G, Kosoff R, Chernoff J. PAK signalling during the development and progression of cancer. Nat Rev Cancer 2014;14:13–25.

[30] Dart AE, Wells CM. P21-activated kinase 4–not just one of the PAK. Eur J Cell Biol 2013; 92:129–38.

[31] Md Hashim NF, Nicholas NS, Dar AE, Kiriakidis S, Paleolog E, Wells CM. Hypoxia-induced invadopodia formation: a role for beta-PIX. Open Biol 2013;3:120159.

[32] Molli PR, Li DQ, Murray BW, Rayala SK, Kumar R. PAK signaling in oncogenesis. Oncogene 2009;28:2545–55.

[33] Minden A. PAK4-6 in cancer and neuronal development. Cell Logist 2012;2:95–104.

[34] Meng J, Meng Y, Hanna A, Janus C, Jia Z. Abnormal long-lasting synaptic plasticity and cognition in mice lacking the mental retardation gene Pak3. J Neurosci 2005;25:6641–50.

[35] Satoh T, Endo M, Nakafuki M, Akiyama T, Yamamoto T, Kaziro Y. Accumulation of p21ras. GTP in response to stimulation with epidermal growth factor and oncogene products with tyrosine kinase activity. Proc Natl Acad Sci USA 1990;87:7926–9.

[36] Leevers SJ, Marshall CJ. Activation of extracellular signal-regulated kinase, ERK2, by p21ras oncoprotein. EMBO J 1992;11:569–74.

[37] Bekri S, Adelaide J, Merscher S, Grosgeorge J, Caroli-Bosc F, Perucca-Lostanlen D, et al. Detailed map of a region commonly amplified at 11q13-->q14 in human breast carcinoma. Cytogenet Cell Genet 1997;79:125–31.

[38] Tang Y, Chen Z, Ambrose D, Liu J, Gibbs JB, Chernoff J, et al. Kinase-deficient Pak1 mutants inhibit Ras transformation of Rat-1 fibroblasts. Mol Cell Biol 1997;17:4454–64.

[39] Adam L, Vadlamudi R, Mandal M, Chernoff J, Kumar R. Regulation of microfilament reorganization and invasiveness of breast cancer cells by kinase dead p21-activated kinase-1. J Biol Chem 2000;275:12041–50.

[40] Vadlamudi RK, Adam L, Wang RA, Mandal M, Nguyen M, Nguyen D, et al. Regulatable expression of p21-activated kinase-1 promotes anchorage-independent growth and abnormal organization of mitotic spindles in human epithelial breast cancer cells. J Biol Chem 2000;275:36238–44.

[41] Bagheri-Yarmand R, Mandal M, Taludker AH, Wang RA, Vadlamudi RK, Kung HJ, et al. Etk/Bmx tyrosine kinase activates Pak1 and regulates tumorigenicity of breast cancer cells. J Biol Chem 2001;276:29403–9.

[42] Li F, Adam L, Vadlamudi RK, Zhou H, Sen S, Chernoff J, et al. p21-activated kinase 1 interacts with and phosphorylates histone H3 in breast cancer cells. EMBO Rep 2002;3:767–73.

[43] Wang RA, Zhang H, Balasenthil S, Medina D, Kumar R. PAK1 hyperactivation is sufficient for mammary gland tumor formation. Oncogene 2006;25:2931–6.

[44] Yoon CH, Hyun KH, Kim RK, Lee H, Lim EJ, Chung HY, et al. The small GTPase Rac1 is involved in the maintenance of stemness and malignancies in glioma stem-like cells. FEBS Lett 2011;585:2331–8.

[45] Coniglio SJ, Zavarella S, Symons MH. Pak1 and Pak2 mediate tumor cell invasion through distinct signaling mechanisms. Mol Cell Biol 2008;28:4162–72.

[46] Flate E, Stalvey JR. Motility of select ovarian cancer cell lines: effect of extra-cellular matrix proteins and the involvement of PAK2. Int J Oncol 2014;45:1401–11.

[47] Sato M, Matsuda Y, Wakai T, Kubota M, Osawa M, Fujimaki S, et al. P21-activated kinase-2 is a critical mediator of transforming growth factor-beta-induced hepatoma cell migration. J Gastroenterol Hepatol 2013;28:1047–55.

[48] Benitah SA, Watt FM. Epidermal deletion of Rac1 causes stem cell depletion, irrespective of whether deletion occurs during embryogenesis or adulthood. J Invest Dermatol 2007;127:1555–7.

[49] Li X, Wen W, Liu K, Zhu F, Malakhova M, Peng C, et al. Phosphorylation of caspase-7 by p21-activated protein kinase (PAK) 2 inhibits chemotherapeutic drug-induced apoptosis of breast cancer cell lines. J Biol Chem 2011;286:22291–9.

[50] Yan BX, Ma JX, Zhang J, Guo Y, Mueller MD, Remick SC, et al. Prostasin may contribute to chemoresistance, repress cancer cells in ovarian cancer, and is involved in the signaling pathways of CASP/PAK2-p34/actin. Cell Death Dis 2014;5:e995.

[51] Cho HJ, Baek KE, Kim IK, Park SM, Choi YL, Nam IK, et al. Proteomics-based strategy to delineate the molecular mechanisms of RhoGDI2-induced metastasis and drug resistance in gastric cancer. J Proteome Res 2012;11:2355–64.

[52] Gopal SK, Greening DW, Hanssen EG, Zhu HJ, Simpson RJ, Mathais RA. Oncogenic epithelial cell-derived exosomes containing Rac1 and PAK2 induce angiogenesis in recipient endothelial cells. Oncotarget 2016, http://dx.doi.org/10.18632/oncotarget.7573.

[53] Liu RX, Wang WQ, Ye L, Bi YF, Fang H, Cui B, et al. p21-activated kinase 3 is overexpressed in thymic neuroendocrine tumors (carcinoids) with ectopic ACTH syndrome and participates in cell migration. Endocrine 2010;38:38–47.

[54] Holderness PN, Donninger H, Birrer MJ, Leaner VD. p21-activated kinase 3 (PAK3) is an AP-1 regulated gene contributing to actin organisation and migration of transformed fibroblasts. PLoS One 2013;8:e66892.

[55] Baldwin A, Grueneberg DA, Hellner K, Sawyer J, Grace M, Li W, et al. Kinase requirements in human cells: V. Synthetic lethal interactions between p53 and the protein kinases SGK2 and PAK3. Proc Natl Acad Sci USA 2010;107:12463–8.

[56] Zhou N, Ding B, Agler M, Cockett M, McPhee F. Lethality of PAK3 and SGK2 shRNAs to human papillomavirus positive cervical cancer cells is independent of PAK3 and SGK2 knockdown. PLoS One 2015;10:e0117357.

[57] Jaffer ZM, Chernoff J. p21-activated kinases: three more join the Pak. Int J Biochem Cell Biol 2002;34:713–7.

[58] Abo A, Qu J, Cammarano MS, Dan C, Fritsch A, Baud V, et al. PAK4, a novel effector for Cdc42Hs, is implicated in the reorganization of the actin cytoskeleton and in the formation of filopodia. EMBO J 1998;17:6527–40.

[59] Yang F, Li X, Sharma M, Zarnegar M, Lim B, Sun Z. Androgen receptor specifically interacts with a novel p21-activated kinase, PAK6. J Biol Chem 2001;276:15345–53.

[60] Pandey A, Dan I, Kristiansen TZ, Watanabe NM, Voldby J, Kajikawa E, et al. Cloning and characterization of PAK5, a novel member of mammalian p21-activated kinase-II subfamily that is predominantly expressed in brain. Oncogene 2002;21:3939–48.

[61] Qu J, Cammarano MS, Shi Q, Ha KC, de Lanerolle P, Minden A. Activated PAK4 regulates cell adhesion and anchorage-independent growth. Mol Cell Biol 2001;21:3523–33.

[62] Callow MG, Clairvoyant F, Zhu S, Schryver B, Shyte DB, Bischoff JR, et al. Requirement for PAK4 in the anchorage-independent growth of human cancer cell lines. J Biol Chem 2002;277:550–8.

[63] Zhang H, Li Z, Viklund EK, Stromblad S. P21-activated kinase 4 interacts with integrin alpha v beta 5 and regulates alpha v beta 5-mediated cell migration. J Cell Biol 2002;158:1287–97.

[64] Ahmed T, Shea K, Masters JR, Jones GE, Wells CM. A PAK4-LIMK1 pathway drives prostate cancer cell migration downstream of HGF. Cell Signal 2008;20:1320–8.

[65] Siu MK, Chan HY, Kong DS, Wong ES, Wong OG, Ngan HY, et al. p21-activated kinase 4 regulates ovarian cancer cell proliferation, migration, and invasion and contributes to poor prognosis in patients. Proc Natl Acad Sci USA 2010;107:18622–7.

[66] Li Z, Zhang H, Lundin L, Thullberg M, Liu Y, Wang Y, et al. p21-activated kinase 4 phosphorylation of integrin beta5 Ser-759 and Ser-762 regulates cell migration. J Biol Chem 2010;285:23699–710.

[67] Kesanakurti D, Chetty C, Rajasekhar MD, Gujrati M, Rao JS. Functional cooperativity by direct interaction between PAK4 and MMP-2 in the regulation of anoikis resistance, migration and invasion in glioma. Cell Death Dis 2012;3:e445.

[68] Lu W, Xia YH, Qu JJ, He YY, Li BL, Lu C, et al. p21-activated kinase 4 regulation of endometrial cancer cell migration and invasion involves the ERK1/2 pathway mediated MMP-2 secretion. Neoplasma 2013;60:493–503.

[69] Eswaran J, Soundararajan M, Knapp S. Targeting group II PAKs in cancer and metastasis. Cancer Metastasis Rev 2009;28:209–17.

[70] Bhardwaj A, Srivastava SK, Singh S, Arora S, Tyagi N, Andrews J, et al. CXCL12/CXCR4 signaling counteracts docetaxel-induced microtubule stabilization via p21-activated kinase 4-dependent activation of LIM domain kinase 1. Oncotarget 2014;5:11490–500.

[71] Fu X, Feng J, Zeng D, Ding Y, Yu C, Yang B. PAK4 confers cisplatin resistance in gastric cancer cells via PI3K/Akt- and MEK/Erk-dependent pathways. Biosci Rep 2014 [Epub ahead of print].

[72] Cai S, Ye Z, Wang X, Pan Y, Weng Y, Lao S, et al. Overexpression of P21-activated kinase 4 is associated with poor prognosis in non-small cell lung cancer and promotes migration and invasion. J Exp Clin Cancer Res 2015;34:48.

[73] Li D, Zhang Y, Li Z, Wang X, Qu X, Liu Y. Activated Pak4 expression correlates with poor prognosis in human gastric cancer patients. Tumour Biol 2015;36:9431–6.

[74] Cotteret S, Jaffer ZM, Beeser A, Chernoff J. p21-Activated kinase 5 (Pak5) localizes to mitochondria and inhibits apoptosis by phosphorylating BAD. Mol Cell Biol 2003;23:5526–39.

[75] Cotteret S, Chernoff J. Nucleocytoplasmic shuttling of Pak5 regulates its antiapoptotic properties. Mol Cell Biol 2006;26:3215–30.

[76] Gong W, An Z, Wang Y, Pan X, Fang W, Jiang B, et al. P21-activated kinase 5 is overexpressed during colorectal cancer progression and regulates colorectal carcinoma cell adhesion and migration. Int J Cancer 2009;125:548–55.

[77] Wang XX, Cheng Q, Zhang SN, Qian HY, Wu JX, Tian H, et al. PAK5-Egr1-MMP2 signaling controls the migration and invasion in breast cancer cell. Tumour Biol 2013;34:2721–9.

[78] Han ZX, Wang XX, Zhang SN, Wu JX, Qian HY, Wen YY, et al. Downregulation of PAK5 inhibits glioma cell migration and invasion potentially through the PAK5-Egr1-MMP2 signaling pathway. Brain Tumor Pathol 2014;31:234–41.

[79] Li Y, Ke Q, Shao Y, Zhu G, Li Y, Geng N, et al. GATA1 induces epithelial-mesenchymal transition in breast cancer cells through PAK5 oncogenic signaling. Oncotarget 2015;6:4345–56.

[80] Zhu G, Li X, Guo B, Ke Q, Dong M, Li F. PAK5-mediated E47 phosphorylation promotes epithelial-mesenchymal transition and metastasis of colon cancer. Oncogene 2015;35(15): 1943–54. http://dx.doi.org/10.1038/onc.2015.259.

[81] Li D, Yao X, Zhang P. The overexpression of P21-activated kinase 5 (PAK5) promotes paclitaxel-chemoresistance of epithelial ovarian cancer. Mol Cell Biochem 2013;383:191–9.

[82] Chen H, Miao J, Li H, Wang C, Li J, Zhu Y, et al. Expression and prognostic significance of p21-activated kinase 6 in hepatocellular carcinoma. J Surg Res 2014;189:81–8.

[83] Cai S, Chen R, Li X, Cai Y, Ye Z, Li S, et al. Downregulation of microRNA-23a suppresses prostate cancer metastasis by targeting the PAK6-LIMK1 signaling pathway. Oncotarget 2015;6:3904–17.

[84] Liu W, Liu Y, Liu H, Zhang W, Fu Q, Xu J, et al. Tumor suppressive function of p21-activated kinase 6 in hepatocellular carcinoma. J Biol Chem 2015;290:28489–501.

[85] Mahlamaki EH, Kauraniemi P, Monni O, Wolf M, Hautaniemi S, Kallioniemi A. High-resolution genomic and expression profiling reveals 105 putative amplification target genes in pancreatic cancer. Neoplasia 2004;6:432–9.

[86] Chen S, Auletta T, Dovirak O, Hutter C, Kuntz K, El-ftesi S, et al. Copy number alterations in pancreatic cancer identify recurrent PAK4 amplification. Cancer Biol Ther 2008;7:1793–802.

[87] Kimmelman AC, Hezel AF, Aguirre AJ, Zheng H, Paik JH, Ying H, et al. Genomic alterations link Rho family of GTPases to the highly invasive phenotype of pancreas cancer. Proc Natl Acad Sci USA 2008;105:19372–7.

[88] Tyagi N, Bhardwaj A, Singh AP, McClellan S, Carter JE, Singh S. p-21 activated kinase 4 promotes proliferation and survival of pancreatic cancer cells through AKT- and ERK-dependent activation of NF-kappaB pathway. Oncotarget 2014;5:8778–89.

[89] Vaz AP, Ponnusamy MP, Seshacharyulu P, Batra SK. A concise review on the current understanding of pancreatic cancer stem cells. J Cancer Stem Cell Res 2014;2.

[90] Moon SU, Kim JW, Sung JH, Kang MH, Kim SH, Chang H, et al. p21-Activated kinase 4 (PAK4) as a predictive marker of gemcitabine sensitivity in pancreatic Cancer cell lines. Cancer Res Treat 2015;47:501–8.

[91] Tyagi N, Marimuthu S, Bhardwaj A, Deshmukh SK, Srivastava SK, Singh AP, et al. p-21 activated kinase 4 (PAK4) maintains stem cell-like phenotypes in pancreatic cancer cells through activation of STAT3 signaling. Cancer Lett 2016;370:260–7.

[92] Yeo D, Huynh N, Beutler JA, Christophi C, Shulkes A, Baldwin GS, et al. Glaucarubinone and gemcitabine synergistically reduce pancreatic cancer growth via down-regulation of P21-activated kinases. Cancer Lett 2014;346:264–72.

[93] Rudolph J, Crawford JJ, Hoeflich KP, Wang W. Inhibitors of p21-activated kinases (PAKs). J Med Chem 2015;58:111–29.

[94] Hsu YH, Johnson DA, Traugh JA. Analysis of conformational changes during activation of protein kinase Pak2 by amide hydrogen/deuterium exchange. J Biol Chem 2008;283:36397–405.

[95] Zhao ZS, Manser E. Do PAKs make good drug targets? F1000 Biol Rep 2010;2:70.

[96] Zhang J, Wang J, Guo Q, Wang Y, Zhou Y, Peng H, et al. LCH-7749944, a novel and potent p21-activated kinase 4 inhibitor, suppresses proliferation and invasion in human gastric cancer cells. Cancer Lett 2012;317:24–32.

[97] Ryu BJ, Kim S, Min B, Kim KY, Lee JS, Park WJ, et al. Discovery and the structural basis of a novel p21-activated kinase 4 inhibitor. Cancer Lett 2014;349:45–50.

[98] Wenthur CJ, Lindsley CW. Classics in chemical neuroscience: clozapine. ACS Chem Neurosci 2013;4:1018–25.

Index

'*Note*: Page numbers followed by "f" indicate figures and "t" indicate tables.'

S

T

U

V

W

Z